Pitman Research Notes in Math Series

Submission of proposals for consideration

Suggestions for publication, in the form of outlines and representative samples, are invited by the Editorial Board for assessment. Intending authors should approach one of the main editors or another member of the Editorial Board, citing the relevant AMS subject classifications. Alternatively, outlines may be sent directly to the publisher's offices. Refereeing is by members of the board and other mathematical authorities in the topic concerned, throughout the world.

Preparation of accepted manuscripts

On acceptance of a proposal, the publisher will supply full instructions for the preparation of manuscripts in a form suitable for direct photo-lithographic reproduction. Specially printed grid sheets can be provided and a contribution is offered by the publisher towards the cost of typing. Word processor output, subject to the publisher's approval, is also acceptable.

Illustrations should be prepared by the authors, ready for direct reproduction without further improvement. The use of hand-drawn symbols should be avoided wherever possible, in order to maintain maximum clarity of the text.

The publisher will be pleased to give any guidance necessary during the preparation of a typescript, and will be happy to answer any queries.

Important note

In order to avoid later retyping, intending authors are strongly urged not to begin final preparation of a typescript before receiving the publisher's guidelines. In this way it is hoped to preserve the uniform appearance of the series.

Longman Scientific & Technical
Longman House
Burnt Mill
Harlow, Essex, CM20 2JE
UK
(Telephone (0279) 426721)

Titles in this series. A full list is available from the publisher on request.

C Bandle

University of Basel, Switzerland

J Bemelmans

RWTH Aachen, Germany

M Chipot
J Saint Jean Paulin
I Shafrir

University of Metz, France

(Editors)

Elliptic and parabolic problems

Pont-à-Mousson 1994

CRC Press
Taylor & Francis Group
Boca Raton London New York

CRC Press is an imprint of the
Taylor & Francis Group, an **informa** business

A CHAPMAN & HALL BOOK

First published 1995 by Longman Group Limited

Published 2019 by CRC Press
Taylor & Francis Group
6000 Broken Sound Parkway NW, Suite 300
Boca Raton, FL 33487-2742

First issued in paperback 2019

No claim to original U.S. Government works

ISBN-13: 978-0-367-44912-4 (pbk)
ISBN-13: 978-0-582-23961-6 (hbk)

Visit the Taylor & Francis Web site at
http://www.taylorandfrancis.com

and the CRC Press Web site at
http://www.crcpress.com

AMS Subject Classifications: (Main) 35Jxx, 35Kxx, 49xx
(Subsidiary) 65xx, 73xx, 65Mxx

British Library Cataloguing in Publication Data

A catalogue record for this book is
available from the British Library

Library of Congress Cataloging-in-Publication Data

A catalog record for this book is available

ISSN 0269-3674

Copublished in the United States with
John Wiley & Sons Inc., 605 Third Avenue, New York, NY 10158

Contents

Preface

This volume collects lectures given during the second European Conference on Elliptic and Parabolic Problems (Pont-à-Mousson, June 1994).

The subjects addressed in this volume include Equations and Systems of Elliptic and Parabolic type and various applications in Physics, Mechanics and Engineering.

We would like to thank all the participants to this meeting for their help in making it succesful. Special thanks go to the contributors for this volume.

The meeting has been made possible by grants from the "Caisse d'épargne de Lorraine Nord", the C.N.R.S., the "Region de Lorraine", the A.A.U.L., the RWTH Aachen, the University of Basel, the University of Metz and the "Gesellschaft von Freuden der Aachener Hochschule e.V.". We express our deep appreciation to them.

Finally, we wish to express our special thanks to Longman for helping us to publish these proceedings.

C. Bandle, J. Bemelmans, M. Chipot, J. Saint Jean Paulin and I. Shafrir

C AMROUCHE, V GIRAULT AND J GIROIRE

Calderón–Zygmund inequalities, Riesz potentials and Riesz transforms in weighted Sobolev spaces

Abstract. In a previous work, we established a complete set of isomorphisms for the Laplace operator in weighted Sobolev spaces. We derive here some applications of these results. In particular, we obtain inequalities similar to those of Calderón-Zygmund and continuity properties of Riesz potentials and transforms in weighted Sobolev spaces. Finally, we give an application to the asymptotic behaviour of the solution of the heat equation.

1. Introduction

For any integer q, we denote by $\mathcal{P}_q$ (respectively $\mathcal{P}_q^\Delta$) the space of polynomials (respectively harmonic polynomials) of degree less than or equal to q, with the convention that $\mathcal{P}_q = \{0\}$ when q is negative. Let $r = |x|$ be the distance to the origin and let $\rho = (1 + r^2)^{1/2}$ and $\lg r = \ln(2 + r^2)$ be two basic weights. For any non negative integer m and real numbers α, β and p with $1 < p < \infty$, we define the weighted Sobolev space

$$W_{\alpha,\beta}^{m,p}(R^n) = \{u \in \mathcal{D}'(R^n)\,;\ \forall \lambda \in N^n\ : 0 \le |\lambda| \le k,\, \rho^{\alpha-m+|\lambda|}(\lg r)^{\beta-1}D^\lambda u \in L^p(R^n)\,;$$

$$k+1 \le |\lambda| \le m,\, \rho^{\alpha-m+|\lambda|}(\lg r)^\beta D^\lambda u \in L^p(R^n)\},$$

where

$$k = k(m,n,p,\alpha) = \begin{cases} m - \dfrac{n}{p} - \alpha & \text{if } \quad \dfrac{n}{p} + \alpha \in \{1,\dots,m\}, \\[2mm] -1 & \text{otherwise,} \end{cases}$$

and for the sake of simplicity, we denote it by $W_\alpha^{m,p}(R^n)$ when $\beta = 0$. It is a Banach space for its graph norm and its dual space is denoted by $W_{-\alpha}^{-m,p'}(R^n)$ where p' is the dual exponent of p: $1/p + 1/p' = 1$. This weighted space is well-adapted to the solution of the Laplace equation in R^n. More precisely, the weights in this definition are chosen so that an inequality similar to Poincaré's inequality holds and when, $k = -1$, they are derived from a standard Hardy's inequality. However, when k is non negative, there are a number of exceptional cases depending on the dimension n and the exponent p where this standard Hardy's inequality does not hold and it is necessary to use a generalized Hardy's inequality (cf. [1], [2], [3]) in which the logarithmic weights are compulsory. Besides this, some critical cases require other spaces, that are not Sobolev spaces, and that we define as follows. For any integers m in Z and l in N and real number p with $1 < p < \infty$, we set:

$$X_l^{m+l,p}(R^n) = \{ u \in W_0^{m,p}(R^n) \, ;$$
$$\forall \lambda \in N^n \, : \, 0 \le |\lambda| \le l, \, x^\lambda u \in W_0^{m+|\lambda|,p}(R^n) \, ; \, u \in W_{\text{loc}}^{m+l,p}(R^n) \},$$

(the index 'loc' refers to neighbourhoods of the origin), equipped with the norm

$$\|u\|_{X_l^{m+l,p}(R^n)} = \Big(\sum_{0 \le |\lambda| \le l} \|x^\lambda u\|_{W_0^{m+|\lambda|,p}(R^n)}^p + \|u\|_{W^{m+l,p}(B(0,1))}^p \Big)^{\frac{1}{p}},$$

where $B(0,1)$ denotes the ball of R^n centered at the origin and with radius one. Its dual space is denoted by $X_{-l}^{-m-l,p'}(R^n)$. The spaces W and X are closely related to each other: they coincide for all values of n/p or n/p', except for some critical ratios depending on m and l. The W spaces without logarithmic weights are widely used but they cannot handle a number of interesting cases. That is why the full W spaces or the X spaces are vital.

In this work, we propose to show how sharp results for the Laplace operator allow to prove, in a straightforward manner, important inequalities and properties of some other operators. In Section 2, we derive a family of Calderón-Zygmund type inequalities under necessary and sufficient conditions (cf. Theorem 2.1). We also establish some equivalent norms for the spaces $W_\alpha^{m,p}(R^n)$, in the same spirit as Nirenberg-Walker [11], but for a larger class of weights (cf. Proposition 2.2). Section 3 is devoted to some continuity properties of Riesz potentials of order one or two (cf. Theorems 3.2–3.4). In Section 4, we determine the domain of continuity of Riesz transforms in adequate weighted subspaces. The technique developed here yields alternate proofs and extensions of some well-known results of harmonic analysis. We end this paper by applications to the time asymptotic behaviour of the solution of the heat equation (cf. Theorems 5.2–5.4). These last results complete the work of Duoandikoetxea-Zuazua [6].

In the sequel, we adopt the convention that the set $\{1, \ldots, q\}$ is empty if the integer q is non positive. The letter C will denote positive constants that may depend on the dimension n, the exponent p and possibly other parameters, but never on the functions under consideration.

2

2. Weighted Calderón-Zygmund inequalities

Recall the definition of Riesz transforms (*cf.* [13]):

$$R_j(f) = c_n \, v.p.(f \star \frac{x_j}{|x|^{n+1}}), \ j = 1, \ldots, n \, ,$$

where

$$c_n = \frac{\Gamma(\frac{n+1}{2})}{\pi^{\frac{n+1}{2}}} \, ,$$

and recall that their Fourier transforms satisfy $\widehat{R_j(f)} = i\frac{\xi_j}{|\xi|}\widehat{f}$. Besides this, as the mappings

$$R_j : L^p(R^n) \mapsto L^p(R^n)$$

are continuous (*cf.* [13]) and satisfy $R_j \circ R_k(\Delta u) = -\frac{\partial^2 u}{\partial x_j \partial x_k}$, we deduce the following Calderón-Zygmund inequality [5] (*cf.* also [13]) for all functions u in $\mathcal{D}(R^n)$:

$$||\frac{\partial^2 u}{\partial x_j \partial x_k}||_{L^p(R^n)} \le C||\Delta u||_{L^p(R^n)}. \tag{2.1}$$

We are going to prove that similar inequalities hold with weighted norms. Indeed, owing to the theory of singular integral operators, we know that for any non negative integer l, the following Riesz transforms:

$$R_j : W_l^{0,p}(R^n) \mapsto W_l^{0,p}(R^n), \tag{2.2}$$

$$R_j : W_{-l}^{0,p}(R^n) \mapsto W_{-l}^{0,p}(R^n), \tag{2.3}$$

are continuous, the first one if and only if $n/p' > l$ and the second one if and only if $n/p > l$ (*cf.* for instance [7]). These two conditions are guaranteed as soon as the corresponding weights satisfy condition A_p of Muckenhoupt (*cf.* [10], [7]). Obviously, under the same conditions, (2.2) and (2.3), imply Calderón-Zygmund inequalities with corresponding weights. However, we are going to prove in the next theorem that $n/p' > l$ is not a necessary condition, in contrast to the condition $n/p > l$.

Theorem 2.1. *For any non negative integer l, there exists a constant $C > 0$, such that the following inequalities hold for all functions u in $\mathcal{D}(R^n)$:*

$$||\frac{\partial^2 u}{\partial x_j \partial x_k}||_{W_l^{0,p}(R^n)} \le C||\Delta u||_{W_l^{0,p}(R^n)} \ iff \ n/p' \notin \{1, \ldots, l\}, \tag{2.4}$$

$$||\frac{\partial^2 u}{\partial x_j \partial x_k}||_{W_l^{0,p}(R^n)} \le C(||\Delta u||_{W_l^{0,p}(R^n)} + ||\Delta u||_{W_0^{-l,p}(R^n)}) \ if \ n/p' \in \{1, \ldots, l\}, \tag{2.5}$$

3

$$\left\|\frac{\partial^2 u}{\partial x_j \partial x_k}\right\|_{W^{0,p}_{-l}(R^n)} \le C\|\Delta u\|_{W^{0,p}_{-l}(R^n)} \text{ iff } n/p > l, \tag{2.6}$$

$$\inf_{q \in \mathcal{P}^{\Delta}_{[l+2-n/p]}} \left\|\frac{\partial^2 (u+q)}{\partial x_j \partial x_k}\right\|_{W^{0,p}_{-l}(R^n)} \le C\|\Delta u\|_{W^{0,p}_{-l}(R^n)} \text{ iff } n/p \notin \{1,\dots,l\}, \tag{2.7}$$

$$\inf_{q \in \mathcal{P}^{\Delta}_{l+2-n/p}} \left\|\frac{\partial^2 (u+q)}{\partial x_j \partial x_k}\right\|_{X^{0,p}_{-l}(R^n)} \le C\|\Delta u\|_{X^{0,p}_{-l}(R^n)} \text{ iff } n/p \in \{1,\dots,l\}. \tag{2.8}$$

Proof.

i) Estimates (2.4) and (2.5) stem from the fact that the Laplace operators

$$\Delta : W^{2,p}_l(R^n)/\mathcal{P}_{[2-l-n/p]} \mapsto W^{0,p}_l(R^n) \perp \mathcal{P}_{[l-n/p']} \text{ iff } n/p' \notin \{1,\dots,l\} \tag{2.9}$$

$$\Delta : W^{2,p}_l(R^n)/\mathcal{P}_{[2-l-n/p]} \mapsto X^{0,p}_l(R^n) \perp \mathcal{P}_{l-n/p'} \text{ otherwise} \tag{2.10}$$

are isomorphisms and from properties of the spaces $X^{0,p}_l(R^n)$ (cf. [1], [2], [3]). In particular, as far as the spaces X are concerned, it can be proved that

$$X^{0,p}_l(R^n) = W^{0,p}_l(R^n) \text{ if } n/p' \notin \{1,\dots,l\},$$
$$X^{0,p}_l(R^n) = W^{0,p}_l(R^n) \cap W^{-l,p}_0(R^n) \text{ otherwise},$$

and more precisely

$$W^{0,p}_{l,1}(R^n) \subset X^{0,p}_l(R^n) \subset W^{0,p}_l(R^n) \text{ if } n/p' \in \{1,\dots,l\},$$
$$W^{0,p}_{-l}(R^n) \subset X^{0,p}_{-l}(R^n) \subset W^{0,p}_{-l,-1}(R^n) \text{ if } n/p \in \{1,\dots,l\}.$$

ii) Estimates (2.6)–(2.8) are a consequence of the isomorphisms

$$\Delta : W^{2,p}_{-l}(R^n)/\mathcal{P}^{\Delta}_{[l+2-n/p]} \mapsto W^{0,p}_{-l}(R^n) \text{ if } n/p \notin \{1,\dots,l\},$$
$$\Delta : X^{2,p}_{-l}(R^n)/\mathcal{P}^{\Delta}_{l+2-n/p} \mapsto X^{0,p}_{-l}(R^n) \text{ otherwise}.$$

The 'only if' parts are essentially related to polynomial properties. For instance, if (2.6) were valid for $n/p \le l$, by density, it would still hold for all u in $W^{2,p}_{-l}(R^n)$ and in particular for all harmonic polynomials of $W^{2,p}_{-l}(R^n)$, which is obviously incorrect. □

In the same spirit, recall a result due to Nirenberg-Walker [11], valid for all u in $W^{2,p}(R^n)$ with u in $W^{0,p}_{l-2}(R^n)$ and Δu in $W^{0,p}_l(R^n)$, where l is an integer:

$$\|u\|_{W^{2,p}_l(R^n)} \le C(\|\Delta u\|_{W^{0,p}_l(R^n)} + \|u\|_{W^{0,p}_{l-2}(R^n)}) \text{ if } n/p + l \notin \{1,2\}, \tag{2.11}$$

$$\sum_{0 \le |\lambda| \le 2} \|\rho^{l-2+|\lambda|} D^\lambda u\|_{L^p(R^n)} \le C(\|\Delta u\|_{W^{0,p}_l(R^n)} + \|u\|_{W^{0,p}_{l-2}(R^n)}) \text{ otherwise}. \tag{2.12}$$

The next proposition shows that (2.12) may be improved by the introduction of logarithmic weights.

4

Proposition 2.2. *Let l be a positive integer.*

i) If $n/p' \notin \{1, \ldots, l\}$ and $n/p + l \in \{1, 2\}$, the following equivalence holds:

$$\|u\|_{W_l^{2,p}(R^n)} \simeq \|\Delta u\|_{W_l^{0,p}(R^n)} + \|u\|_{W_{l-2,-1}^{0,p}(R^n)} .$$

ii) If $n/p \notin \{1, \ldots, l\}$ and $n/p - l \in \{1, 2\}$, the following equivalence holds

$$\|u\|_{W_{-l}^{2,p}(R^n)} \simeq \|\Delta u\|_{W_{-l}^{0,p}(R^n)} + \|u\|_{W_{-l-2,-1}^{0,p}(R^n)} .$$

3. Riesz potentials

For any real number α with $0 < \alpha < n$ and any function f defined in R^n, we introduce formally the convolution operation

$$I_\alpha(f) = F_\alpha \star f = (-\Delta)^{-\alpha/2}(f) \quad \text{where} \quad F_\alpha(x) = \frac{1}{\gamma(\alpha)} |x|^{\alpha-n},$$

and

$$\gamma(\alpha) = \frac{\pi^{\frac{n}{2}} 2^\alpha \Gamma(\frac{\alpha}{2})}{\Gamma(\frac{n}{2} - \frac{\alpha}{2})} .$$

Observe that for $n \geq 3$, F_2 is the fundamental solution of the Laplace operator. This suggests to take also for F_2 the fundamental solution of the Laplace operator when $n = 2$, *i.e.*

$$F_2(x) = \frac{1}{2\pi} \log \frac{1}{|x|} .$$

If it exists, the function $I_\alpha(f)$ is called the Riesz potential (or Newtonian potential) of f of order α. We can extend formally the definition of F_2 to the case where f is a distribution of $W_l^{-1,p}(R^n)$:

$$\forall \varphi \in \mathcal{D}(R^n) , \ < F_2 \star f, \varphi > := \ < f, F_2 \star \varphi >_{W_l^{-1,p}(R^n) \times W_{-l}^{1,p'}(R^n)} .$$

Theorem 3.2 will specify under what conditions this last definition makes sense. More generally, this section will derive continuity properties of Riesz potentials I_1 and I_2 in several weighted spaces. Beforehand, recall the Hardy-Littlewood-Soboleff-Thorin Theorem (*cf.* [12] and [13]):

5

Theorem 3.1. *For any real number α in $]0, n[$, the following operators are continuous:*

$$I_\alpha : L^p(R^n) \mapsto L^r(R^n), \;\; for \;\; 1/r = 1/p - \alpha/n \;\; if \; n/p > \alpha,$$
$$I_\alpha : L^p(R^n) \mapsto \dot{C}^{k,\alpha-n/p-k}(R^n), \;\; for \;\; k = [\alpha - n/p] \;\; if \; n/p < \alpha,$$
$$I_\alpha : L^p(R^n) \mapsto B.M.O \;\; if \; n/p = \alpha.$$

Remark. The first statement can be found in Peetre (*cf.* [12]) and Stein (*cf.* [13]). The second statement can be found in Peetre (*cf.* [12]). The third one can be found in Stein and Weiss (*cf.* [14]), where it is expressed in its dual form with the help of the Hardy space $\mathcal{H}^1$.

Recall the definition of the space B.M.O. (*cf.* [8]) (standing for " bounded mean oscillation ") : a locally integrable function f belongs to B.M.O. if

$$\|f\|_{\text{B.M.O.}} =: \sup_Q \frac{1}{|Q|} \int_Q |f(x) - f_Q| dx < \infty,$$

where the supremum is taken over all cubes and $f_Q = (1/|Q|) \int_{R^n} f(x)dx$ is the mean of f over Q.

The following theorem establishes some properties of I_2.

Theorem 3.2. *Let l be a positive integer and p be a real number with $1 < p < \infty$ and such that $n/p' > l$: Then the following operators I_2 are continuous:*

$$I_2 : W_l^{-1,p}(R^n) \perp \mathcal{P}_{[l+1-n/p']} \mapsto W_l^{1,p}(R^n), \tag{3.1}$$
$$I_2 : W_{l+1}^{0,p}(R^n) \perp \mathcal{P}_{[l+1-n/p']} \mapsto W_{l-1}^{0,p}(R^n) \;\; if \; n/p' \neq l+1. \tag{3.2}$$

Proof.

i) Let us prove (3.1). It can be readily checked that if φ tends to zero in $\mathcal{D}(R^n)$, then $F_2 \star \varphi$ also tends to zero in $W_{-l}^{1,p'}(R^n)$, so that $I_2(f)$ is indeed a distribution.

1) For any f in $W_l^{-1,p}(R^n) \perp \mathcal{P}_{[l+1-n/p']}$, there exists u in $W_l^{0,p}(R^n)$ such that (*cf.* [3])

$$\text{div } u = f, \;\; \|u\|_{W_l^{0,p}(R^n)} \le C\|f\|_{W_l^{-1,p}(R^n)}.$$

Next, for any φ in $\mathcal{D}(R^n)$, since $F_2 \star \varphi \in W_{-l}^{2,p'}(R^n)$, we have:

$$< \frac{\partial}{\partial x_j} I_2(f), \varphi > = - < I_2(f), \frac{\partial \varphi}{\partial x_j} > = - < \text{div } u, I_2(\frac{\partial \varphi}{\partial x_j}) > = < u_k, \frac{\partial^2}{\partial x_j \partial x_k} I_2(\varphi) > .$$

6

Then, in view of Calderón-Zygmund's inequality (2.6) and owing that $n/p' > l$, we immediately derive

$$\|\nabla(F_2 \star f)\|_{W_l^{0,p}(R^n)} \leq C\|f\|_{W_l^{-1,p}(R^n)}.$$

On the other hand,

$$< (F_2 \star f), \varphi > = - < u, \nabla(F_2 \star \varphi) > .$$

But $F_2 \star \varphi$ also belongs to $W_{-l+1}^{2,p'}(R^n)$ and there are no polynomials in this space. Therefore, it follows from [3] that:

$$\|\nabla(F_2 \star \varphi)\|_{W_{-l}^{0,p'}(R^n)} \leq C\|\nabla^2(F_2 \star \varphi)\|_{W_{-l+1}^{0,p'}(R^n)}$$

and (2.6) yields, as $n/p' > l$,

$$
\begin{aligned}
| < F_2 \star f, \varphi > | &\leq \|u\|_{W_l^{0,p}(R^n)}\|\nabla(F_2 \star \varphi)\|_{W_{-l}^{0,p'}(R^n)} \\
&\leq C\|u\|_{W_l^{0,p}(R^n)}\|\nabla^2(F_2 \star \varphi)\|_{W_{-l+1}^{0,p'}(R^n)} \\
&\leq C\|u\|_{W_l^{0,p}(R^n)}\|\varphi\|_{W_{-l+1}^{0,p'}(R^n)} \leq C\|f\|_{W_l^{-1,p}(R^n)}\|\varphi\|_{W_{-l+1}^{0,p'}(R^n)},
\end{aligned}
$$

whence the continuity of the mapping (3.1).

2) The continuity of the mapping (3.2) follows immediately from (3.1) and the continuous imbeddings of $W_l^{1,p}(R^n)$ in $W_{l-1}^{0,p}(R^n)$ and $W_{l+1}^{0,p}(R^n)$ in $W_l^{-1,p}(R^n)$. The last one holds when $n/p' \neq l+1$. □

Remark. When $l < n/p' < l + 1$, by applying the preceding argument with f replaced by $f - \Phi < f, 1 >_{W_l^{-1,p}(R^n) \times W_{-l}^{1,p'}(R^n)}$, where Φ is a fixed function of $\mathcal{D}(R^n)$ with mean-value one, we can prove the continuity of the operator:

$$J_2 : W_l^{-1,p}(R^n) \mapsto W_l^{1,p}(R^n), \; J_2(f) = I_2(f) - F_2 \star \Phi < f, 1 >_{W_l^{-1,p}(R^n) \times W_{-l}^{1,p'}(R^n)} .$$

Remark. More generally, for any integer l and any non negative integer m, we can establish the continuity of the operator I_2

$$I_2 : W_l^{-1-m,p}(R^n) \mapsto W_l^{1-m,p}(R^n) \text{ if } n/p' > l+1+m \text{ and } n/p > -l+1-m.$$

In particular, when $m = 0$, the operators I_2

$$I_2 : W_l^{-1,p}(R^n) \mapsto W_l^{1,p}(R^n) \text{ and } I_2 : W_{l+1}^{0,p}(R^n) \mapsto W_{l-1}^{0,p}(R^n)$$

are continuous provided $n/p' > l+1$ and $n/p > -l+1$.

Now, we turn to the operator I_1.

Theorem 3.3. *Let l be a positive integer and p a real number with $1 < p < \infty$ and such that $n/p' > l$. Then the operator I_1 is continuous:*

$$I_1 : W_l^{0,p}(R^n) \mapsto W_l^{1,p}(R^n). \tag{3.3}$$

Proof.

Recall that the operator I_1 is continuous, if $n/p' > l$ (*cf.* for instance [9]):

$$I_1 : W_l^{0,p}(R^n) \mapsto W_{l-1}^{0,p}(R^n), \tag{3.4}$$

and recall the equality (*cf.* [13]):

$$\forall \psi \in \mathcal{D}(R^n), \ R_j(\psi) = -\frac{\partial}{\partial x_j} I_1(\psi). \tag{3.5}$$

As $n/p' > l$, the Riesz transform R_j is a continuous mapping from $W_l^{0,p}(R^n)$ into $W_l^{0,p}(R^n)$ and therefore (3.4) implies that (3.5) still holds for ψ in $W_l^{0,p}(R^n)$:

$$\forall \psi \in W_l^{0,p}(R^n), \ R_j(\psi) = -\frac{\partial}{\partial x_j} I_1(\psi), \tag{3.6}$$

(in fact (3.6) is valid even if $n/p' \leq l$). Then the desired continuity of I_1 follows from (3.4) and the fact that there exists a positive constant C such that

$$\forall f \in W_l^{0,p}(R^n), \ \|\frac{\partial}{\partial x_j} I_1(f)\|_{W_l^{0,p}(R^n)} = \|R_j(f)\|_{W_l^{0,p}(R^n)} \leq C\|f\|_{W_l^{0,p}(R^n)}.$$

$\square$

Theorem 3.4. *Let l be a positive integer and p a real number with $1 < p < \infty$ and such that $n/p' > l$. Then the following operator I_1 is continuous:*

$$I_1 : W_l^{1,p}(R^n) \mapsto W_l^{2,p}(R^n). \tag{3.7}$$

Proof.

By applying the previous theorem with $l - 1$ instead of l, we see that it suffices to prove the estimate

$$\|\frac{\partial^2}{\partial x_j \partial x_k} I_1(f)\|_{W_l^{0,p}(R^n)} \leq C\|f\|_{W_l^{1,p}(R^n)}.$$

We start by showing that I_1 and the derivative operators commute in $W_l^{1,p}(R^n)$. As we know that they commute in $\mathcal{D}(R^n)$, let f_m be a sequence of functions of $\mathcal{D}(R^n)$ that converge to f in $W_l^{1,p}(R^n)$. Since

$$\frac{\partial f_m}{\partial x_j} \longrightarrow \frac{\partial f}{\partial x_j} \text{ dans } W_l^{0,p}(R^n),$$

we infer from Theorem 3.3 that

$$I_1(\frac{\partial f_m}{\partial x_j}) \longrightarrow I_1(\frac{\partial f}{\partial x_j}) \text{ in } W_l^{1,p}(R^n).$$

On the other hand,

$$f_m \longrightarrow f \text{ in } W_{l-1}^{0,p}(R^n),$$

and applying again Theorem 3.3, we have

$$\frac{\partial}{\partial x_j} I_1(f_m) \longrightarrow \frac{\partial}{\partial x_j} I_1(f) \text{ dans } W_{l-1}^{0,p}(R^n);$$

hence they also commute in $W_l^{1,p}(R^n)$.

Now, for any f in $W_l^{1,p}(R^n)$ and φ in $\mathcal{D}(R^n)$, we can write

$$< \frac{\partial^2}{\partial x_j \partial x_k} I_1(f), \varphi > = - < \frac{\partial}{\partial x_k} I_1(f), \frac{\partial \varphi}{\partial x_j} > = - < I_1(\frac{\partial f}{\partial x_k}), \frac{\partial \varphi}{\partial x_j} >$$
$$= - < \frac{\partial f}{\partial x_k}, I_1(\frac{\partial \varphi}{\partial x_j}) > = - < \frac{\partial f}{\partial x_k}, \frac{\partial}{\partial x_j} I_1(\varphi) > .$$

Therefore

$$| < \frac{\partial^2}{\partial x_j \partial x_k} I_1(f), \varphi > | \leq C \|\frac{\partial f}{\partial x_k}\|_{W_l^{0,p}(R^n)} \|\frac{\partial}{\partial x_j} I_1(\varphi)\|_{W_{-l}^{0,p'}(R^n)}$$
$$\leq C \|\frac{\partial f}{\partial x_k}\|_{W_l^{0,p}(R^n)} \|R_j(\varphi)\|_{W_{-l}^{0,p'}(R^n)} \leq C \|f\|_{W_l^{1,p}(R^n)} \|\varphi\|_{W_{-l}^{0,p'}(R^n)},$$

by virtue of (2.3). $\qquad\qquad\qquad\qquad\qquad\qquad\qquad\qquad\qquad\qquad\qquad\qquad\qquad\square$

4. Riesz transforms

The next theorem extends (2.2) and (2.3).

Theorem 4.1. *If $n/p' > l$, the following Riesz transforms are continuous :*

$$R_j : W_l^{1,p}(R^n) \mapsto W_l^{1,p}(R^n). \tag{4.1}$$

Proof.

Since $n/p' > l$, we can apply (3.6) with $l-1$ instead of l, and since $W_l^{1,p}(R^n) \subset W_{l-1}^{0,p}(R^n)$, we have

$$\forall f \in W_l^{1,p}(R^n), \ \ R_j(f) = -\frac{\partial}{\partial x_j} I_1(f). \tag{4.2}$$

The theorem is an easy consequence of (4.2) and the continuity of I_1 proved in Theorem 3.4. $\quad\square$

The next theorem proves that we can suppress the condition $n/p' > l$ if we work in adequate subspaces.

Theorem 4.2. *The following composed Riesz transforms are continuous:*

$$R_j \circ R_k : W_l^{0,p}(R^n) \bot \mathcal{P}_{[l-n/p']} \mapsto W_l^{0,p}(R^n) \ \ if \, n/p' \notin \{1,\dots,l\}, \tag{4.3}$$

$$R_j \circ R_k : X_l^{0,p}(R^n) \bot \mathcal{P}_{l-n/p'} \mapsto W_l^{0,p}(R^n) \ \ if \, n/p' \in \{1,\dots,l\}, \tag{4.4}$$

$$R_j \circ R_k : W_{-l}^{0,p}(R^n) \mapsto W_{-l}^{0,p}(R^n)/\mathcal{P}_{[l-n/p]} \ \ if \, n/p \notin \{1,\dots,l\}. \tag{4.5}$$

Proof.

Properties (4.3) and (4.4) are an immediate consequence of isomorphisms (2.9) and (2.10). Property (4.5) is the dual statement of (4.3). $\quad\square$

5. Application to the solution of the heat equation

The heat equation:

$$u_t - \Delta u = 0, \ \text{in } R^n \times \,]0, \infty[, \tag{5.1}$$

$$u(\cdot, 0) = f, \ \text{in } R^n, \tag{5.2}$$

has a unique solution u obtained by taking the convolution of the heat kernel G with the initial data f :

$$u(\cdot, t) = G(\cdot, t) \star f, \ \text{ where } \ G(x, t) = (4\pi t)^{-\frac{n}{2}} e^{-\frac{|x|^2}{4t}}. \tag{5.3}$$

For any non negative integer l, we set: $K(x) = e^{-|x|^2/4t}$, $K_l(x) = |x|^l K(x)$.

The following lemma is fundamental.

Lemma 5.1. *Let α, β and p be three real numbers with $1 < p < \infty$. Then for any real number q such that $q \geq p$ and for any function v in $W^{0,p}_{\alpha,\beta}(R^n)$, we have for all $t > 0$:*

$$\|K \star v\|_{W^{0,q}_{\alpha,\beta}(R^n)} \leq Ct^{\frac{n}{2}(\frac{1}{p'}+\frac{1}{q})}\|v\|_{W^{0,p}_{\alpha,\beta}(R^n)}, \tag{5.4}$$

$$\|K_l \star v\|_{W^{0,q}_{\alpha,\beta}(R^n)} \leq Ct^{\frac{n}{2}(\frac{1}{n}+\frac{1}{p'}+\frac{1}{q})}\|v\|_{W^{0,p}_{\alpha,\beta}(R^n)}, \tag{5.5}$$

where the constant C is independent of v.

Proof.

For v in $W^{0,p}_{\alpha,\beta}(R^n)$, let us define:

$$w(y) = (1 + |y|^2)^{\frac{\alpha}{2}}(\lg \rho)^\beta v(y),$$

and set

$$Tw(x) = \int_{R^n} K(x-y)(1+|y|^2)^{-\frac{\alpha}{2}}(\lg \rho)^{-\beta}w(y)dy = K \star v(x),$$

where $\lg \rho$ is defined as $\lg r$, but with respect to y: $\lg \rho = \ln(2 + |y|^2)$. Observe that if v belongs to $W^{0,p}_{\alpha,\beta}(R^n)$ then w belongs to $L^p(R^n)$.

We set $\delta = \frac{1}{p} - \frac{1}{q}$ and $\frac{1}{r} = 1 - \delta$. Next, let us write

$$K\rho^{-\alpha}(\lg \rho)^{-\beta}w = K^{\frac{r}{q}}w^{\frac{p}{q}}K^{\frac{r}{p'}}w^{p\delta}\rho^{-\alpha}(\lg \rho)^{-\beta}.$$

Then Hölder's inequality yields

$$|Tw(x)|^q \leq (\int_{R^n} K^r(x-y)|w(y)|^p dy)\times$$

$$\times (\int_{R^n} K^r(x-y)(1+|y|^2)^{-\frac{\alpha p'}{2}}(\lg \rho)^{-p'\beta}dy)^{\frac{q}{p'}}\|w\|^{\delta qp}_{L^p(R^n)}.$$

Next, for $|x|$ sufficiently large, we can show that the integral function

$$I(x) = \int_{R^n} K^r(x-y)(1+|y|^2)^{-\frac{\alpha p'}{2}}(\lg \rho)^{-p'\beta}\,dy$$

satisfies the estimate:

$$I(x) \leq C(1+|x|^2)^{-\frac{\alpha p'}{2}}(\text{Log}\,(2+|x|^2))^{-p'\beta}\int_{R^n} e^{-r|z|^2/4t}\,dz$$

$$\leq C(1+|x|^2)^{-\frac{\alpha p'}{2}}(\text{Log}\,(2+|x|^2))^{-p'\beta}t^{\frac{n}{2}}.$$

Therefore

$$\|Tw\|_{W^{0,q}_{\alpha,\beta}(R^n)} \le Ct^{\frac{n}{2p'}}\|w\|^{\delta p}_{L^p(R^n)}\Big(\int_{R^n}\int_{R^n} K^r(x-y)|w(y)|^p\,dy\,dx\Big)^{\frac{1}{q}}$$

$$\le Ct^{\frac{n}{2p'}}\|w\|^{\delta p}_{L^p(R^n)} t^{\frac{n}{2q}}\|w\|^{\frac{p}{q}}_{L^p(R^n)},$$

thus establishing (5.4). A similar argument yields (5.5). □

Now, for any distribution f in $W^{-1,p}_0(R^n)$, we define the kernel

$$S^{1,p}_0(f) = G(\cdot,t) < f,1 >_{W^{-1,p}_0 \times W^{1,p'}_0} \quad \text{if } n/p' < 1,$$

$$S^{1,p}_0(f) = (G(\cdot,t) \star \Phi) < f,1 >_{W^{-1,p}_0 \times W^{1,p'}_0} \quad \text{if } n/p' = 1,$$

where Φ is again a fixed function of $\mathcal{D}(R^n)$ with mean-value one. In the sequel, q denotes a real number greater than or equal to p.

Theorem 5.2. *There exists a constant C, independent of u, such that, for all $t > 0$ and for all initial distribution f of (5.2) in $W^{-1,p}_0(R^n)$, we have*

$$\|u(\cdot,t) - S^{1,p}_0(f)\|_{L^q(R^n)} \le Ct^{-\frac{n}{2}(\frac{1}{n}+\frac{1}{p}-\frac{1}{q})}\|f\|_{W^{-1,p}_0(R^n)} \quad \text{if } n/p' \le 1, \tag{5.6}$$

$$\|u(\cdot,t)\|_{L^q(R^n)} \le Ct^{-\frac{n}{2}(\frac{1}{n}+\frac{1}{p}-\frac{1}{q})}\|f\|_{W^{-1,p}_0(R^n)} \quad \text{if } n/p' > 1. \tag{5.7}$$

Proof.

For any f in $W^{-1,p}_0(R^n)$, there exists v in $W^{1,p}_0(R^n)$ (*cf.* [1], [2], [3]) such that

$$f = \Delta v + < f,1 > \delta \quad \text{if } n/p' < 1,$$

$$f = \Delta v + < f,1 > \Phi \quad \text{if } n/p' = 1,$$

$$f = \Delta v \quad \text{if } n/p' > 1,$$

and

$$\|v\|_{W^{1,p}_0(R^n)} \le C\|f\|_{W^{-1,p}_0(R^n)}.$$

Consider for example the case where $n/p' \le 1$; we infer from (5.3) and Young's inequality that

$$\|u(\cdot,t) - S^{1,p}_0(f)\|_{L^q(R^n)} \le \|\nabla G\|_{L^r(R^n)}\|\nabla v\|_{L^p(R^n)},$$

12

with $\frac{1}{q} = \frac{1}{p} + \frac{1}{r} - 1$ and $1 \le r \le p'$. The estimate (5.6) follows finally from the fact that

$$||\nabla G||_{L^r(R^n)} \le Ct^{-\frac{n}{2}(1-\frac{1}{r})-\frac{1}{2}} = Ct^{-\frac{n}{2}(\frac{1}{n}+\frac{1}{p}-\frac{1}{q})}.$$

The same argument can be applied to establish (5.7). □

Remark. Choose R sufficiently large for the support of Φ to be contained in the ball of radius R and centered at the origin. All $|x| \ge 2R$ and $|y| \le R$ satisfy $|x - y| \ge \frac{1}{2}|x|$ and in addition

$$\forall |x| \ge R, \ |G(\cdot, t) \star \Phi(x)| \le (4\pi t)^{-\frac{n}{2}} \int_{|x|<R} K(x, y)|\Phi(y)| \, dy \le Ct^{-\frac{n}{2}} e^{-\frac{|x|^2}{16t}}.$$

This shows that $G(\cdot, t) \star \Phi(x)$ has the same asymptotic behaviour in x and t as $G(x, t)$.

Theorem 5.3. *There exists a constant C, independent of u, such that, for any $t > 0$ and for all initial distribution f of (5.2) in $W_0^{-1,p}(R^n)$, we have*

i) if $n/p \ne 1$, then

$$||u(\cdot, t) - S_0^{1,p}(f)||_{W_{-1}^{0,q}(R^n)} \le Ct^{-\frac{n}{2}(\frac{2}{n}+\frac{1}{p}-\frac{1}{q})}||f||_{W_0^{-1,p}(R^n)} \quad if \ n/p' \le 1, \qquad (5.8)$$

$$||u(\cdot, t)||_{W_{-1}^{0,q}(R^n)} \le Ct^{-\frac{n}{2}(\frac{2}{n}+\frac{1}{p}-\frac{1}{q})}||f||_{W_0^{-1,p}(R^n)} \quad if \ n/p' > 1; \qquad (5.9)$$

ii) if $n/p = 1$, then

$$||u(\cdot, t) - S_0^{1,p}(f)||_{W_{-1,-1}^{0,q}(R^n)} \le Ct^{-\frac{n}{2}(\frac{2}{n}+\frac{1}{p}-\frac{1}{q})}||f||_{W_0^{-1,p}(R^n)} \quad if \ n/p' = 1, \quad (5.10)$$

$$||u(\cdot, t)||_{W_{-1,-1}^{0,q}(R^n)} \le Ct^{-\frac{n}{2}(\frac{2}{n}+\frac{1}{p}-\frac{1}{q})}||f||_{W_0^{-1,p}(R^n)} \quad if \ n/p' > 1. \qquad (5.11)$$

Proof.

   i) Let us prove (5.8), the proof for (5.9) being essentially the same. Let v be associated with f as in the proof of Theorem 5.2. We can write

$$u(x, t) - S_0^{1,p}(f) = \Delta G(\cdot, t) \star v(x),$$

and observe that since $n/p \ne 1$, $W_0^{1,p}(R^n) \subset W_{-1}^{0,p}(R^n)$. Then, as

$$\Delta G = \frac{1}{t}(-\frac{n}{2} + \frac{|x|^2}{4t})G,$$

13

the estimates (5.8) and (5.9) follow from

$$\|u(\cdot,t) - S_0^{1,p}(f)\|_{W_{-1}^{0,q}(R^n)} \le Ct^{-\frac{n}{2}-2}(t\|K \star v\|_{W_{-1}^{0,q}(R^n)} + \|K_2 \star v\|_{W_{-1}^{0,q}(R^n)}),$$

and Lemma 5.1.

ii) The proof of (5.10) and (5.11) is much the same, but observe that if $p = n$, then v belongs only to $W_{-1,-1}^{0,p}(R^n)$, thus explaining the difference between (5.8) and (5.9) on one hand and (5.10) and (5.11) on the other hand. □

Theorem 5.4. *For all $t > 0$ and for all initial data f of (5.2) in $W_1^{0,p}(R^n)$, we have*

$$\|u(\cdot,t)\|_{W_1^{0,q}(R^n)} \le Ct^{-\frac{n}{2}(\frac{1}{p}-\frac{1}{q})}\|f\|_{W_1^{0,p}(R^n)}, \tag{5.12}$$

$$|u(\cdot,t)|_{W_1^{2,q}(R^n)} \le Ct^{-\frac{n}{2}(\frac{2}{n}+\frac{1}{p}-\frac{1}{q})}\|f\|_{W_1^{0,p}(R^n)}. \tag{5.13}$$

Proof.

The estimate (5.12) is an easy consequence of Lemma 5.1 and the equality

$$u(x,t) = G(\cdot,t) \star f(x) = (4\pi t)^{-\frac{n}{2}} K(\cdot,t) \star f(x).$$

For the second bound, it suffices to observe that

$$\frac{\partial^2 u}{\partial x_i \partial x_j} = \frac{\partial^2 G}{\partial x_i \partial x_j} \star f = \frac{1}{2t}[(-\delta_{ij} + \frac{x_i x_j}{2t})G] \star f,$$

where δ_{ij} is Kroneker's symbol. □

References

[1] Amrouche, C. , Girault, V. and Giroire, J. Espaces de Sobolev avec poids et équation de Laplace dans R^n, Partie I, *C. R. Acad. Sci. Paris*, série I, 315 (1992), 269–274.

[2] Amrouche, C. , Girault, V. and Giroire, J. Espaces de Sobolev avec poids et équation de Laplace dans R^n, Partie II, *C. R. Acad. Sci. Paris*, série I, 315 (1992), 889–894.

[3] Amrouche, C. , Girault, V. and Giroire, J. Weighted Sobolev Spaces for the Laplace equation in R^n, to appear in *J. Math. Pures et Appl.*

[4] Calderón, A.P. Intermediate spaces and interpolation, the complex method, *Studia Math.*, 24 (1964), 113–190.

[5] Calderón, A.P. and Zygmund, A. Singular integral operators and differential equations, *Amer. J. Math.*, 79 (1957), 901-921.

[6] Duoandikoetxea, J. and Zuazua, E. Moments, masses de Dirac et décomposition de fonctions, *C. R. Acad. Sci. Paris* série I, 315 (1992) , 693–698.

[7] Garcia-Cuerva, J. and Rubio de Francia, J.L. Weighted norm inequalities and related topics, *North-Holland* Amsterdam, (1985).

[8] John, F., and Nirenberg, L. On functions of bounded mean oscillation, *Comm. Pure Appl. Math.*, 14 (1961), 415–426.

[9] Lockhart, R.B. Fredholm properties of a class of elliptic operators on non-compact manifolds, *Duke Math. J.*, 48, 1 (1981), 289–312.

[10] Muckenhoupt, B. Weighted norm inequalities for the Hardy maximal function, *Trans. Amer. Math. Soc.*, 165 (1972), 207–226.

[11] Nirenberg, L. and Walker, H. The null spaces of elliptic partial differential operators in R^n, *J. Math. Anal. Appl.*, 42 (1973), 271–301.

[12] Peetre, J. Espaces d'interpolation et théorème de Soboleff, *Ann. Inst. Fourier, Grenoble*, 16 (1966), 276–317.

[13] Stein, E.M. Singular Integrals and Differentiability Properties of Functions, *Princeton University Press*, Princeton, New Jersey, (1970).

[14] Stein, E.M., and Weiss, G. On the theory of harmonic functions of several variables. I. The Theory of H^p spaces, *Acta Math.*, 103 (1960), 25–62.

Chérif AMROUCHE
Centre Benjamin Franklin
Université de Technologie de Compiègne
60206 Compiègne Cedex, France

Vivette GIRAULT
Laboratoire d'Analyse Numérique
Université Pierre et Marie Curie
75252 Paris Cedex 05, France

Jean GIROIRE
Département de Mathématiques,
Université de Nantes
44072 Nantes Cedex 03, France

N ANDRE

Nonuniqueness results for quasilinear elliptic systems

1 Introduction

In this note we would like to consider systems of the type

$$\begin{cases} -\big(a_i(u_1,...,u_n)u_i'\big)' = f_i & \text{in} \quad I = (a,b) \quad \forall i = 1,2,.,n, \\ u_i \in H^1((a,b)) \quad \forall i = 1,2,.,n, \\ u_i(a) = A_i \quad \text{et} \quad u_i(b) = B_i \quad \forall i = 1,2,.,n, \end{cases} \tag{1}$$

where f_i are in $H^{-1}((a,b))$ and a_i are continuous functions verfiying for some constants α, β

$$0 < \alpha \le a_i \le \beta \quad \forall i = 1,...,n. \tag{2}$$

We will set $u = (u_1....u_n)$.

Some of these systems were studied in [C.F.M.] and [G.].

First we will show existence of a solution; then we will study the question of uniqueness. Except for the case where $f_i = 0 \quad \forall i$, uniqueness fails in general even for smooth a_i's.

Let us first remark that, using the Schauder fixed point theorem [G.T.], we have :

Theorem 1

The system (1) has always at least one weak solution in $\Big(H^1(I) \Big)^n$.

2 Uniqueness and nonuniqueness results.

In this section we would like to show that (1) may have more than one solution even if the a_i's are very regular. Indeed we have :

Theorem 2

If $a_i(u) = A(u) \quad \forall i = 1, ... n$ and if A is a C^∞ function, the system (1) may have more than one solution.

Proof of theorem 2

We construct a counter-example. Let $n = 2$ and $I = (0, 1)$. We show that the system

$$
\begin{cases}
-(A(u_1, u_2)u_i')' = f_i \text{ on } (0, 1), \quad 0 < \alpha \le A(u_1, u_2) \le \beta \\
u_1(0) = a, \quad u_1(1) = b, \quad u_2(0) = a', \quad u_2(1) = b',
\end{cases} \tag{3}
$$

may have more than one solution even if A is in $C^\infty([a, b] \mathrm{x} \mathbf{R})$.

We choose u_1

$$
\begin{cases}
u_1 \in C^\infty([0, 1]) \quad \text{increasing} \\
u_1(0) = a, \quad u_1(1) = b, \quad u_1' > 0 \ (u_1 \text{ is one-to-one from } [0, 1] \text{ onto } [a, b]).
\end{cases} \tag{4}
$$

Next we choose θ such that

$$
\begin{cases}
\theta \in C^\infty([0, 1]) \text{ with values in } [\tau, 1]) \text{ where } \tau > \frac{1}{2}, \\
\theta(0) = \theta(1) = 1, \\
\theta \text{ such that } u_1'(1 - \theta) \text{ has exactly one maximum in } [0, 1].
\end{cases} \tag{5}
$$

(for instance, if u_1' is constant, we can choose θ convex. Note that $u_1'(1 - \theta) \ge 0$.)
Finally, we choose $\psi \in C^\infty([a, b])$ such that

$$
\begin{cases}
\psi : [a, b] \to \mathbf{R}, \\
\psi' \text{ increasing}, \\
\int_0^1 \psi'(u_1(t))u_1'(t)(1 - \theta(t))dt = \int_0^1 u_1'(t)(1 - \theta(t))dt.
\end{cases} \tag{6}
$$

For instance if $a > 0$, we can choose $\psi = \frac{1}{2b}x^2 + hx$ so that $\psi' = \frac{1}{b}x + h$. Then

$$
S : h \to \int_0^1 (\psi'(u_1(t)) - 1).u_1'(t).(1 - \theta(t))dt
$$

is continuous from $[0, 1]$ into R. $S(0) < 0$ and $S(1) > 0$, so for some h, S is equal to 0. We can take ψ corresponding to that h.

We select a' and b' by setting $a' = \psi(a)$ $b' = \psi(b)$.

Now, we define u_2 using the preceeding functions ie, we set

$$u_2 = \psi(u_1). \tag{7}$$

We have $u_2' = \psi'(u_1)u_1'$ and $u_2(0) = a'$, $u_2(1) = b'$.

We are now going to construct v_1 and v_2 such that (u_1, u_2), (v_1, v_2) are both solutions of (3).

Set

$$v_1(t) = a + \int_0^t u_1'(t)\theta(t)dt + ct \tag{8}$$

where (cf.(4))

$$c = b - a - \int_0^1 u_1'(t)\theta(t)dt = \int_0^1 u_1'(t)(1 - \theta(t))dt > 0. \tag{9}$$

Thus $v_1(0) = a$ $v_1(1) = b$ and

$$v_1'(t) = u_1'(t)\theta(t) + c > 0. \tag{10}$$

So v_1 is increasing. Next, set now

$$v_2(t) = a' + \int_0^t u_2'(t)\theta(t)dt + et \tag{11}$$

where

$$e = b' - a' - \int_0^1 u_2'(t)\theta(t)dt = \int_0^1 u_2'(t)(1 - \theta(t))dt.$$

By (7), we have

$$e = \int_0^1 \psi'(u_1(t))u_1'(t)(1 - \theta(t))dt = c. \tag{12}$$

So we have:
$$v_2(0) = a' \quad v_2(1) = b' \text{ et } v_2'(t) = u_2'(t)\theta(t) + c > 0. \tag{13}$$

We now verify that we never have $v_2 = \psi(v_1)$ except at 0 and 1.

We have

$$(v_2(t) - \psi(v_1(t)))' = u_1'(t)\psi'(u_1(t))\theta(t) - \psi'(v_1(t))u_1'(t)\theta(t)$$
$$= (\psi'(u_1(t)) - \psi'(v_1(t))u_1'(t)\theta(t).$$

18

$u_1' > 0$, $\theta > \tau$ and ψ' is increasing then $(v_2 - \psi(v_1))' = 0$ only when $v_1 = u_1$.
Moreover

$$(u_1 - v_1)'(t) = u_1'(t) - u_1'(t)\theta(t) - c$$
$$= u_1'(t)(1 - \theta(t)) - \int_0^1 u_1'(t)(1 - \theta(t))dt.$$

Since $u_1'(1 - \theta)$ admits only one maximum $u_1' - v_1'$ vanishes exactly twice on $[0, 1]$. We know $u_1'(0) - v_1'(0) = -c < 0$, thus $v_2 - \psi(v_1) = 0$ only at 0 and 1.

Set

$$\varphi(z) = v_2(v_1^{-1}(z)) \iff v_2 = \varphi(v_1),$$

it is clear that $\varphi - \psi = 0$ only at a and b and that $\varphi - \psi < 0$ on (a, b).

Using these functions, we are going to define A such that $A(v_1(x), v_2(x)) = 1$ and $A(u_1(x), u_2(x)) = \theta(x)$.

First we set :

$$B(z_1, z_2) = \begin{cases} \theta(u_1^{-1}(z_1)) & \text{if } z_2 = \psi(z_1), \\ \\ 1 & \text{if } z_2 = v_2(v_1^{-1}(z_1)) = \varphi(z_1), \\ \\ (1 - t) + t.\theta(u_1^{-1}(z_1)) = 1 - t(1 - \theta(u_1^{-1}(z_1))) \\ \quad \text{if } z_2 = t\psi(z_1) + (1 - t)\varphi(z_1), \quad t = t(z_1, z_2). \end{cases} \tag{14}$$

B is clearly continuous on $[a, b]\mathbf{xR}$. Then, define

$$D = \{(z_1, z_2) \mid \varphi(z_1) \le z_2 \le \psi(z_1) \}.$$

We have

$$\tau \le B(z_1, z_2) \le 1 \quad \forall (z_1, z_2) \in D.$$

Since B is continuous on $[a, b]\mathbf{xR}$, we can find $\delta > 0$ such that

$$\frac{\tau}{2} \le B(z_1, z_2) \le 2 \quad \forall z = (z_1, z_2), \ dist(z, D) \le \delta,$$

and $\chi \in C^\infty([a, b]\mathbf{xR})$ such that

$$\chi = \begin{cases} 1 & \text{if} \quad dist(z, D) \le \frac{\delta}{3} \\ 0 & \text{if} \quad dist(z, D) \ge \frac{\delta}{2} \\ 0 \le \chi \le 1. \end{cases}$$

Now, we set
$$A = \chi.B + (1 - \chi)$$
We now remark that if z verifies $dist(z, D) \geq \frac{\delta}{2}$ then $A = 1$ and if $dist(z, D) \leq \frac{\delta}{2}$ then
$$\frac{\tau}{2} \leq inf(B, 1) \leq A \leq sup(B, 1) \leq 2.$$
It is clear that $A \in C^{\infty}([a, b]\mathrm{x}\mathbf{R})$ if and only if $B \in C^{\infty}([a, b]\mathrm{x}\mathbf{R})$. Moreover, (u_1, u_2) and (v_1, v_2) are solutions of (3) where $f_i = (u_i'.\theta)'$. Indeed using (13) and (10), we have
$$\begin{cases} A(u_1, u_2).u_i' = u_i'.\theta & \forall i = 1, 2 \\ A(v_1, v_2).v_i' = v_i' = u_i'.\theta + c & \forall i = 1, 2. \end{cases}$$

We can now prove that if u_1, ψ and θ are C^{∞} then B is too.

We kwow (4) that $u_1' > 0$ on $[0, 1]$ then u_1^{-1} is C^{∞} and so are u_2, v_1, and v_2. $v_1' > 0$ on $[0, 1]$, so we know that v_1^{-1} and φ are C^{∞}.

$$B(z_1, z_2) = B(z_1, t.\psi(z_1) + (1 - t).\varphi(z_1)) = 1 + t.(1 - \theta(u_1^{-1}(z_1)))$$

where
$$t = \frac{z_2 - \psi(z_1)}{\varphi(z_1) - \psi(z_1)}. \tag{15}$$

Then we can now prove that B is C^{∞} using its asymptotic expansion near a and b.

We would like now to complete this note to take a look to the case where $f_i = 0 \quad \forall i$. In this case we have first :

Theorem 3

Assume that $a_i(u_1...u_n)$ does not depend on i, i.e. $a_i(u_1...u_n) = A(u)$, then, when the f_i's are identically equal to 0 the problem (1) has only one solution.

Proof of theorem 3

In this case the problem becomes
$$A(u_1...u_n).u_i' = C_i \quad \forall i = 1, 2...n. \tag{16}$$

$A(u) \geq \alpha > 0$ so we have
$$u_i' = \frac{C_i}{A(u)} \tag{17}$$

and by integration

$$B_i - A_i = u_i(b) - u_i(a)$$
$$= C_i. \int_a^b \frac{dx}{A(u)}.$$

There exist $k_1, .., k_n$ not all equal to 0 such that

$$\sum_1^n k_i(B_i - A_i) = 0$$

where the k_i's depend only on the A_i, B_i's. So we have $\sum_1^n k_i.C_i = 0$ and

$$A(u). \sum_1^n k_i.u_i' = 0 \text{ or } \sum_1^n k.u_i' = 0.$$

We can find i such that

$$u_i = \sum_{j \neq i} k_j'(u_j) + c_i.$$

The proof can be terminated by induction, since we ending up to a single quasilinear equation.

In order for the above result to hold we have to suppose that the $a_i(u_1...u_n)$ are all the same. Indeed we have

Theorem 4

If $a_i(u_1...u_n)$ depend on i, (1) may have more than one solution.

Proof of theorem 4:

We construct a counter-example. We choose $n = 2$ and $(a, b) = (0, 1)$. Let (1) be the system :

$$\begin{cases} (a_i(u_1, u_2)u_i')' = 0 & \text{on } (0, 1) \quad 0 < \alpha \leq a_i(u_1, u_2) \leq \beta \\ u_i(0) = A_i \quad u_i(1) = B_i & \text{where} \quad B_i > A_i. \end{cases} \tag{18}$$

Let u_i and v_i $(i = 1, 2)$, be two functions in $C^1([0, 1])$, increasing such that

$$u_i(0) = A_i \quad u_i(1) = B_i \quad u_i'(x) > 0 \ \forall x \in [0, 1].$$

Set

$$u_i'(0) = r_i > 0 \quad u_i'(1) = r_i' > 0 \tag{19}$$

and suppose that v_i $(i = 1, 2)$ are two functions in $C^1([0,1])$, increasing, such that

$$\begin{cases} v_i(0) = A_i \quad v_i(1) = B_i \\ v_i'(x) > 0 \quad \forall x \in [0,1] \\ v_i'(0) = r_i \quad v_i'(1) = r_i' \\ u_1 < v_1 \text{ on } (0,1) \\ u_2 > v_2 \text{ on } (0,1). \end{cases} \tag{20}$$

Clearly, u_1 et v_1 are one-to-one from $[0,1]$ onto $[A_1, B_1]$. On $(0,1)$, $u_1(x) < v_1(x)$, so on (A_1, B_1), we have $u_1^{-1}(t) > v_1^{-1}(t)$. $\forall x \in (A_1, B_1)$ we have

$$v_2(v_1^{-1}) < v_2(u_1^{-1}) < u_2(u_1^{-1}) \tag{21}$$

It is clear that for some $k > 0$, we have $u_i' > k$.

We now construct a_i $(i = 1, 2)$ in using u_i and v_i $(i = 1, 2)$.
If $y \in [v_2(v_1^{-1}(x)), u_2(u_1^{-1}(x))]$ set (for $x \in (A_1, B_1)$)

$$t = t(y) = \frac{y - v_2(v_1^{-1}(x))}{u_2(u_1^{-1}(x)) - v_2(v_1^{-1}(x))} \in [0,1]$$

and

$$a_i(x, y) = \begin{cases} \frac{1}{u_i'(u_1^{-1}(B_1))} = \frac{1}{u_i'(0)} \text{ if } x \leq A_1 \\[2mm] \frac{1}{u_i'(u_1^{-1}(B_1'))} = \frac{1}{u_i'(1)} \text{ if } x \geq B_1 \\[2mm] \frac{1}{u_i'(u_1^{-1}(x))} \text{ if } y \geq u_2(u_1^{-1}(x)) \\[2mm] \frac{1}{v_i'(v_1^{-1}(x))} \text{ if } y \leq v_2(v_1^{-1}(x)) \\[2mm] t.\frac{1}{v_i'(v_1^{-1}(x))} + (1 - t).\frac{1}{u_i'(u_1^{-1}(x))} \\[2mm] \text{if } y \in [v_2(v_1^{-1}(x)), u_2(u_1^{-1}(x))], y = t.v_2(v_1^{-1}(x)) + (1 - t).u_2(u_1^{-1}(x)). \end{cases}$$

The functions a_i are continuous on $(C, D) \times R$ where $A_1 < C < D < B_1$ because u_i', v_i', u_1^{-1} and v_1^{-1} are continuous in x.

We know also that $u'_i(u_1^{-1}(A_1)) = v'_i(v_1^{-1}(A_1)) = u'_i(0)$ and that $u'_i(u_1^{-1}(B_1)) = v'_i(v_1^{-1}(B_1)) = u'(1)$, so the functions a_i are continuous everywhere.

It is clear that

$$a_i(u_1, u_2).u'_i = 1 = a_i(v_1, v_2).v'_i \quad \forall i = 1, 2.$$

Thus (u_1, u_2) et (v_1, v_2) are solutions of (18). This completes the proof.

REFERENCES

[C.F.M.] M. Chipot, A. Feggous and G. Michaille : Monotonicity properties for variational inequalities associated with nonlinear diagonal systems.
Proceedings of the Nancy Meeting, 1988.
Pitman Research notes # 208, P. Bénilan, M. Chipot, L. C. Evans,
M. Pierre Edts., Longman, (1989).

[G.] M. Giaquinta : Multiple Integrals in the Calculus of Variations and Nonlinear Elliptic Systems
Princeton University Press, Princeton, (1983).

[G.T.] D. Gilbarg, N.S. Trudinger : Elliptic Partial Differential Equations of Second Order.
Springer Verlag, Berlin, (1985).

N. André,
Centre d'Analyse Non Linéaire
Université de Metz
Ile du Saulcy 57045 Metz-cedex 01.

S N ANTONTSEV, J I DIAZ AND S I SHMAREV

The support shrinking in solutions of parabolic equations with non-homogeneous absorption terms

1. Introduction

1.1 STATEMENT OF THE PROBLEM. This paper deals with the propagation and vanishing properties of local weak solutions of nonlinear parabolic equations. Let $\Omega \subset \mathbf{R}^N$, $N = 1, 2, \ldots$, be an open connected domain with the smooth boundary $\partial\Omega$, and $T > 0$. We consider the problem

$$\left. \begin{array}{c} \frac{\partial}{\partial t}\left(|u|^{\alpha-1}u\right) = \operatorname{div}\left(\vec{A}(x,t,u,\nabla u)\right) - B(x,t,u) + f(x,t) \\ \text{in} \quad Q = \Omega \times (0,T), \\ u(x,0) = u_0(x) \quad \text{in} \quad \Omega \end{array} \right\} \tag{1}$$

assuming that the functions $\vec{A}$ and B are subject to the following structural conditions: there exist constants $\lambda > 0$ and $p > 1$ such that

$$\forall (x,t,s,\rho) \in \Omega \times \mathbf{R}^+ \times \mathbf{R} \times \mathbf{R}^N \quad M_1|\rho|^p \le (\vec{A}(x,t,s,\rho),\rho) \le M_2|\rho|^p, \tag{2}$$

$$\forall (x,t,s) \in \Omega \times \mathbf{R}^+ \times \mathbf{R} \quad sB(x,t,s) \ge M_3\, a(x,t)\, |s|^{\lambda+1}, \tag{3}$$

with $a(x,t) \ge 0$ a given measurable bounded function satisfying

$$a^{-1} \in L^{(1+\lambda)/(\tilde{\lambda}-\lambda)}(Q), \qquad 0 < \lambda < \tilde{\lambda}. \tag{4}$$

In (2)-(3) M_i, $i = 1, 2, 3$, are positive constants. The additional (and crucial) assumption in all further consideration is:

$$\lambda < \alpha. \tag{5}$$

The right-hand side $f(x,t)$ of equation (1) and the initial data $u_0(x)$ are assumed to satisfy

$$u_0 \in L^{\alpha+1}(\Omega), \quad f \in L^{(1+\lambda)/\lambda}(Q), \quad f\, a^{-1/(1+\tilde{\lambda})}(Q). \tag{6}$$

We are interested in the qualitative properties of solutions of problem (1), understood in the following sense.

Definition 1 *A measurable in Q function $u(x,t)$ is said to be a weak solution of problem (1) if*

a) $u \in L^p\left(0, T; W^{1,p}(\Omega)\right) \cap L^\infty\left(0, T; L^{\alpha+1}(\Omega)\right)$;

b) $\lim_{t \to 0} \|u(x, t) - u_0(x)\|_{L^{\alpha+1}(\Omega)} = 0$;

c) *for any test function* $\zeta(x, t) \in W^{1,\infty}(0, T : W_0^{1,p}(\Omega))$, *vanishing at* $t = T$, *the integral identity holds*

$$\int_Q \left\{ |u|^{\alpha-1} u \zeta_t - (\vec{A}, \nabla \zeta) - B\zeta + f\zeta \right\} dx dt + \int_\Omega |u_0|^{\alpha-1} u_0 \zeta(x, 0) dx = 0. \quad (7)$$

So far, the theory of problems of the type (1) already accounts for a number of existence results. We refer the reader to papers [1, 9, 13, 18] and their references.

The class of equations of (1) includes, in particular, the following model equation

$$v_t = \Delta\left(|v|^{m-1} v\right) - M_3 |v|^{\gamma-1} v + f(x, t). \quad (8)$$

To pass to an equation of the form (1) with the parameters $\alpha = 1/m$, $\lambda = \gamma/m$, $p = 2$ amounts to introduce the new unknown $v := |u|^{\frac{1}{m}} sign\ u$. Assumption (5) holds if, for instance,

$$m \geq 1, \quad \gamma \in (0, 1).$$

In this choice of the exponents of nonlinearity the disturbances originated by data propagate with *finite speed*, (see [17] and references therein). Moreover, it is known [20, 21, 16] that in this range of the parameters the supports of nonnegative weak solutions to equation (8) may shrink as t grows. It is known also, [8, 12], that solutions of the Cauchy problem and the Cauchy-Dirichlet problem for equation (8) may even vanish on some subset of the problem domain Q despite of the fact that u_0 and the boundary data are strictly positive. These properties were derived by means of comparison of solutions of (8) with suitable sub and supersolutions of these problems.

It is to be pointed out here that in our formulation the function $\vec{A}(x, t, s, \rho)$ is not subject to any monotonicity assumptions neither in s nor in ρ. Next, we are not constrained by any special boundary conditions. Lastly, as follows from Definition 1.1 the solutions of problem (1) are not supposed to have a definite sign.

Our purpose is to generalize the referred results and to describe the dynamics of the supports of solutions of equation (1) without having recourse to any comparison method. We propose certain refinement of the energy methods in the literature [10, 24, 25, 5, 6, 7, 4, 3]. This techniques allows us to make certain conclusions about the properties of the supports of local weak solutions to problem (1) which rely only on some assumptions about the properties of initial data or even use only the information on the character of the nonlinearity of the equation in (1).

The results we obtain below may be illustrated by the following simplified description. Let $v(x, t)$ be a local weak solution of the model equation

$$v_t = \Delta_p(|v|^{m-1} v) - |v|^{\gamma-1} v,$$

where $\Delta_p(\cdot)$ denotes the p-Laplace operator given by

$$\Delta_p v \equiv \mathrm{div}\left(|\nabla v|^{p-2}\nabla v\right), \quad p > 1.$$

Then:

(i) if $0 < \gamma < 1$, $m(p-1) \geq 1$ and $v(x,0)$ is flat enough near the boundary of its support, the so-called "waiting time" of v is complete, i.e., $\forall \ t \in [0,T] \quad supp\ v(\cdot,t) \subseteq supp\ v(\cdot,0)$ (we also may say that there is no dilatation of the initial support);

(ii) if γ, m, and p additionally satisfy the relation $m + \gamma \leq p/(p-1)$, we have shrinking of the initial support, i.e., the above inclusion is strict: $supp\ v(\cdot,t) \subset\subset supp\ v(\cdot,0)$ for $t > 0$ small enough;

(iii) under the assumptions of item (ii) on the exponents but without any assumption on the initial datum, a null-set with nonempty interior (or dead core) is formed, i.e., $\exists t^* > 0 : \quad \forall t > t^* \quad \overline{\Omega} \setminus \{supp\ v(\cdot,t)\} \neq \emptyset.$

In order to compare these results and the theorems below, recall that $\alpha = \frac{1}{m}$ and $\lambda = \frac{\gamma}{m}$. We also remark that the above results remain true when the diffusion is linear, i.e., $p = 2$ and $m = 1$, but in the presence of the non-homogeneous strong absorption term: $\gamma \in (0,1)$, $a(x,t)$ is admitted to vanish at some set set of zero measure.

1.2. FORMULATION OF RESULTS. Let us introduce the following notation: given $T > 0$, $t \in [0,T)$, $x_0 \in \Omega$, $\rho \geq 0$, and nonnegative parameters σ and μ,

$$P(t,\rho) \equiv \{(x,s) \in Q : |x - x_0| < \rho(s) \equiv \rho + \sigma(s-t)^\mu, \ s \in (t,T)\} \equiv P(t,\rho;\sigma,\mu).$$

It is clear that the choice of the parameters σ, μ, ρ, T determines the shape of the domains $P(t,\rho)$. We distinguish three cases.

a) $\sigma = 0$, $\mu = 0$, $\rho > 0$; in this case $P(t,\rho)$ is a cylinder $B_\rho(x_0) \times (t,T)$;

b) $\sigma > 0$, $\mu = 1$, $\rho > 0$; $P(0,\rho)$ renders a truncated cone centered in the point $x_0 \in \Omega$ and with the base $B_\rho(x_0) := \{x \in \Omega : |x - x_0| < \rho\}$ on the plane $t = 0$;

c) $\sigma > 0$, $0 < \mu < 1$, $\rho = 0$; then $P(t,0)$ becomes a paraboloid.

To simplify the notation we will omit the arguments of P wherever possible. Treating separately cases a), b), c) we indicate specially which of the parameters are essential and which are not. The domains of the type $P(t,\rho)$ will play the fundamental role in the definition of the local energy functions

$$E(P) := \int_{P(t,\rho)} |\nabla u(x,\tau)|^p dx d\tau, \qquad C(P) := \int_{P(t,\rho)} |u(x,\tau)|^{\lambda+1} dx d\tau,$$

$$C_a(P) := \int_{P(t,\rho)} a(x,t)\,|u(x,\tau)|^{\tilde{\lambda}+1} dx d\tau,$$

$$b(T) := ess \sup_{s \in (t,T)} \int_{|x-x_0| < \rho + \sigma(s-t)^\mu} |u(x,s)|^{\alpha+1} dx,$$

associated to any of local weak solutions of problem (1).

Let us assume that

$$\left\| \frac{1}{a} \right\|_{L^{(1+\lambda)/(\tilde\lambda - \lambda)}(Q)} \le K, \quad K = const, \tag{9}$$

whence

$$C^{(1+\tilde\lambda)/(1+\lambda)} \le C_a \left\| \frac{1}{a} \right\|_{L^{(1+\lambda)/(\tilde\lambda-\lambda)}(Q)} \le K\, C_a. \tag{10}$$

We now pass to the precise statement of our results. The only global information we need will be formulated in terms of *the global energy function*

$$D(u(\cdot,\cdot)) := b(T,\Omega) + \int_Q \left(|\nabla u|^p + a\, |u|^{\tilde\lambda+1} \right) dx\, dt,$$

where

$$b(T,\Omega) := ess \sup_{t \in (0,T)} \int_\Omega |u(x,t)|^{\alpha+1} dx.$$

Our first result referres to the situation when the support of u (an arbitrary local weak solution of (1)) does not display the property of dilatation with respect to the initial support $supp\ u_0$ and the support of the forcing term $supp\ f(\cdot,t)$. Assume that

$$u_0 \equiv 0 \quad \text{in} \quad B_{\rho_0}(x_0) \quad \text{for some } x_0 \in \Omega \text{ and } \rho_0 > 0 \tag{11}$$

$$f \equiv 0 \quad \text{in the cylinder} \quad P = P(0,\rho_0) = P(0,\rho_0 : 0,0) \quad (= B_{\rho_0}(x_0) \times (0,T)). \tag{12}$$

and claim the convergence (near $\rho = \rho_0$) of the auxiliary integral

$$I := \int_{\rho_0+0} (\rho - \rho_0)^\beta \left[\|u_0\|^{\alpha+1}_{L^{\alpha+1}(B_\rho(x_0))} + \left\| f\, a^{-1/(1+\tilde\lambda)} \right\|^{(1+\tilde\lambda)/\tilde\lambda}_{L^{(1+\tilde\lambda)/\lambda}(P(0,\rho))} \right]^{p/(p-1)} d\rho < \infty, \tag{13}$$

where

$$\beta = (1 - \delta\tilde\theta)(1 + \kappa), \quad \delta = -\left(1 + \frac{p-1-\lambda}{p(1+\lambda)} N\right), \quad \tilde\theta = \frac{pN - r(N-1)}{(N+1)p - Nr}, \tag{14}$$

with some

$$\kappa \in \left(0, \frac{p(1+\alpha)}{(p-1-\lambda)(1-\tilde\theta)}\right), \tag{15}$$

Note that condition (13) implies certain restrictions on the vanishing rates of the functions $\|u_0\|_{L^{\alpha+1}(B_\rho(x_0))}$ and $\|f(\cdot,t)\, a^{-1/(1+\tilde\lambda)}\|_{L^{(\lambda+1)/\lambda}(B_\rho(x_0))}$ as $\rho \to \rho_0$.

27

Theorem 1 *Assume* (2), (3), (9) *and*

$$\lambda < \alpha \leq p - 1. \tag{16}$$

Let u_0 and f satisfy (11), (12) *and* (13). *Then there exists positive constants M (depending only on the constants in* (2), (3), ρ_0, $dist(x_0, \partial\Omega)$ *and the difference $\tilde{\lambda} - \lambda$) such that any weak solution of* (1) *with bounded global energy, $D(u) \leq M$, possesses the property*

$$u(x, t) \equiv 0 \quad in \quad B_{\rho_0}(x_0) \times (0, T).$$

Under some additional assumptions on the structural exponents α, λ, p and the function f one may get a stronger result which means that the support of $u(\cdot, t)$ shrinks strictly with respect to the initial support.

Theorem 2 *Assume* (2) $-$ (5), (16), (9) *and let*

$$1 + \lambda \leq \alpha \frac{p}{p-1}. \tag{17}$$

Let u_0 satisfy (11). *Assume*

$$f \equiv 0 \quad in \ the \ truncated \ cone \ P \equiv P(0, \rho_0 : \sigma, 1) \ for \ some \ \sigma > 0 \tag{18}$$

and let (13) *be true. Then there exist positive constants M, $\tilde{\lambda} > \lambda$ and t^* such that each weak solution of problem* (1) *with global energy satisfying the inequality $D(u) \leq M$, possesses the property*

$$u(x, t) \equiv 0 \quad in \ P(0, \rho_0 : \sigma, 1) \bigcap \{t \leq t^*\}.$$

Remark 1 *It is curious to observe that the assertion of Theorem 2 has a local character in the sense that different parts of the boundary of supp u_0 may originate pieces of the boundary of the null-set of $u(x, t)$, which display different shrinking properties. Having a possibility to control the rate of vanishing of u_0 and $f(x, t)$, one may design solutions of problem* (1) *which have prescribed shapes of supports. For the model equation* (8) *this phenomenon is already known as "the heat cristall"* [23, Ch.3, Sec.3].

The last of our main results refers to the case when the initial datum need not vanish, that is, the parameter ρ_0 in the conditions of Theorems 1 and 2 is assummed to be zero. Assuming $f \equiv 0$ we show how the strong absorption term causes the formation of the null-set of the solution.

Theorem 3 *Assume* (2) $-$ (5), (16) $-$ (17), (9). *Let $f \equiv 0$. Then there exist positive constants M, t^*, and $\mu \in (0, 1)$ such that any weak solution of problem* (1) *satisfying the inequality $D(u) \leq M$ possesses the property*

$$u(x, t) \equiv 0 \quad in \ P(t^*, 0 : 1, \mu).$$

2. Differential Inequalities

2.1 FORMULA OF INTEGRATION BY PARTS. It follows from results of [] that for local weak solutions of equation (1) the following formula of integration by parts holds:

$$
\begin{aligned}
i_1 + i_2 + i_3 \;=&: \; \frac{\alpha}{\alpha+1} \int_{P \cap \{t=T\}} |u|^{\alpha+1} dx + \int_P \left(\vec{A}, \nabla u\right) dx d\theta + \int_P u\, B dx d\theta \\
=& \; \int_{\partial_l P} \left(\vec{n}_x, \vec{A}\right) u d\Gamma d\theta + \frac{\alpha}{\alpha+1} \int_{\partial_l P} n_\tau |u|^{\alpha+1} d\Gamma d\theta \\
&+ \; \frac{\alpha}{\alpha+1} \int_{P \cap \{t=0\}} |u_0|^{\alpha+1} dx + \int_P u f dx d\theta \\
:=& \; j_1 + j_2 + j_3 + j_4.
\end{aligned} \tag{1}
$$

Here $d\Gamma$ is the differential form on the hypersurface $\partial_l P \cap \{t = const\}$, $\vec{n}_x$ and n_τ are the components of the unit normal vector to $\partial_l P$, $|\vec{n}_x|^2 + |n_\tau|^2 = 1$.

2.2. THE ENERGY DIFFERENTIAL INEQUALITIES. DOMAINS OF THE TYPE C). Now we derive differential inequalities for the energy function $E + C$ which later on will be utilized for the proofs of Theorems 1-3. We begin with the most complicated case c) where the domain P is a paraboloid determined by the parameters $\mu \in (0,1)$, $\sigma > 0$, and t:

$$
P = P(t) = \{(x, \tau): \; |x - x_0| \equiv \rho(\tau) \le \sigma(\tau - t)^\mu, \; \tau \in (t, T)\}, \; t \in (0, T).
$$

We assume that $f \equiv 0$ and that P does not touch the initial plane $\{t = 0\}$. These assumptions simplify the basic energy equality (1) $i_1 + i_2 + i_3 = j_1 + j_2$.

Let us estimate the first term j_1. It is easy to see that

$$
\vec{n} \equiv (\vec{n}_x, n_\tau) = \frac{1}{\left(\sigma^2 \mu^2 + (\theta - t)^{2(1-\mu)}\right)^{1/2}} \left((\theta - t)^{1-\mu} \vec{e}_x - \mu \sigma \vec{e}_\tau\right)
$$

where $\vec{e}_\tau$ and $\vec{e}_x$ are unit vectors orthogonal to the hyperplane $t = 0$ and the axis t respectively.

Let (ρ, ω), $\rho > 0$, $\omega \in \partial B_1$, be the polar coordinate system in $\mathbf{R}^N$. Given an arbitrary function $F(x, t)$, we use the notation $x = (\rho, \omega)$ and $F(x, t) = \Phi(\rho, \vec{\omega}, t)$. There holds the equality

$$
I(t) := \int_P F(x, \theta) dx d\theta \equiv \int_t^T d\theta \int_0^{\rho(\theta, t)} \rho^{N-1} d\rho \int_{\partial B_1} \Phi(\rho, \vec{\omega}, \theta) |J| d\omega,
$$

where J is the Jacobi matrix and, due to the definition of P, $\rho(\theta, t) = \sigma(\theta - t)^\mu$. It is easy to check that:

$$\frac{dI(t)}{dt} = -\int_0^{\rho(\theta,t)} \rho^{N-1} d\rho \int_{\partial B_1} \Phi(\rho,\vec{\omega},\theta)|J|d\omega \bigg|_{\theta=t}$$

$$+ \int_t^T \rho_t(\theta,t)\rho^{N-1}(\theta,t)d\theta \int_{\partial B_1} \Phi(\rho,\vec{\omega},t)|J|d\omega$$

$$= \int_{\partial_t P} \rho_t F(x,\theta)d\Gamma d\theta. \tag{2}$$

Treating the energy function E as a function of t, with the use of (2), (2), and the Hölder inequality, we have now:

$$|j_1| \leq \left|\int_{\partial_t P} \left(\vec{n}_x, \vec{A}\right) u d\Gamma d\theta\right| \leq M_2 \int_{\partial_t P} |\vec{n}_x||\nabla u|^{p-1}|u|d\Gamma d\theta$$

$$= M_2 \left(-\frac{dE}{dt}\right)^{(p-1)/p} \left(\int_t^T \frac{|\vec{n}_x|^p}{|\rho_t|^{p-1}} \left(\int_{\partial B_{\rho(\theta,t)}} |u|^p d\Gamma\right) d\theta\right)^{1/p}. \tag{3}$$

To estimate the right-hand side of (3) we use the following interpolation inequality: given $v \in W^{1,p}(B_\rho)$ and $\lambda \leq p-1$,

$$\|v\|_{p,S_\rho} \leq L_0 \left(\|\nabla v\|_{p,B_\rho} + \rho^\delta\|v\|_{\lambda+1,B_\rho}\right)^{\tilde{\theta}} \cdot \left(\|v\|_{r,B_\rho}\right)^{1-\tilde{\theta}} \tag{4}$$

with a universal constant $L_0 > 0$ not depending on $v(x)$ and the exponents

$$r \in [1, Np/(N-1)], \quad \tilde{\theta} = \frac{pN - r(N-1)}{(N+r)p - Nr} \in (0,1), \quad \delta = -\left(1 + \frac{p-1-\lambda}{p(1+\lambda)}N\right)$$

(see, e.g., Diaz-Veron [14]). Let us introduce the notation

$$E_*(t,\rho) := \int_{B_\rho} |\nabla u|^p dx, \quad C_*(t,\rho) := \int_{B_\rho} |u|^{\lambda+1} dx,$$

so that

$$E = \int_t^T E_*(\theta, \rho(\theta,t))d\theta, \quad C = \int_t^T C_*(\theta, \rho(\theta,t))d\theta,$$

and make use of the Hölder inequality

$$\left(\int_{B_\rho} |u|^r dx\right)^{1/r} \leq \left(\int_{B_\rho} |u|^{1+\lambda} dx\right)^{(1+\lambda)/qr} \cdot \left(\int_{B_\rho} |u|^{\alpha+1} dx\right)^{(1+\alpha)(q-1)/qr},$$

where

$$q = \frac{\alpha - \lambda}{\alpha - r + 1}, \quad r \in [1+\lambda, 1+\alpha].$$

To estimate the second factor in the right-hand side of (3), we choose r satisfying the inequalities

$$1 < 1 + \lambda < \frac{p(1+\alpha)}{\alpha - \lambda + p} < r < 1 + \alpha < \frac{pN}{N-1}.$$

It is easy to check then

$$p < qr = \frac{(\alpha - \lambda)r}{\alpha + 1 - r}, \qquad \gamma = 1 - \frac{\tilde{\theta}}{p} - \frac{1-\tilde{\theta}}{qr} \in \left(1 - \frac{1}{qr}, 1 - \frac{1}{p}\right),$$

$$\frac{\gamma p}{p-1} \in \left(\frac{qr-1}{qr}\frac{p}{p-1}, 1\right), \qquad (1-\gamma)p \in (0,1). \tag{5}$$

Then, by virtue of (4),

$$\int_{\partial B_\rho} |u|^p d\Gamma \;\leq\; L_0 \left(\int_{B_\rho} |\nabla u|^p dx + \rho^{\delta p}\left(\int_{B_\rho} |u|^{\lambda+1} dx\right)^{p/(\lambda+1)}\right)^{\tilde{\theta}} \left(\int_{B_\rho} |u|^r dx\right)^{p(1-\tilde{\theta})/r}$$

$$\leq\; K \rho^{\delta\tilde{\theta}p}\left(\int_{B_\rho} |\nabla u|^p dx + \int_{B_\rho} |u|^{\lambda+1} dx\right)^{\tilde{\theta}}$$

$$\times \left(\int_{B_\rho} |u|^{\lambda+1} dx\right)^{p(1-\tilde{\theta})/qr} \left(\int_{B_\rho} |u|^{\alpha+1} dx\right)^{p(q-1)(1-\tilde{\theta})/qr}$$

$$\leq\; K\rho^{\delta\tilde{\theta}p}(E_* + C_*)^{\tilde{\theta}+(1-\tilde{\theta})p/qr}\, b^{(q-1)(1-\tilde{\theta})p/qr}, \tag{6}$$

where

$$K \;=\; L_0 \max\left(\rho_0^{-\delta p}, \left(\operatorname*{ess\,sup}_{(t,T)} \int_{B_{\rho(\theta)}} |u|^{\lambda+1} dx\right)^{\frac{p}{\lambda+1}-1}\right)^{\tilde{\theta}}$$

$$\leq\; L_0 \max\left(\rho_0^{-\delta p}, \left(\operatorname{meas} B_{\rho(T)}\right)^{\frac{\alpha-\lambda}{\alpha+1}\left(\frac{p}{\lambda+1}-1\right)}(b(T))^{\frac{\lambda+1}{\alpha+1}\left(\frac{p}{\lambda+1}-1\right)}\right)^{\tilde{\theta}}, \qquad \rho \leq \rho_0.$$

Returning to (3) and applying once again the Hölder inequality, we have from (6)

$$|j_1| \;\leq\; L\left(-\frac{dE}{dt}\right)^{(p-1)/p}\left(\int_t^T \frac{|\vec{n}_x|^p}{|\rho_t|^{p-1}} K\rho^{\delta\tilde{\theta}p}(E_* + C_*)^{p(1-\gamma)}\, b^{(q-1)(1-\tilde{\theta})p/qr} d\tau\right)^{1/p}$$

$$\leq\; L\Lambda(t)\left(-\frac{d(E+C)}{dt}\right)^{(p-1)/p} \cdot b^{(q-1)(1-\tilde{\theta})/qr}\,(E+C)^{1-\gamma} \tag{7}$$

for a suitable positive constant L and the exponent $\mu = 1/[1 - p(1-\gamma)]$.

$$\Lambda(t) := \left(\int_t^T \left(\frac{1}{|\rho_t|^{p-1}}\rho^{\delta\tilde{\theta}p}(\tau)\right)^\mu d\tau\right)^{1/\mu p} < \infty. \tag{8}$$

To satisfy (8) one has to take μ small enough, since the condition of convergence of the integral $\Lambda(t)$ is:

$$(1 - \mu)(2p - 1) + \mu\delta\tilde{\theta}p > -(1 - \tilde{\theta})\left(1 - \frac{p(\alpha - r + 1)}{(\alpha - \lambda)r}\right).$$

So, we have obtained the estimate of the following type:

$$|j_1| \leq L_1 \Lambda(t) D(u)^{(q-1)(1-\tilde{\theta})/qr} (E + C)^{1-\gamma} \left(-\frac{d(E + C)}{dt}\right)^{(p-1)/p} \tag{9}$$

where L_1 is a universal positive constant, $D(u)$ is *the total energy* of the solution under investigation.

Let us estimate j_2. For this purpose we use the interpolation inequality

$$\|v\|_{\alpha+1,\partial B_\rho} \leq L_0 \left(\|\nabla v\|_{p,B_\rho} + \rho^\delta \|v\|_{\lambda+1,B_\rho}\right)^s \cdot \|v\|_{r,B_\rho}^{1-s} \tag{10}$$

with a universal positive constant $L_0 > 0$, the exponent

$$s = \frac{(\alpha + 1)N - r(N - 1)}{(N + r)p - Nr} \cdot \frac{p}{\alpha + 1},$$

and δ from (4), which holds for each $v \in W^{1,p}(B_\rho)$. Now we choose the exponent r in (4) as follows

$$1 + \lambda < r < \frac{\alpha p}{p - 1},$$

whence,

$$\kappa = (1 + \alpha)\left(\frac{s}{p} + \frac{1 - s}{qr}\right) \in (0, 1), \qquad \eta = \kappa + \frac{(q - 1)(1 - s)}{qr}(1 + \alpha) = \kappa + \xi > 1.$$

Similarly to the previous estimate, using (10) we have:

$$\int_{\partial B_\rho} |u|^{\alpha+1} dx \leq L \left(\int_{B_\rho} |\nabla u|^p dx + \int_{B_\rho} |u|^{\lambda+1} dx\right)^{s(\alpha+1)/p}$$

$$\times \left[\left(\int_{B_\rho} |u|^{\lambda+1} dx\right)^{1/qr} \left(\int_{B_\rho} |u|^{\alpha+1} dx\right)^{(q-1)/qr}\right]^{(1-s)(\alpha+1)} K^{s(\alpha+1)/\tilde{\theta}p}.$$

Here K is defined in (6). Using the Hölder inequality and reminding that always $|n_\tau| \leq 1$, we arrive to the inequality

$$|j_2| = \left|\int_t^T |n_\tau| d\tau \int_{\partial B_{\rho(\tau)}} |u|^{\alpha+1} d\Gamma\right|$$

$$\leq L (E + C + b(T, \Omega)) (b(T, \Omega))^{\eta-1} \left(\int_t^T \left(K^{s(\alpha+1)/\tilde{\theta}p}\right)^{1/(1-\kappa)} d\tau\right)^{1-\kappa}. \tag{11}$$

We now turn to estimating the left-hand side of (1). By (2)-(3) we have at once that

$$i_1 + i_2 + i_3 \geq i_1 + M_1 E + M_3 C_a \geq M_4 D^{1-m} (E + C + i_1)^m, \qquad m = \frac{1+\tilde{\lambda}}{1+\lambda} > 1, \quad (12)$$

$M_4 = M_4(M_1, M_3, m)$.

Since the right-hand side of (1) is an increasing function of T, we may always replace i_1 by $\frac{\alpha}{\alpha+1} b(T)$ in the left-hand side of (1). Gathering now (1) with $j_3 = j_4 = 0$ and (11), (12), we get:

$$M_4 D^{1-m} \left(E + C + \frac{\alpha}{1+\alpha} b\right)^m \leq L(E + C + b) b^{\eta-1} \left(\int_t^T K^{s(1+\alpha)/(\tilde{\theta}p(1-\kappa))} d\tau\right)^{1-\kappa}$$

$$+ L_1 \Lambda(t) D^{(q-1)(1-\tilde{\theta})/qr} (E + C) \left(-\frac{d(E+C)}{dt}\right)^{(p-1)/p},$$

Let us now choose $\tilde{\lambda}$ satisfying the inequality

$$1 < m = \frac{1+\tilde{\lambda}}{1+\lambda} \leq \eta(\alpha, \lambda, p), \qquad (13)$$

and assume $T - t$ and $D(u)$ be so small that

$$L(b(T, \Omega))^\kappa \int_t^T \left(K^{s(\alpha+1)/\tilde{\theta}p}\right)^\varepsilon d\tau < \frac{M_4}{2}.$$

The we arrive at the inequality

$$(E+C)^m \leq (E+C+b(T,\Omega))^m \leq L_2 \Lambda(t) D(u)^{(m+1)(q-1)(1-\tilde{\theta})/qr} (E+C)^{1-\gamma} \left(-\frac{d(E+C)}{dt}\right)^{(p-1)/p},$$

whence we get the desired differential inequality for the energy function $Y(t) := E + C$:

$$Y^\nu(t) \leq c(t) (-Y(t))', \qquad \nu = \frac{(m-1+\gamma)}{p-1}, \qquad (14)$$

with

$$c(t) = \left(L_1 (M_*)^{(q-1)(1-\tilde{\theta})/qr} \Lambda(t)\right)^{p/(p-1)}, \qquad L_1 = const > 0$$

for $M_* := D(u)$. Note that $c(t) \to 0$ as $t \to T$. According to (5) $p\gamma/(p-1) < 1$; thus we may take, additionally to (13),

$$\nu \left(\frac{p(m-1)}{p-1} + \frac{\gamma p}{p-1}\right) < 1.$$

2.3. The energy differential inequalities. Domains of the types A)-B).

In these cases the differential inequality for the energy function $E + C$ is derived in same way that in the case c) but with certain simplifications due to the choice of the domain P.

Let us begin with the case b). Let

$$P = \{(x,t) : |x - x_0| < \rho + \sigma\theta, \ \sigma \in (0,T)\}, \quad \rho \geq \rho_0 > 0.$$

The unit outer normal to $\partial_l P$ has the form

$$\vec{n} = \frac{1}{\sqrt{1 + \sigma^2}}(1, -\sigma)$$

and if we treat now the energy function $Y := E + C$ as a function of ρ, we have:

$$
\begin{aligned}
\frac{dY(\rho)}{d\rho} &= \frac{d}{d\rho}\left\{\int_0^T d\theta \int_0^{\rho+\sigma\theta} \tau^{N-1}d\tau \int_{\partial B_1} |J|\left(|\nabla u|^p + |u|^{\lambda+1}\right)\Big|_{x=(\tau,\omega)} d\omega\right\} \\
&= \int_0^T d\theta \int_{\partial B_1}\left\{(\rho+\sigma\theta)^{N-1}|J|\left(|\nabla u|^p + |u|^{\lambda+1}\right)\Big|_{x=(\rho+\sigma\theta,\omega)}\right\} d\omega \\
&= \int_{\partial_l P}\left(|\nabla u|^p + |u|^{\lambda+1}\right) d\Gamma d\theta.
\end{aligned}
\tag{15}
$$

Following the above scheme for estimating the term j_1 in (1) and applying (15), we arrive at the following inequality

$$|j_1| \leq \frac{K}{\sqrt{1+\sigma^2}}\left(\frac{dE}{d\rho}\right)^{(p-1)/p} \rho^{\delta\tilde{\theta}}\left(b(T)\right)^{(q-1)(1-\tilde{\theta})/qr}\left(\int_0^T (E_*) + C_*\right)^{\tilde{\theta}+p(1-\tilde{\theta})/qr} d\theta\right)^{1/p}.$$

Let r be such that $\tilde{\theta} + (1-\tilde{\theta})p/qr = 1$. Such a choice is always possible, since

$$\tilde{\theta} + \frac{(1-\tilde{\theta})p}{qr} = 1 \quad \Leftrightarrow \quad r = \frac{p(1+\alpha)}{p+\alpha-\lambda},$$

and the last equality is compatible with the conditions $p > 1 + \lambda$, $\alpha > \lambda$, and the starting choice of r: $r \in [1 + \lambda, \nu + \alpha]$. The estimate for j_1 then takes the form

$$|j_1| \leq \frac{K\rho^{\delta\tilde{\theta}}}{\sqrt{1+\sigma^2}}\left(\frac{dE}{d\rho}\right)^{(p-1)/p}\left(b(T)\right)^{(q-1)(1-\tilde{\theta})/qr-\varepsilon}\left(E + C\right)^{\varepsilon+1/p}.$$

with an arbitrary $\varepsilon \in \left(0, (q-1)(1-\tilde{\theta})/qr\right)$.

The estimate for j_2 is the same that of the case c). The only difference is that now we need not claim that T is small. The value of the coefficient in the estimate for j_2 is controlled now by the choice of σ, since $n_\tau = -\sigma/\sqrt{1+\sigma^2}$. Due to (11) we have $j = 0$. At last, we estimate j_4 with the help of the Hölder and Young inequality

$$j_4 \leq \tau C + L(\tau) \int_P a^{-1/\tilde{\lambda}} |f|^{(\tilde{\lambda}+1)/\tilde{\lambda}} dx d\theta.$$

Gathering these estimates with (1), (12), we arrive to the inequality

$$Y(\rho) \leq c(\rho) Y^{\varepsilon+1/p}(\rho) (Y'(\rho))^{(p-1)/p} + F(\rho), \quad \rho > \rho_0$$

with the coefficient $c(\rho) = \rho^{\delta\tilde{\theta}} K \left(D(u) \right)^{(q-1)(1-\tilde{\theta})/qr-\varepsilon}$ and the right-hand side term

$$F(\rho) = \frac{\alpha}{\alpha+1} \int_{R_\rho(x_0)} |u_0|^{\alpha+1} dx + L(\tau) \int_P a^{-1/\tilde{\lambda}} |f|^{(\tilde{\lambda}+1)/\tilde{\lambda}} dx d\theta.$$

It is easy to see now that the function

$$Z := Y^{p(1+\varepsilon)/(p-1)}(\rho)$$

satisfies the inequality

$$Z^\gamma(\rho) \leq \frac{p-1}{p(1+\varepsilon)} c^{p/(p-1)} Z'(\rho) + F^{p/(p-1)}(\rho), \qquad \rho > \rho_0, \qquad \gamma = \frac{m}{1+\varepsilon} < 1. \quad (16)$$

In the case a), the desired inequality (16) for the energy function $Z(\rho) := (E + C)^{p(1+\varepsilon)/(p-1)}$ defined on the cylinders $P = \{(x,t): |x - x_0| < \rho, \ t > 0\}$ is a by-product of the previous consideration, since the term j_2 of the right-hand side of (1) vanishes.

3. Analysis of the Differential Inequalities

3.1 THE MAIN LEMMA.

Lemma 1 *Let a function $U(\rho)$ be defined for $\rho \in (\rho_0, R)$, $\rho_0 \geq 0$ and possesses the properties: $0 \leq U(\rho) \leq M = const.$, $U'(\rho) \geq 0$ and*

$$AU^s(\rho) \leq G\rho^{-\delta} U'(\rho) + \varphi(\rho) \quad as \ \rho \in (\rho_0, R) \quad (1)$$

where $R < \infty$, $s \in (0,1)$, A, G, δ are finite positive constants, and $\varphi(\rho)$ is a given function. If the integral

$$i(\rho) := \int_{\rho_0}^\rho \sigma^\delta (\sigma - \rho_0)^{-(1+\delta)/(1-s)} \varphi(\sigma) d\sigma$$

converges and the equation

$$(\rho - \rho_0)^{(1+\delta)/(1-s)} \left\{ \left(\frac{A(1-s)}{G(1+\delta)} \right)^{1/(1-s)} - \frac{1}{G} i(\rho) \right\} = M \quad (2)$$

has a root $\rho_* \in (\rho_0, R)$, then $U(\rho_0) = 0$.

Proof. Let us consider the function

$$z(\rho) = \left(\frac{A(1-s)}{G(1+\delta)}\right)^{1/(1-s)} (\rho - \rho_0)^{(1+\delta)/(1-s)},$$

satisfying the conditions

$$Az^s = G\rho^{-\delta}z' \quad \text{as } \rho \in (\rho_0, R), \quad z(\rho_0) = 0. \tag{3}$$

Introduce the function

$$\Phi(\rho) := \exp\left(-\frac{sA}{G}\int_{\rho_0}^{\rho}\sigma^\delta d\sigma \int_0^1 (\theta U + (1-\theta z))^{s-1}\, d\theta\right)$$

and observe that always

$$U^s - z^s \equiv \int_0^1 (\theta U + (1-\theta z))^{s-1}\, d\theta (U - z).$$

Subtracting now termwise equality (3) from inequality (1) and multiplying the result by the function $\rho^{-\delta}G^{-1}\Phi(\rho)$, we get:

$$\frac{d}{d\rho}\left\{(u-z)\,\Phi\right\} \geq -\frac{\rho^\delta}{B}\Phi\varphi. \tag{4}$$

Integrate inequality (4) over the interval (ρ_0, ρ):

$$U(\rho) \geq z(\rho) + \frac{1}{\Phi(\rho)}U(\rho_0) - \frac{1}{G\Phi(\rho)}\int_{\rho_0}^{\rho}\sigma^\delta\Phi(\sigma)\varphi(\sigma)d\sigma. \tag{5}$$

Let us relax (5), having rewritten it in the form

$$\begin{aligned}
M \;\geq\; & U(\rho_0) - \frac{1}{G}\int_{\rho_0}^{\rho}\sigma^\delta\varphi(\sigma) \\
& \times \exp\left(\frac{sA}{G}\int_\sigma^\rho \tau^\delta d\tau \int_0^1 (\theta U(\tau) + (1-\theta)z(\tau))^{s-1}\, d\theta\right) d\sigma
\end{aligned}$$

and then make use of the following relations:

$$\begin{aligned}
&\exp\left(\frac{sA}{G}\int_\sigma^\rho \tau^\delta d\tau \int_0^1 (\theta U(\tau) + (1-\theta)z(\tau))^{s-1}\, d\theta\right) \\
&\leq \exp\left(\frac{sA}{G}\int_0^1 (1-\theta)^{s-1}d\theta \int_\sigma^\rho \tau^\delta z^{s-1}(\tau)d\tau\right) \\
&= \exp\left(\int_\sigma^\rho \frac{d(z(\tau))}{z(\tau)}\right) = \exp\left(\ln\left(\frac{z(\rho)}{z(\sigma)}\right)\right) = \frac{z(\rho)}{z(\sigma)}.
\end{aligned}$$

36

In the result we have

$$0 \le U(\rho_0) \le M - z(\rho)\left\{ 1 - \frac{1}{G}\int_{\rho_0}^{\rho} \sigma^\delta \frac{\varphi(\sigma)}{z(\sigma)} d\sigma \right\}$$

$$\equiv M - (\rho - \rho_0)^{(1+\delta)/(1-s)}\left\{ \left(\frac{A(1-s)}{G(1+\delta)} \right)^{1/(1-s)} - \frac{i(\rho)}{G} \right\} := F(\rho). \quad (6)$$

Assuming existence of some $\rho_* \in (\rho_0, R)$ such that $F(\rho_*) = 0$, we get: $U(\rho_0) = 0$.
$\square$

3.2. PROOFS OF THEOREMS 1-3.

We begin with the proof of Theorem 1. One has just to verify that the conditions of Lemma 1 are fulfilled. Assume $u_0(x) \equiv 0$ in a ball $B_{\rho_0}(x_0)$ and $f \equiv 0$ in the cylinder $P(0, \rho)$, having this ball as the down-base. Let $R > 0$ be such that $P(0, R) \subset Q$, and the integral I defined in the conditions of Theorem 1 is convergent. Assuming the restrictions on the structural constants listed in the conditions of Theorem 1, we derive for the corresponding energy function inequality (16). By Lemma 1, we see that it is sufficient to point out a treshold value of the total energy $M_*^{p(1+\varepsilon)/(p-1)}$ such that equation (2) would have a solution ρ_*. Recall that in the case of inequality (16) the coefficient G of inequality (1) depends only on structural constants and the energy $M_*^{p(1+\varepsilon)/(p-1)}$, but does not depend on t. So, for the function $F(\rho)$ defined in (6) satisfies $F(\rho) \to -\infty$ as $M \to 0$ for each $\rho \in (\rho_0, R)$ fixed. Further, G is a linear function of the argument $M_*^{p(1+\varepsilon)/(p-1)}$ so that $G \to \infty$ as $M \to \infty$, $G \to 0$ as $M \to 0$. Then from (6), having just compared the orders of M of positive and negative terms of $F(\rho)$, that $F(\rho) > 0$ for large M. This means that $F(R)$, being viewed as a function of M, is always nonnegative for small M, which proves the theorem. $\square$

The proof of Theorem 2 literally repeats the arguments just presented. The only difference is that now one has to add condition (11), needed for the derivation of (16).

For the proof of Theorem 3 we assume that the value of T is taken so as to satisfy $P \subset Q$. Remind that the coefficient $c(t)$ in inequality (14) may be estimated from above by $l := c(0)$. Introduce the function $z(T - t) := Y(t)$. Since it satisfies the inequality

$$z^{\gamma p/(p-1)}(t) \le l z'(t) \quad \text{as } t \in (0, T), \quad z(0) = 0, \quad z(t) \in [0, d(u)],$$

there remains to apply Lemma 1 with $i(\rho) \equiv 0$ to complete the proof of Theorem 3.

References

[1] H.W.ALT, S.LUKHAUS Quasilinear elliptic-parabolic differential equations. *Math.Z.*, 1983, v.183, pp.311-341.

[2] S.N.ANTONTSEV On the localization of solutions of nonlinear degenerate elliptic and parabolic equations. *Dokl. Akad. Nauk. SSSR*, 1981, 260, n.6, pp.1289-1293. (in Russian; English translation in *Soviet Math. Dokl.*, 1981, 24. n.2, pp.420-424).

[3] S.N.ANTONTSEV, J.I.DIAZ, S.I.SHMAREV New applications of energy methods to parabolic free boundary Problems. *Uspekhi Mat. Nauk*, 1991, 46, issue 6 (282), pp. 181-182. (in Russian; English translation: in *Russian Mathematical Surveys*, 1991, 46, n.6, pp.193-194).

[4] S.N.ANTONTSEV, J.I.DIAZ, S.I.SHMAREV New results of the character of localization of solutions for elliptic and parabolic equations. *Birkhauser-Verlag*, International Series of Numerical Mathematics, 1992, 106, pp. 59-66.

[5] S.N.ANTONTSEV, S.I. SHMAREV Localization of solutions of nonlinear parabolic equations with linear sources of general type. *Dinamica Sploshnoi Sredy*, Novosibirsk, 1989, 89, pp. 28-42.

[6] S.N.ANTONTSEV, S.I.SHMAREV Local energy method and vanishing of weak solutions of nonlinear parabolic equations. *Dokl. Akad. Nauk SSSR*, 1991, 318, n.4. pp. 777-781. (in Russian; English translation: in *Soviet Math. Doklady*, 1991, 43).

[7] S.N.ANTONTSEV, S.I.SHMAREV Local energy method and vanishing properties of weak solutions of quasilinear parabolic equations. In *Free boundary problems involving solids*. Proceedings of the Internatinal J.Chadam and H.Rasmussen eds., Pitman Res. Notes in Math. ser., 281, Longman. 1993, pp.2-5.

[8] C.BANDLE, I.STAKGOLD The formation of the dead core in parabolic reaction-diffusion problems. *Trans. Amer. Math. Soc.*, 1984, V.286, pp.275-293.

[9] F.BERNIS Existence results for double nonlinear higher order parabolic equations on unbounded domains. *Math Annalen*, 1988, v.279, pp. 373-394.

[10] M.CHIPOT, T.SIDELIS An upper bound for the waiting time for nonlinear degenerate parabolic equations. *Trans. Amer. Math. Soc.*, 1985, v.288, n.1, pp.423-427.

[11] J.I.DIAZ, J.HERNANDEZ Some results on the existence of free boundaries for parabolic reaction-diffusion systems. In *Trends in theory and practice of nonlinear differential equations*, V.Lakshmikantham ed., Marcel Dekker, 1984, pp. 149-156.

[12] J.I.DIAZ, J.HERNANDEZ Qualitative properties of free boundaries for some nonlinear degenerate parabolic equations. In *Nonlinear parabolic equations: qualitative properties of solutions*, L.Boccardo and A.Tesei eds., Pitman No 149, 1987, pp.85-93.

[13] J.I.DIAZ, DE THÉLIN On a nonlinear parabolic problem arising in some models related to turbulent flows. To appear in *SIAM J. Math. Analysis*.

[14] J.I.DIAZ, L.VERON Local vanishing properties of solutions of elliptic and parabolic quasilinear equations. *Trans. Amer. Math. Soc.*, 1985, 290, n.2, pp. 787-814.

[15] L.C.EVANS, B.KNERR Instantaneous shrinking of the support of nonnegative solutions to certain nonlinear parabolic equations and variational inequalities. *Illinois J. Math.*, 1979, V.23, pp.153-166.

[16] V.A.GALAKTIONOV, J.L.VAZQUEZ Extinction for a quasilinear heat equation with absorption II. A dynamical systems approach, *to appear.*

[17] B.GILDING, R.KERSNER The characterization of reaction-convection-diffusion processes by travelling waves *Faculty of Applied Mathematics, University of Twente,* Memorandum No.1111, 1993.

[18] J.KAČUR On a solution of degenerate elliptic-parabolic systems in Orlicz-Sobolev spaces II. *Math.Z.*, 1990, V.203, p.569-579.

[19] A.S.KALASHNIKOV Some questions of the qualitative theory of nonlinear degenerate second-order parabolic equations. *Russian Math Surveys*, 1987, V.42, n.2, pp.169-222.

[20] R.KERSNER On the behavior of temperature fronts in nonlinear heat-conductive media with absorption. *Vestnik MGU*, ser.1, Mathematics and Mechanics, 1978, n.5, pp.44-51.

[21] R.KERSNER Nonlinear heat conduction with absorption: space localization and extinction in finite time. *SIAM J.Appl. Math.*, 1983, 43, pp. 1274-1285.

[22] I.PALIMSKII Some qualitative properties of solutions of equations of nonlinear heat conductivity with absorption. *Chislennye metody mekhaniki sploshnoi sredy,* Novosibirsk, 1985, 16, n.1, pp.136-145. (in Russian)

[23] A.A.SAMARSKY,V.A.GALAKTIONOV, S.P.KURDUMOV, A.M.MIKHAYLOV *Blow-up in Problems for Quasilinear Parabolic Problems*, Nauka, Moscow, 1987 (in Russian; English translation: Walter de Groyter, Berlin, to appear).

[24] S.I.SHMAREV Local properties of solutions of nonlinear equation of heat conduction with absoption. *Dinamika Sploshnoi Sredy*, Novosibirsk, 1990, 95, pp. 28-42. (in Russian)

[25] S.I.SHMAREV The effect of support shrinking for solutions of a degenerate parabolic equation. *Dinamica sploshnoi Sredy*, 1990, 97, pp.194-200. (in Russian)

S.N.Antontsev: *Lavrentiev Institute of Hydrodynamics, Novosibirsk, RUSSIA and Departamento de Matemáticas, Universidad de Oviedo SPAIN; supported by FICYT (SPAIN)*

J.I.Diaz: *Departamento de Matemática Aplicada Universidad Complutense de Madrid, SPAIN; partially sponsored by the DGICYT (SPAIN), project PB90/0620.*

S.I.Shmarev: *as the first author*

D Z AROV, I P GAVRILYUK AND V L MAKAROV

Representation and approximation of solutions of initial value problems for differential equations in Hilbert space based on the Cayley transform

The Cayley transform

$$T_\gamma = (\gamma I + A)(\gamma I - A)^{-1}, \quad \gamma > 0$$

is frequently used to switch from the study of closed but in general unbounded linear operators with dense domain $\mathfrak{D}(A)$ in some Hilbert space H to that of the bounded operators T_γ [1, 6, 10]. This transform has been used in prediction theory of stationary stochastic processes in order to turn processes with continuous time parameter into such with discrete time [7]. In [2] it was proposed to change passive linear stationary systems with continuous time parameter into equivalent ones with discrete time by the Cayley transform, which preserves the main global properties of the system.

In the present paper we give a new application of the Cayley transform to get a new representation of the solution of initial value problem

$$\dot{x}(t) = Ax(t), \qquad x(0) = x_0 \tag{1}$$

under the assumption that A is the generator of a C_0-semigroup $T(t)$ in the Hilbert space H with the scalar product $\langle \cdot, \cdot \rangle$ and the associated norm $|\langle \cdot \rangle|$ and

$$Re\langle Ax, x \rangle \leq \beta \langle x, x \rangle, \qquad x \in H, \qquad \beta < 0, \tag{2}$$

so that

$$|\langle T(t) \rangle| \leq Me^{\beta t}. \tag{3}$$

By a solution of (1) we mean an H-valued function $x(t)$ such that $x(t)$ is continuous for $t \geq 0$, continuously differentiable and $x(t) \in \mathfrak{D}(A)$ for $t > 0$ and (1) is satisfied.

If A is the infinitesimal generator of a C_0 semigroup $T(t)$, the abstract Cauchy problem for A has a solution, namely $x(t) = T(t)x_0$, for every $x_0 \in \mathfrak{D}(A)$, which is the only solution. If A is the infinitesimal generator of a C_0 semigroup which is not differentiable, then, in general, if $x_0 \notin \mathfrak{D}(A)$, the initial value problem (1) does not have a solution. The function $t \to T(t)x_0$ is then a "generalized solution" which is also called a mild solution. There are many different ways to define generalized solutions of the problem (1). One such way is the following [8, p. 105]: a continuous function x on $[0, \infty)$ is a generalized solution of (1) if there are $x_n \in \mathfrak{D}(A)$ such that $x_n \to x(0)$ as $n \to \infty$ and $T(t)x_n \to x(t)$ uniformly on bounded intervals. It is obvious that the generalized solution thus defined is indepent of the sequence $\{x_n\}$, is unique, and if $x(0) \in \mathfrak{D}(A)$, it gives the solution of (1). With this definition (1) has a generalized solution for every $x_0 \in H$ and this generalized solution is $T(t)x_0$.

Using the Cayley transform we translate the problem (1), the so-called continuous problem, into the discrete initial value problem

$$y_{\gamma,n+1} = T_\gamma y_{\gamma,n}, \qquad n = 0, 1, ..., \qquad y_{\gamma,0} = x_0 \tag{4}$$

which will be also called the discrete problem. Using the discrete semigroup T_γ^n the solution of problem (4) can be given in the form

$$y_{\gamma,n} = T_\gamma^n y_0 \equiv T_\gamma^n x_0, \qquad . n \geq 0. \tag{5}$$

We justify the simple formulas connecting the semigroups $T(t)$ and T_γ^n and the solutions of (1) and (4). These formulas for a bounded operator A were derived in [3], and for an unbounded operator in [5].

We choose a real number $\gamma > 0$ such that

$$\beta < 0 < \gamma. \tag{6}$$

It was shown in [3] that

$$|\langle T_\gamma \rangle| \leq \begin{cases} 1 & \text{if A is unbounded} \\ q_\gamma < 1 & \text{if A is bounded} \end{cases} \tag{7}$$

where

$$q_\gamma = \sqrt{1 - \frac{4\gamma|\beta|}{s^2}}, \qquad s = |\langle \gamma I - A \rangle|. $$

We first assume that A is a bounded operator.
If $\mathrm{Re}\ \lambda > \beta$, the resolvent of the operaor A has the following representation:

$$R_\lambda(A) = (\lambda I - A)^{-1} = \int_0^\infty e^{-\lambda t} T(t) dt \tag{8}$$

and for the solution $x(t)$ of the initial value problem (1) the Laplace image is defined by

$$\hat{x}(\lambda) = \int_0^\infty e^{-\lambda t} x(t) dt. \tag{9}$$

Moreover, since $\beta < 0$ and

$$|\langle x(t) \rangle| = |\langle T(t)x_0 \rangle| \le e^{\beta t} |\langle x_0 \rangle| \tag{10}$$

this image can be considered on the imaginary axis, i.e.

$$\hat{x}(i\mu) = \int_0^\infty e^{-i\mu t} x(t) dt \tag{11}$$

and the Fourier reconversion

$$x(t) = \frac{1}{2\pi} \int_{-\infty}^\infty e^{i\mu t} \hat{x}(i\mu) d\mu \tag{12}$$

are defined.

Let us consider the conformal mapping

$$z = \frac{\gamma - \lambda}{\gamma + \lambda} \tag{13}$$

sending the right half-plane ($Re\ \lambda > 0$) onto the unit circle ($|z| < 1$) and set

$$\hat{y}(z) = (I - zT_\gamma)^{-1} x_0. \tag{14}$$

It is easy to verify that

$$\hat{y}(z) = \frac{\gamma + \lambda}{2\gamma} x_0 + \frac{\gamma^2 - \lambda^2}{2\gamma} \hat{x}(\lambda). \tag{15}$$

On the other hand we have for $|z| \le 1$

$$\hat{y}(z) = \sum_{n=0}^\infty z^n T_\gamma^n x_0 = \sum_{n=0}^\infty z^n y_{\gamma,n}$$

$$|\langle \hat{y}(z) \rangle| \le \sum_{n=0}^\infty q_\gamma^n |\langle x_0 \rangle| = \frac{1}{1 - q_\gamma} |\langle x_0 \rangle|. \tag{16}$$

From (15), (16) we obtain the following relation between the function $\hat{x}$ and the sequence $\{y_{\gamma,n}\}$:

$$
\begin{aligned}
\hat{x}(i\mu) &= \frac{2\gamma}{\gamma^2 + \mu^2} \sum_{n=0}^{\infty} \left(\frac{\gamma - i\mu}{\gamma + i\mu}\right)^n y_{\gamma,n} - \frac{1}{\gamma - i\mu} y_{\gamma,0} \\
&= \frac{1}{\gamma + i\mu} y_{\gamma,0} + \frac{2\gamma}{\gamma^2 + \mu^2} \sum_{n=1}^{\infty} (-1)^n \left(\frac{\mu + i\gamma}{\mu - i\gamma}\right)^n y_{\gamma,n}.
\end{aligned}
\tag{17}
$$

By substituting (17) into (12) we get

$$
\begin{aligned}
x(t) &= \frac{1}{2\pi i} \int_{-\infty}^{\infty} e^{i\mu t} \frac{d\mu}{\mu - i\gamma} x_0 \\
&+ \sum_{n=1}^{\infty} (-1)^n \frac{1}{2\pi i} \int_{-\infty}^{\infty} e^{i\mu t} \frac{2\gamma i}{(\mu + i\gamma)(\mu - i\gamma)} \left(\frac{\mu + i\gamma}{\mu - i\gamma}\right)^n d\mu \, y_{\gamma,n}.
\end{aligned}
\tag{18}
$$

Calculating the integrals in (18), we receive the following relation between solutions of the continuous and discrete initial value problems:

$$
x(t) = \sum_{n=0}^{\infty} \varphi_n(2\gamma t) y_{\gamma,n}
\tag{19}
$$

where

$$
\begin{aligned}
\varphi_0(t) &= e^{-\frac{t}{2}}, \\
\varphi_n(t) &= (-1)^{n+1} e^{-\frac{t}{2}} \frac{t}{n} L_{n-1}^{(1)}(t) = (-1)^n e^{-\frac{t}{2}} \left[L_n^{(0)}(t) - L_{n-1}^{(0)}(t) \right],
\end{aligned}
\tag{20}
$$

and

$$
L_n^{(\alpha)}(t) = \sum_{j=0}^{n} \binom{n+\alpha}{n-j} \frac{(-t)^j}{j!}, \qquad L_n^{(\alpha)}(0) = \binom{n-\alpha}{n} \qquad \alpha > -1
\tag{21}
$$

are the Laguerre polynomials. It is important to remark that

$$
\begin{aligned}
|\varphi_0(t)| &= e^{-\frac{t}{2}} \leq 1, \\
|\varphi_n(t)| &= \left| (-1)^n \frac{1}{2\pi i} \int_{-\infty}^{\infty} \frac{2i\gamma}{\mu^2 + \gamma^2} e^{i\mu t/(2\gamma)} \left(\frac{\mu + i\gamma}{\mu - i\gamma}\right)^n d\mu \right| \\
&\leq \frac{\gamma}{\pi} \int_{-\infty}^{\infty} \frac{d\mu}{\mu^2 + \gamma^2} = \frac{\gamma}{\pi} \frac{1}{\gamma} \arctan \frac{\mu}{\gamma} \Big|_{-\infty}^{\infty} = 1, \qquad n \geq 1, \qquad t \in [0, \infty).
\end{aligned}
\tag{22}
$$

From (19) we also have

$$T\left(\frac{t}{2\gamma}\right) = \sum_{n=0}^{\infty} \varphi_n(t) T_\gamma^n. \tag{23}$$

In order to get the inversion formula we note that

$$R_\lambda(A) = (\lambda I - A)^{-1} = \int_0^\infty e^{-\lambda t} T(t)dt,$$

$$(k-1)!(\lambda I - A)^{-k} = \int_0^\infty e^{-\lambda t} t^{k-1} T(t)dt$$

what yields

$$\begin{aligned}
T_\gamma^n &= (\gamma I + A)^n (\gamma I - A)^{-n} \\
&= [2\gamma I - (\gamma I - A)]^n (\gamma I - A)^{-n} \\
&= (-1)^n \sum_{k=0}^n \binom{n}{k} (-2\gamma)^k (\gamma I - A)^{-k} \\
&= (-1)^n \left[I + \sum_{k=1}^n \binom{n}{k} \frac{(-2\gamma)^k}{(k-1)!} \int_0^\infty e^{-\gamma t} t^{k-1} T(t)dt \right] \\
&= (-1)^n \left[I + \int_0^\infty \Psi_n(t) T\left(\frac{t}{2\gamma}\right) dt \right]
\end{aligned} \tag{24}$$

where

$$\Psi_n(t) = e^{-\frac{t}{2}} \frac{d}{dt} L_n^{(0)}(t) = -e^{-\frac{t}{2}} L_{n-1}^{(1)}(t). \tag{25}$$

As an approximation to the solution of the problem (1) one can take truncated sum

$$x^N(t) = \sum_{k=0}^N \varphi_k(2\gamma t) y_{\gamma,k}. \tag{26}$$

It follows from (5), (7), (19) and (22) that

$$|\langle x(t) - x^N(t)\rangle| \leq |\langle x_0\rangle| \sum_{n=N+1}^\infty q_\gamma^n = \frac{q_\gamma^{N+1}}{1-q_\gamma} |\langle x_0\rangle|, \qquad 0 < q_\gamma < 1. \tag{27}$$

We are now in the position to consider the case A is an unbounded selfadjoint operator with dense domain $\mathfrak{D}(A)$ in the separable Hilbert space, $-A$ is positiv. We will show that the representation (19) which one can write as

$$x(t) = e^{-\gamma t} \sum_{k=0}^\infty (-1)^k L_k^{(0)}(2\gamma t)(y_{\gamma,k} + y_{\gamma,k+1}) \tag{28}$$

holds also in this case. The approximate solution is of the form

$$x^N(t) = e^{-\gamma t} \sum_{k=0}^{N} (-1)^k L_k^{(0)}(2\gamma t)(y_{\gamma,k} + y_{\gamma,k+1}). \tag{29}$$

We consider together with (28) the series

$$\tilde{x}(t) = -\gamma e^{-\gamma t} \sum_{k=0}^{\infty} (-1)^k L_k^{(0)}(2\gamma t)(y_{\gamma,k} - y_{\gamma,k+1}) \tag{30}$$

which one obtains by formal differentiation of (29).

We shall use the spectral representation

$$f(-A)x = \int_{\lambda_0}^{\infty} f(\lambda) dE_\lambda x, \qquad 0 < \lambda_0 < -\beta \tag{31}$$

where $f(\lambda)$ is a continuous function and E_λ is the spectral family associated with $-A$. The domain of the operator $f(-A)$ is the set of vectors $x \in H$ such that $f(\lambda)$ is measurable with respect to the measure $d_\lambda \langle E_\lambda x, x \rangle = d_\lambda |\langle E_\lambda x \rangle|^2$ and such that

$$\int_{\lambda_0}^{\infty} |f(\lambda)|^2 d|\langle E_\lambda x \rangle|^2 < \infty. \tag{32}$$

For example, the powers of the operator $-A$ can be defined by

$$(-A)^\sigma x = \int_{\lambda_0}^{\infty} \lambda^\sigma dE_\lambda x, \ \sigma \geq 0.$$

Lemma 1. *Let $x_0 \in \mathfrak{D}^\sigma \equiv \mathfrak{D}((-A)^\sigma)$, $-A$ be a densely defined selfadjoint positive operator. Then*

1) the series (28) converges in H uniformly in $t \geq 0$ and $x(t)$ is continuous on $[0, \infty)$ provided that $\sigma > 0$;

2) the series (30) converges in H uniformly in $t \in [0, \infty)$, $\tilde{x}(t)$ is continuous on $[0, \infty)$ and $\tilde{x}(t) = \dot{x}(t)$ provided that $\sigma > 1$;

3) $x(t) \in \mathfrak{D}(A)$ for all $t \geq 0$, if $\sigma > 1$.

Proof. We have for $0 < \lambda_0 < -\beta$

$$y_{\gamma,k} + y_{\gamma,k+1} = T_\gamma^k(I + T_\gamma)x_0 = 2\gamma \int_{\lambda_0}^{\infty} \left(\frac{\gamma - \lambda}{\gamma + \lambda} \right)^k \frac{1}{\gamma + \lambda} \frac{1}{\lambda^\sigma} dE_\lambda x_0^\sigma \tag{33}$$

$$y_{\gamma,k} - y_{\gamma,k-1} = T_\gamma^k(I - T_\gamma)x_0 = 2\int_{\lambda_0}^\infty \left(\frac{\gamma - \lambda}{\gamma + \lambda}\right)^k \frac{\lambda}{\gamma + \lambda} \frac{1}{\lambda^\sigma} dE_\lambda x_0^\sigma \tag{34}$$

where $x_0^\sigma = (-A)^\sigma x_0$. The function

$$\varphi(\lambda) = \frac{1}{\lambda^{2\rho}} \left(\frac{\gamma - \lambda}{\gamma + \lambda}\right)^{2k} \tag{35}$$

for large k has its maximum at the point $\lambda^* = \frac{\gamma}{\rho}\left(k + \sqrt{k^2 + \rho^2}\right)$ and

$$\varphi_{max} \equiv \varphi(\lambda^*) = \frac{\rho^{2\rho}}{\gamma^{2\rho}(k + \sqrt{k^2 + \rho^2})^{2\rho}} \left(1 - \frac{2\rho}{\rho + k + \sqrt{k^2 + \rho^2}}\right)^{2k},$$

i.e.

$$\varphi(\lambda) \leq \frac{c(\rho)}{k^{2\rho}}, \tag{36}$$

where the constant c is independent of k. Now from (33), (34) we get

$$|\langle y_{\gamma,k} + y_{\gamma,k+1}\rangle|^2 = 4\gamma^2 \int_{\lambda_0}^\infty \left(\frac{\gamma - \lambda}{\gamma + \lambda}\right)^{2k} \frac{1}{(\gamma + \lambda)^2} \frac{1}{\lambda^{2\sigma}} d|\langle E_\lambda x_0^\sigma\rangle|^2$$
$$\leq \frac{c(\sigma)}{k^{2(1+\sigma)}} |\langle x_0^\sigma\rangle|^2 \tag{37}$$

and

$$|\langle y_{\gamma,k} - y_{\gamma,k+1}\rangle|^2 = 4 \int_{\lambda_0}^\infty \left(\frac{\gamma - \lambda}{\gamma + \lambda}\right)^{2k} \left(\frac{\lambda}{\gamma + \lambda}\right)^2 \frac{1}{\lambda^{2\sigma}} d|\langle E_\lambda x_0^\sigma\rangle|^2$$
$$\leq \frac{c(\sigma)}{k^{2\sigma}} |\langle x_0^\sigma\rangle|^2. \tag{38}$$

The estimates (37), (38) and the inequality [4]

$$e^{-\frac{t}{2}} \left| L_k^{(0)}(t)\right| \leq 1 \tag{39}$$

imply that the series (29) and (30) are majorized by the number series

$$c\sum_{k=1}^\infty k^{-(1+\sigma)} \quad \text{and} \quad c\sum_{k=1}^\infty k^{-\sigma}, \quad c = const > 0. \tag{40}$$

Therefore the series (29) converges uniformly in $t \in [0, \infty)$ provided that $\sigma > 0$ and the series (30) converges uniformly in $t \in [0, \infty)$ provided that $\sigma > 1$. That means the continuity of $x(t)$, $\tilde{x}(t)$ and $\tilde{x}(t) = \dot{x}(t)$.

46

If $x_0 \in \mathfrak{D}^\sigma$, $\sigma > 1$, then we have

$$Ax^N(t) = e^{-\gamma t} \sum_{k=0}^{N} (-1)^k L_k^{(0)}(2\gamma t) T_\gamma^k (I + T_\gamma) A x_0. \tag{41}$$

Let us consider the series

$$x_A(t) = e^{-\gamma t} \sum_{k=0}^{\infty} (-1)^k L_k^{(0)}(2\gamma t) T_\gamma^k (I + T_\gamma) A x_0 = \lim_{N \to \infty} A x^N(t). \tag{42}$$

Analogously to (33) we get

$$T_\gamma^k (I + T_\gamma) A x_0 = 2\gamma \int_{\lambda_0}^{\infty} \left(\frac{\gamma - \lambda}{\gamma + \lambda} \right)^k \frac{1}{\gamma + \lambda} \frac{1}{\lambda^{\sigma-1}} dE_\lambda x_0^\sigma$$

and further making use of (36) we have

$$\|T_\gamma^k (I + T_\gamma) A x_0\|^2 = 4\gamma^2 \int_{\lambda_0}^{\infty} \left(\frac{\gamma - \lambda}{\gamma + \lambda} \right)^{2k} \frac{1}{(\gamma + \lambda)^2} \frac{1}{\lambda^{2(\sigma-1)}} d|\langle E_\lambda x_0^\sigma \rangle|^2$$
$$\leq \frac{c(\sigma)}{k^{2\sigma}} |\langle x_0^\sigma \rangle|^2. \tag{43}$$

One can see that the series (42) converges uniformly in $t \in [0, \infty)$ provided that $\sigma > 1$. From (41), (42) and from the closedness of A we get $x(t) = \lim_{N \to \infty} x^N(t) \in \mathfrak{D}(A)$. The proof is complete.

The assumptions of Lemma 1 can be weakened if we consider a finite interval $[\varepsilon, \omega] \subset (0, \infty)$.

Lemma 2. *Let $x_0 \in \mathfrak{D}^\sigma$, $-A$ be a densely defined selfadjoint positive operator and $[\varepsilon, \omega]$ an arbitrary interval, $0 < \varepsilon < \omega < \infty$.*

1) The series (28) converges in H uniformly in $t \in [\varepsilon, \omega]$ and $x(t)$ is continuous on $[\varepsilon, \omega]$ provided that $\sigma \geq 0$;

2) The series (30) converges in H uniformly in $t \in [\varepsilon, \omega]$, $\tilde{x}(t)$ is continuous on $[\varepsilon, \omega]$ and $\tilde{x}(t) = \dot{x}(t)$ provided that $\sigma > \frac{3}{4}$;

3) $x(t) \in \mathfrak{D}(A)$ for all $t \in [\varepsilon, \omega]$, if $\sigma > \frac{3}{4}$.

Proof. The proof is similar to that of Lemma 1 if one takes into account the representation [9, p. 193]

$$L_k^{(\alpha)}(t) = \pi^{-\frac{1}{2}} e^{\frac{t}{2}} t^{-\frac{\alpha}{2} - \frac{1}{4}} k^{\frac{\alpha}{2} - \frac{1}{4}} \left\{ \cos \left[2(kt)^{\frac{1}{2}} - \frac{\alpha \pi}{2} - \frac{\pi}{4} \right] + (nt)^{-\frac{1}{2}} O(1) \right\}, \tag{44}$$
$$\alpha > -1, \quad ck^{-1} \leq t \leq \omega, \quad c = const > 0.$$

It follows from (44) that

$$|e^{-\frac{t}{2}}L_k^{(0)}(t)| \le ck^{-\frac{1}{4}} \tag{45}$$

uniformly in $t \in [\varepsilon, \omega]$. Hence the series (29), (30) and (42) are majorized by the number series

$$c\sum_{k=1}^{\infty} k^{-(\frac{5}{4}+\sigma)} \quad \text{and} \quad c\sum_{k=1}^{\infty} k^{-(\sigma+\frac{1}{4})} \qquad c = const > 0$$

uniformly on $[\varepsilon, \omega]$ from where the statements of the lemma follow.

Theorem 1. *Let the assumptions of Lemma 1 hold, A be the infinitesimal generator of a C_0 semigroup $T(t)$ and $\sigma > \frac{3}{4}$. Then the function $x(t)$ given by (19) is the only solution of the Cauchy problem (1) and the formulas (23), (24) for the continuous semigroup $T(t)$ and the discrete semigroup T_γ^n hold.*

Proof. Taking into account Lemma 1, 2 it is only necessary to show that $x(t)$ satisfies the equation (1). We have

$$\dot{x}(t) - Ax(t) = e^{-\gamma t}\sum_{k=0}^{\infty}(-1)^k L_k^{(0)}(2\gamma t)\int_{\lambda_0}^{\infty}\left(\frac{\gamma-\lambda}{\gamma+\lambda}\right)^k \lambda^{-\sigma}$$

$$\times\left[\gamma - \gamma\frac{\gamma-\lambda}{\gamma+\lambda} - \lambda - \lambda\frac{\gamma-\lambda}{\gamma+\lambda}\right]dE_\lambda x_0^\sigma = 0.$$

The uniqueness follows from Theorem 1.4 in [8, p. 104].

Remark 1. The solution $x(t)$ from Theorem 1 for $\sigma \in (0,1)$ is a generalized solution.

We will now study the approximate solution (29) and the convergence of $x^N(t)$ to the exact solution $x(t)$ as $N \to \infty$.

Theorem 2. *Let $x_0 \in \mathfrak{D}^\sigma$, $\sigma > 0$, $-A$ be a densely defined selfadjoint positive operator. Then*

$$|\langle x^N(t) - x(t)\rangle| \le cN^{-\sigma}|\langle x_0^\sigma\rangle| \tag{46}$$

uniformly in $t \in [0, \infty)$, where $x_0^\sigma = (-A)^\sigma x_0$, c is independent of N, x_0.

Proof. Making use of (37), (38) we get

$$|\langle x^N(t) - x(t)\rangle| \le c(\sigma)|\langle x_0^\sigma\rangle| \sum_{k=N+1}^{\infty} k^{-(1+\sigma)} \le cN^{-\sigma}|\langle x_0^\sigma\rangle| \tag{47}$$

with some constant c which does not depends on N, x_0. The proof is complete.

The estimate (46) can be improved if one regards (19) and (29) at some fixed point $t \in (0, \infty)$ or on some intervall $[\varepsilon, \omega]$, $0 < \varepsilon < \omega < \infty$. The inequalities (37), (45) yield the next statement.

Theorem 3. *Under assumptions of Theorem 2 the estimate*

$$|\langle x^N(t) - x(t) \rangle| \le cN^{-\sigma - \frac{1}{4}} |\langle x_0^\sigma \rangle|$$

holds uniformly on each bounded interval $[\varepsilon, \omega] \subset (0, \infty)$.

References

[1] Achieser, N.I., Glasmann, I.M.: *Theorie der linearen Operatoren im Hilbert-Raum.* Berlin: Akademie-Verlag 1975.

[2] Arov, D.Z.: *Passive linear steady-state dynamical systems.* Sibirian Math. J. **20**(1979) 2, 149 – 162.

[3] Arov, D.Z., Gavrilyuk, I.P.: *A method for solving initial value problems for linear differential equations in Hilbert space based on the Cayley transform.* Numer. Funct. Anal. and Optimiz. **14**(1993)5 & 6, 456 – 473.

[4] Bateman, H., Erdelyi, A.: *Higher transcendental functions,* vol. 2. Mc. Graw – Hill Book Company, Inc. 1953.

[5] Gawrilyuk, I.P., Makarov, V.L.: *The Cauley transform and the solution of an initial value problem for a first order differential equation with an unbounded operator coefficient in Hilbert space.* Numer. Funct. Anal. and Optimiz. **15**(1994) 5 & 6, 583 - 598.

[6] Hille, E., Phillips, R.S.: *Functional Analysis and Semi-Groups.* AMS Colloquium publications, Vol. 31. Providence, Rhode Island 1957.

[7] Krein, M.G.: *On fundamental approximating problem in the extrapolation theory and filtration of stationary stochastic processes.* (Russian). Dokl. Akad. Nauk SSSR **94**(1954)1, 13 – 16.

[8] Pazy, A.: *Semigroups of linear Operators and Applications to Partial Differential Equations.* New York – Berlin – Heidelberg : Springer-Verlag 1983.

[9] Szegö, G.: *Orthogonal polynomials.* New York: AMS 1939.

[10] Sz.-Nagy, B., Foiaś, C.: *Harmonic Analysis of Operators in Hilbert space.* Budapest, 1970.

Authors:

Damir Z. Arov: Department of Mathematics, Pedagogical Institute, 270020 Odessa, Ukraine

Iwan P. Gavrilyuk: Institut für Mathematik, Universität Leipzig, Augustusplatz 10, 04109 Leipzig, Germany

Vladimir L. Makarov: Department of Cybernetics, Kiev University "Shevtshenko", 252127 Kiev, Ukraine

J CARRILLO[1] AND A ALONSO

A unified formulation for the boundary conditions in some convection-diffusion problem

1 Introduction

Let Ω be a bounded domain in $\mathbb{R}^n$ with Lipschitz boundary Γ. Ω represents an homogeneous isotropic porous medium separating several reservoirs. The water flows from the reservoirs through the porous medium Ω generating 2 regions: a wet (and unknown) region Ω_W (where the normalized pressure of the fluid is positive), and a dry region $\Omega_D = \Omega \setminus \Omega_W$ (where the normalized pressure vanishes).

We assume that the flow in Ω has reached a steady state.

The classical formulation of the Dam Problem allows the boundary Γ to be divided into three parts: the bottom of the dam which is considered as an impervious part and therefore the flow vanishes at the bottom; the part of the boundary representing the bottom of the reservoirs and the part of the boundary in contact with air. At the part representing the bottom of the reservoirs different boundary conditions can be considered. At this part the classical formulation assumes that the pressure of the fluid is continuous through the boundary (this means that the boundary is infinitely pervious) and then Dirichlet boundary conditions are prescribed (see [Ba, BKS]). However, Bear (1979) (see [Be]) proposed to consider leaky boundary conditions in which the flow at the bottom of the reservoirs is a monotone increasing function of the difference between the pressure of the reservoir and the pressure of the dam (see [CC]).

The goal of this work is to give a unified formulation of the Steady State Dam Problem in the way of Brézis (1972) (see [B1]). An existence result will be proved for this formulation.

1.1 The motion of the fluid into Ω

The Darcy's law states that

$$v = -k\nabla(p + x_n) \text{ in } \Omega_W,$$

[1]Partially supported by grant No. PB90-0245 from DGICYT

where

$$v = \text{velocity of the fluid}$$
$$p = \text{pressure of the fluid}$$
$$k = \text{coefficient of permeability of the medium (k=1).}$$

Since the fluid is incompressible we have

$$\left. \begin{array}{l} div\, v = 0 \\ \Delta p = 0 \end{array} \right\} \text{ in } \Omega_W.$$

At the free boundary $\partial\Omega_W \cap \Omega$ we assume that

$$\left. \begin{array}{l} p = 0 \\ v.\nu = -\dfrac{\partial}{\partial \nu}(p + x_n) = 0 \end{array} \right\} \text{ on } \partial\Omega_W \cap \Omega.$$

Taking into account the above assumptions, for any $\xi \in \mathcal{D}(\Omega)$ we have

$$\int_{\Omega_W} (\nabla p.\nabla\xi + \xi_{x_n}) = \int_{\Omega_W} \nabla(p + x_n).\nabla\xi$$

$$= \int_{\partial\Omega_W \cap \Omega} \frac{\partial(p + x_n)}{\partial \nu} \xi d\sigma = 0$$

whence, since $p = 0$ in Ω_D for p smooth enough we have

$$\int_\Omega (\nabla p.\nabla\xi + \chi(\Omega_W)\xi_{x_n}) = 0.$$

Thus we get the following weak formulation.

Weak formulation:

We look for a pair (p, χ) satisfying

$$\begin{array}{ll} i) & (p, \chi) \in H^1(\Omega) \times L^\infty(\Omega) \\ ii) & p \geq 0 \text{ a.e. in } \Omega, \\ iii) & 0 \leq \chi \leq 1 \text{ a.e. in } \Omega, \\ iv) & (1 - \chi)p = 0 \text{ a.e. in } \Omega, \end{array}$$

$$v) \left\{ \begin{array}{l} \Delta p + \dfrac{\partial \chi}{\partial x_n} = 0 \text{ in } H^{-1}(\Omega) \\ \Longleftrightarrow \\ \displaystyle\int_\Omega (\nabla p.\nabla\xi + \chi\xi_{x_n}) = 0 \,\, \forall \xi \in H_0^1(\Omega). \end{array} \right.$$

1.2 Boundary conditions

In the classical formulation of the Dam Problem ([Ba, Al, BKS]) as well as in the Dam Problem with leaky boundary conditions ([CC]) we have

$$\Gamma = \Gamma_1 \cup \Gamma_2 \cup \Gamma_3$$
$$\Gamma_i \cap \Gamma_j = \emptyset \text{ for } i \neq j,$$

where

Γ_1 is the impervious bottom of the dam,
Γ_2 is the part of Γ in contact with air,
Γ_3 is the part of Γ representing the bottom of the reservoirs.

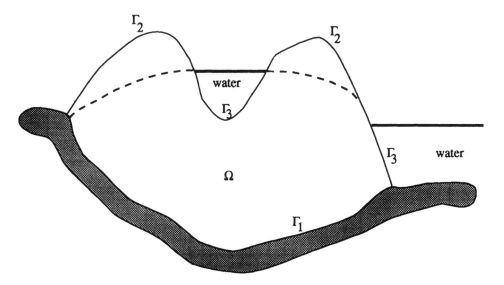

Figure 1: Dam

In the classical formulation the boundary conditions are the following

$$\textbf{DBC} \begin{cases} p = \phi & \text{on } \Gamma_2 \cup \Gamma_3, \\ v.\nu = -\dfrac{\partial p}{\partial \nu} - \chi \, \nu_{x_n} \geq 0 & \text{on } \Gamma_2, \\ v.\nu = -\dfrac{\partial p}{\partial \nu} - \chi \, \nu_{x_n} = 0 & \text{on } \Gamma_1, \end{cases}$$

where $\phi \in \mathcal{C}^{0,1}(\overline{\Omega})$, $\phi \geq 0$, represents the external pressure on $\Gamma_2 \cup \Gamma_3$: $\phi = 0$ on Γ_2 and $\phi > 0$ on Γ_3.

By taking into account this boundary conditions, the classical weak formulation of the Dam Problem is

$$
\textbf{DDP}
\begin{cases}
(p, \chi) \in H^1(\Omega) \times L^\infty(\Omega) \\
p \geq 0 \text{ a.e. in } \Omega, \\
0 \leq \chi \leq 1 \text{ a.e. in } \Omega, \\
(1 - \chi)p = 0 \text{ a.e. in } \Omega, \\
p = \phi \text{ on } \Gamma_2 \cup \Gamma_3, \\
\begin{cases}
\int_\Omega (\nabla p . \nabla \xi + \chi \xi_y) \leq 0 \\
\forall \xi \in H^1(\Omega) \text{ s.t. } \xi = 0 \text{ on } \Gamma_3, \ \xi \geq 0 \text{ on } \Gamma_2.
\end{cases}
\end{cases}
$$

In the Dam Problem with Leaky Boundary Conditions we have

$$
\textbf{LBC}
\begin{cases}
p = \phi = 0 & \text{on } \Gamma_2, \\
v.\nu = -\dfrac{\partial p}{\partial \nu} - \chi \nu_{x_n} \geq 0 & \text{on } \Gamma_2, \\
v.\nu = -\dfrac{\partial p}{\partial \nu} - \chi \nu_{x_n} = -\beta(\phi - p) & \text{on } \Gamma_3, \\
v.\nu = -\dfrac{\partial p}{\partial \nu} + \chi \nu_{x_n} = 0 & \text{on } \Gamma_1,
\end{cases}
$$

where $\phi \in \mathcal{C}^{0,1}(\overline{\Omega})$, $\phi \geq 0$, represents the external pressure on $\Gamma_2 \cup \Gamma_3$: $\phi = 0$ on Γ_2 and $\phi > 0$ on Γ_3; and β is a monotone nondecreasing continuous function.

Then the weak formulation of the Dam Problem with Leaky Boundary Conditions is

$$
\textbf{LDP}
\begin{cases}
(p, \chi) \in H^1(\Omega) \times L^\infty(\Omega) \\
p \geq 0 \text{ a.e. in } \Omega, \\
0 \leq \chi \leq 1 \text{ a.e. in } \Omega, \\
(1 - \chi)p = 0 \text{ a.e. in } \Omega, \\
p = \phi = 0 \text{ on } \Gamma_2, \\
\begin{cases}
\int_\Omega (\nabla p . \nabla \xi + \chi \xi_y) \leq \int_{\Gamma_3} \beta(\phi - p) \xi \, d\sigma \\
\forall \xi \in H^1(\Omega) \text{ s.t. } \xi \geq 0 \text{ on } \Gamma_2.
\end{cases}
\end{cases}
$$

Our goal is to give a formulation of the Dam Problem containing both: **DDP** and **LDP**.

To do this we observe that the boundary conditions **DBC** as well as the boundary conditions **LBC** can be (formally) written as

$$
\frac{\partial p}{\partial \nu} + \chi_{x_n} = \gamma \in B(x, \phi - p) \text{ on } \Gamma,
$$

where, for a.e. $x \in \Gamma$

$$
s \mapsto B(x, s)
$$

is a maximal monotone graph in $\mathbb{R}^2$.

In particular

$$\text{for DBC} \quad \begin{cases} B(x,\cdot) = \begin{cases} \mathbb{R} \times \{0\} \text{ for a.e. } x \in \Gamma_1, \\ \{0\} \times \mathbb{R} \text{ for a.e. } x \in \Gamma_2 \cup \Gamma_3, \end{cases} \\ \gamma \leq 0 \text{ on } \Gamma_2; \end{cases}$$

and

$$\text{for LBC} \quad \begin{cases} B(x,\cdot) = \begin{cases} \mathbb{R} \times \{0\} \text{ for a.e. } x \in \Gamma_1, \\ \beta(\phi - p) \text{ for a.e. } x \in \Gamma_3, \\ \{0\} \times \mathbb{R} \text{ for a.e. } x \in \Gamma_2, \end{cases} \\ \gamma \leq 0 \text{ on } \Gamma_2; \end{cases}$$

2 Statement of the problem

Let us consider B such that:

A_0 for a.e. $x \in \Gamma$ $s \mapsto B(x,s)$
       is a maximal monotone graph of $\mathbb{R}^2$,

A_1 $0 \in B(x,0)$ for a.e. $x \in \Gamma$

A_2 for a.e. $x \in \Gamma$ $D(B(x,\cdot))$ is closed.

Remark 1 *For a.e. $x \in \Gamma$ let $(a(x), b(x)) = Int(D(B(x,\cdot)))$ where*

$$-\infty \leq a(x) \leq 0 \leq b(x) \leq \infty.$$

A_2 means that for a.e. $x \in \Gamma$ there exists a unique pair $B_1(x,\cdot)$ and $B_2(x,\cdot)$ of maximal monotone graphs in $\mathbb{R}^2$ such that

$$\begin{cases} \begin{cases} D(B_2(x,\cdot)) = \mathbb{R} \\ B_2(x,\cdot) = B(x,\cdot) \text{ in } (a(x), b(x)), \\ B_2(x,s) = (s, B_0(x, a(x))) \ \forall s \leq a(x) \\ B_2(x,s) = (s, B_0(x, b(x))) \ \forall s \geq b(x), \end{cases} \\ \begin{cases} D(B_1(x,\cdot)) = D(B(x,\cdot)) \\ B_1(x,\cdot) = 0 \text{ in } Int(D(B(x,\cdot))), \end{cases} \\ B = B_1 + B_2; \end{cases}$$

where B_0 is the minimal section of B (see [B2]).

Moreover we assume

$$A_3 \quad \begin{cases} B_1 \text{ and } B_2 \text{ are Caratheodory Maximal Monotone} \\ \text{Graphs in the following sense} \\ B_i^\varepsilon(x,s) = \dfrac{s - (I + \varepsilon B_i(x,\cdot))^{-1}(s)}{\varepsilon} \\ \text{is a Caratheodory function } \forall i = 1, 2 \ \forall \varepsilon > 0. \end{cases}$$

Remark 2 *In fact B_i^ε is the Yosida approximation of B_i whence it is uniformly Lipschitz continuous with respect to the second variable.*

Moreover

$$
\begin{aligned}
A_4 \quad & \exists \Gamma_0 \subset \Gamma, \ |\Gamma_0| > 0, \\
& \text{s.t. } \forall x \in \Gamma_0, \ 0 \in B(x, s) \Rightarrow s = 0, \\
A_5 \quad & \forall R \ \exists C_R \text{ s.t. } B_2(x, s) \subset (-C_R, C_R), \\
& \forall s \in (-R, R), \text{ for a.e. } x \in \Gamma.
\end{aligned}
$$

Remark 3 *A_4 is an ellipticity assumption.*

Weak formulation of the Dam Problem

Let $\phi \in C^{0,1}(\overline{\Omega})$ such that $\phi \geq 0$ in Ω; and let B satisfy A_0 to A_5. For a.e. $x \in \Gamma$ let

$$
c(x) = \begin{cases} b(x) & \text{if } \phi(x) > 0, \\ +\infty & \text{if } \phi(x) = 0, \end{cases}
$$

then we define $\mathbb{K}$

$$
\mathbb{K} = \{ \xi \in H^1(\Omega) / \, a(x) \leq \phi(x) - \xi(x) \leq c(x) \text{ a.e. in } \Gamma \}
$$

(where $a(x)$ and $b(x)$ have been defined in Remark 1).

Thus the weak formulation of the Dam Problem is: find (p, χ, γ_2) such that

$$
\mathbf{DP} \begin{cases}
i) & (p, \chi, \gamma_2) \in H^1(\Omega) \times L^\infty(\Omega) \times L^2(\Gamma) \\
ii) & \phi(x) - p(x) \in D(B(x, \cdot)) \text{ for a.e. } x \in \Gamma, \\
iii) & p \geq 0 \text{ for a.e. } x \in \Omega, \\
iv) & 0 \leq \chi \leq 1 \text{ for a.e. } x \in \Omega, \\
v) & (1 - \chi)p = 0 \text{ for a.e. } x \in \Omega, \\
vi) & \gamma_2(x) \in B_2(x, \phi(x) - p(x)) \text{ for a.e } x \in \Gamma, \\
vii) & \gamma_2(x) \leq 0 \text{ for a.e. } x \in \Gamma \text{ s.t. } \phi(x) = 0, \\
viii) & \displaystyle\int_\Omega (\nabla p . \nabla (\xi - p) + \chi(\xi - p)_{x_n}) \\
& \geq \displaystyle\int_\Gamma \gamma_2 . (\xi - p) \, d\sigma \ \forall \xi \in \mathbb{K}.
\end{cases}
$$

We check easily that **DDP** and **LDP** are particular cases of **DP**. For instance if we choose B like in **DBC**, we have

$$
D(B(x, \cdot)) = \{0\} \text{ for a.e. } x \in \Gamma_2 \cup \Gamma_3
$$

whence

$$
p = \phi \text{ a.e. on } \Gamma_2 \cup \Gamma_3,
$$

and

$$
B_2(x, \cdot) = \mathbb{R} \times \{0\} \text{ for a.e. } x \in \Gamma_2 \cup \Gamma_3.
$$

Moreover, we have

$$a(x) = \begin{cases} -\infty \text{ for a.e. } x \in \Gamma_1, \\ 0 \text{ for a.e. } x \in \Gamma_2 \cup \Gamma_3, \end{cases}$$

$$b(x) = \begin{cases} +\infty \text{ for a.e. } x \in \Gamma_1, \\ 0 \text{ for a.e. } x \in \Gamma_2 \cup \Gamma_3, \end{cases}$$

$$c(x) = \begin{cases} +\infty \text{ for a.e. } x \in \Gamma_1 \cup \Gamma_2, \\ 0 \text{ for a.e. } x \in \Gamma_3, \end{cases}$$

thus

$$\mathbb{K} = \{\xi \in H^1(\Omega)/ \ xi \leq 0 \text{ on } \Gamma_2, \xi = \phi \text{ on } \Gamma_3.\}$$

Now, since γ_2 vanishes on Γ, the inequality **DP**$viii)$ becomes

$$\int_\Omega (\nabla p.\nabla(\xi - p) + \chi(\xi - p)_{x_n}) \geq 0 \ \forall \xi \in \mathbb{K}$$

or equivalently

$$\int_\Omega (\nabla p.\nabla\zeta + \chi\zeta_{x_n}) \geq 0 \ \forall \zeta \in H^1(\Omega) \text{ s.t. } \zeta = 0 \text{ on } \Gamma_3 \text{ and } \zeta \leq 0 \text{ on } \Gamma_2,$$

then, a solution of **DP** is still a solution of **DDP**.

Now, if we take B like in **LBC** we have

$$D(B(x, \cdot) = \{0\} \text{ for a.e. } x \in \Gamma_2$$

whence

$$p = \phi = 0 \text{ a.e. on } \Gamma_2,$$

and

$$B_2(x, \cdot) = \mathbb{R} \times \{0\} \text{ for a.e. } x \in \Gamma_2;$$

$$D(B(x, \cdot) = \mathbb{R} \text{ for a.e. } x \in \Gamma_3$$

and

$$B_2(x, \cdot) = \{(s, \beta(s))/ s \in \mathbb{R}\} \text{ for a.e. } x \in \Gamma_3,$$

Moreover, we have

$$a(x) = \begin{cases} -\infty \text{ for a.e. } x \in \Gamma_1 \cup \Gamma_3, \\ 0 \text{ for a.e. } x \in \Gamma_2, \end{cases}$$

$$b(x) = \begin{cases} +\infty \text{ for a.e. } x \in \Gamma_1 \cup \Gamma_3, \\ 0 \text{ for a.e. } x \in \Gamma_2, \end{cases}$$

$$c(x) = +\infty \text{ for a.e. } x \in \Gamma;$$

whence

$$\mathbb{K} = \{\xi \in H^1(\Omega)/ \ \xi i \leq 0 \text{ on } \Gamma_2.\}$$

Thus the inequality **DP**$viii$) becomes

$$\int_\Omega (\nabla p.\nabla(\xi - p) + \chi(\xi - p)_{x_n}) \geq \int_{\Gamma_3} \beta(\xi - p)\, d\sigma \ \forall \xi \in \mathbb{K},$$

or, equivalently

$$\int_\Omega (\nabla p.\nabla \zeta + \chi \zeta_{x_n}) \geq \int_{\Gamma_3} \beta \zeta \, d\sigma \ \forall \zeta \in H^1(\Omega) \text{ s.t. } \zeta \leq 0 \text{ on } \Gamma_2.$$

In this case a solution of **DP** is also a solution of **LDP**.

3 A regularized problem

In order to solve the existence of solutions for **DP** we introduce a regularized (or penalized) problem.

For any $\varepsilon > 0$ we look for p_ε such that

$$P_\varepsilon \begin{cases} i) & p_\varepsilon \in H^1(\Omega) \\ ii) & \int_\Omega (\nabla p_\varepsilon.\nabla \xi + H_\varepsilon(p_\varepsilon)\xi_{x_n}) \\ & = \int_\Gamma (B_1^\varepsilon(x, \phi - p_\varepsilon) + B_2^\varepsilon(x, \phi - p_\varepsilon))\xi \, d\sigma \\ & \forall \xi \in H^1(\Omega), \end{cases}$$

where $H_\varepsilon(s) = min(s^+/\varepsilon, 1)$.

First we prove an a priori estimate for the solutions of a family of problems containing P_ε.

Lemma 1 *Let $u \in H^1(\Omega)$ satisfy*

$$\int_\Omega (\nabla u.\nabla \xi + H_\varepsilon(u)\xi_{x_n} + \alpha u\xi)$$

$$= \alpha \int_\Omega f\xi + \int_\Gamma (B_1^\varepsilon(x, \phi - u) + B_2^\varepsilon(x, \phi - u))\xi \, d\sigma$$

for any $\xi \in H^1(\Omega)$, where $\alpha \in \mathbb{R}$, $\alpha \geq 0$ and $f \in L^\infty(\Omega)$, $f \geq 0$. Then u satisfies:

$$0 \leq u(x) \leq H - x_n \text{ a.e. in } \Omega,$$

for some $H \in \mathbb{R}$ large enough.

Proof: We consider H large enough such that

$$H - x_n \geq \varepsilon \text{ in } \Omega,$$
$$\alpha(H - x_n) \geq \alpha f \text{ a.e. in } \Omega,$$
$$H - x_n \geq \phi(x) \text{ on } \Gamma.$$

We use the test function

$$(u - (H - x_n))^+$$

in the above equation. We get

$$0 \leq \int_\Omega |\nabla(u - (H - x_n))^+|^2 \leq$$

$$\leq \int_\Gamma (B_1^\varepsilon(x, \phi - u) + B_2^\varepsilon(x, \phi - u))(u - (H - x_n))^+ \, d\sigma$$

$$\leq \int_\Gamma (B_1^\varepsilon(x, H - x_n - u) + B_2^\varepsilon(x, H - x_n - u))(u - (H - x_n))^+ \, d\sigma \leq 0.$$

Then $(u - (H - x_n))^+ = c$ where c is a nonnegative constant. Then:

$$\int_{\Gamma_0} (B_1^\varepsilon(x, -c) + B_2^\varepsilon(x, -c))c$$

$$= \int_\Gamma (B_1^\varepsilon(x, -c) + B_2^\varepsilon(x, -c))c = 0$$

whence from A_4 we get $c = 0$.

To prove that $u \geq 0$ we use the test function u^- and we take into account the fact that $f \geq 0$.

Theorem 2 *For any $\varepsilon > 0$, (P_ε) has one and only one solution.*

Proof: (we prove only the existence of the the solution, for a complete proof see [AC] .) We take $H \in \mathbb{R}$ such that

$$H - x_n \geq \varepsilon \text{ in } \Omega,$$
$$H - x_n \geq \phi(x) \text{ on } \Gamma.$$

We define the continuous mapping

$$T : L^2(\Omega) \mapsto L^2(\Omega)$$

by: for any $v \in L^2(\Omega)$ $u = Tv$ is the unique solution (see [Li] théorème 2.1, Chap.2) of

$$P_\varepsilon' \begin{cases} i) & u \in H^1(\Omega) \\ ii) & \int_\Omega (\nabla u . \nabla \xi + u\xi) \\ & - \int_\Gamma (B_1^\varepsilon(x, \phi - u) + B_2^\varepsilon(x, \phi - u))\xi \, d\sigma \\ & = \int_\Omega (min(v^+, H - x_n)\xi - H_\varepsilon(v)\xi_{x_n}) \ \forall \xi H^1(\Omega). \end{cases}$$

We deduce easily that T maps $L^2(\Omega)$ into a bounded subset of $H^1(\Omega)$ whence, from the Schauder theorem we deduce the existence of a fixed point, p_ε, of T. From the above lemma we deduce that the fixed point p_ε satisfies

$$0 \leq p_\varepsilon \leq H - x_n$$

whence p_ε is a solution of P_ε.

From lemma 1 and theorem 2 we deduce the following corollary.

Corollary 3 *Let p_ε be the unique solution of P_ε. Then there exists a constant C and a constant H such that*

$$i) \quad \| p_\varepsilon \|_{H^1(\Omega)} \leq C \quad \forall \varepsilon > 0$$
$$ii) \quad 0 \leq p_\varepsilon \leq H - x_n \ a.e. \ in \ \Omega, \ \forall \varepsilon > 0.$$

4 Existence of a solution to DP

Now we can prove the existence result for **DP**.

Theorem 4 *There exists at least one solution to **DP**. Moreover, let $(p_\varepsilon)_{\varepsilon>0}$ be the solutions of $(P_\varepsilon)_{\varepsilon>0}$; then there exists a subsequence of $(p_\varepsilon)_{\varepsilon>0}$, still denoted by $(p_\varepsilon)_{\varepsilon>0}$ such that*

$$i) \quad (p_\varepsilon) \overset{\varepsilon \to 0}{\rightharpoonup} p \ in \ H^1(\Omega) \ weakly,$$
$$ii) \quad (p_\varepsilon) \overset{\varepsilon \to 0}{\to} p \ in \ L^2(\Omega) \ strongly,$$
$$iii) \quad (p_\varepsilon) \overset{\varepsilon \to 0}{\to} p \ in \ L^2(\Gamma) \ strongly,$$
$$iv) \quad H_\varepsilon(p_\varepsilon) \overset{\varepsilon \to 0}{\rightharpoonup} \chi \ in \ L^\infty(\Omega) \ weak\text{-}\star,$$
$$v) \quad B_2^\varepsilon(x, \phi - p_\varepsilon) \overset{\varepsilon \to 0}{\rightharpoonup} \gamma_2 \ in \ L^\infty(\Gamma) \ weak\text{-}\star,$$

*and (p, χ, γ_2) is a solution of **DP**.*

Proof: $i)$ to $v)$ are consequences of the estimates of the above corollary. Obviously we have

$$(p, \chi, \gamma_2) \in H^1(\Omega) \times L^\infty(\Omega) \times L^2(\Gamma)$$

$$p \geq 0 \ \text{for a.e.} \ x \in \Omega,$$

$$0 \leq \chi \leq 1 \ \text{for a.e.} \ x \in \Omega,$$

and

$$\gamma_2(x) \leq 0 \ \text{for a.e.} \ x \in \Gamma \ \text{s.t.} \ \phi(x) = 0.$$

Moreover

$$0 = \int_\Omega (1 - H_e(p_\varepsilon))(p_\varepsilon - \varepsilon)^+ \overset{\varepsilon \to 0}{\to} \int_\Omega (1 - \chi)p,$$

whence

$$(1 - \chi)p = 0 \text{ for a.e. } x \in \Omega.$$

Furthermore

$$\int_\Gamma B_2^\varepsilon(x, \phi - p_\varepsilon)(\phi - p_\varepsilon) \overset{\varepsilon \to 0}{\to} \int_\Gamma \gamma_2(\phi - p)$$

and thus

$$\gamma_2 \in B_2(x, \phi - p).$$

Since

$$B_1^\varepsilon(x, \phi - p_\varepsilon) = \frac{1}{\varepsilon}(min(\phi - p_\varepsilon - a, 0) + max(\phi - p_\varepsilon - b, 0)$$

we have

$$B_1^\varepsilon(x, \phi - p_\varepsilon)(\xi - p_\varepsilon) \geq 0 \ \forall \xi \in I\!\!K,$$

then

$$\int_\Omega (\nabla p_\varepsilon . \nabla(\xi - p_\varepsilon) + H_e(p_\varepsilon)(\xi - p_\varepsilon)_{x_n})$$

$$= \int_\Gamma (B_1^\varepsilon(x, \phi - p_\varepsilon) + B_2^\varepsilon(x, \phi - p_\varepsilon))(\xi - p_\varepsilon) \, d\sigma$$

$$\geq \int_\Gamma B_2^\varepsilon(x, \phi - p_\varepsilon))(\xi - p_\varepsilon) \, d\sigma \ \ \forall \xi \in I\!\!K.$$

By taking into account A_5 we deduce

$$C \geq \frac{1}{\varepsilon}\int_\Gamma (min(\phi - p_\varepsilon - a, 0) + max(\phi - p_\varepsilon - b, 0))(\xi - p_\varepsilon) \geq 0,$$

for any $\xi \in I\!\!K$. We choose $\xi = \phi$, then by letting $\varepsilon \to 0$ we deduce

$$\int_\Gamma (min(\phi - p - a, 0) + max(\phi - p - b, 0))(\phi - p) = 0$$

and thus

$$\phi(x) - p(x) \in D(B(x, \cdot)) \text{ for a.e. } x \in \Gamma,$$

Finally, by letting $\varepsilon \to 0$

$$\int_\Omega \nabla p . \nabla(\xi - p) \geq lim \, sup \int_\Omega \nabla p_\varepsilon . \nabla(\xi - p_\varepsilon)$$

$$\int_\Omega H_e(p_\varepsilon)\xi_{x_n} \to \int_\Omega \chi \xi_{x_n},$$

$$\int_\Omega H_e(p_\varepsilon)(p_\varepsilon)_{x_n} = \int_\Omega G_e(p_\varepsilon)_{x_n}$$

$$= \int_\Gamma G_\varepsilon(p_\varepsilon)\nu_{x_n}\,d\sigma \to \int_\Gamma p\,\nu_{x_n}\,d\sigma$$

$$= \int_\Omega p_{x_n} = \int_\Omega \chi\,p_{x_n},$$

where

$$G_\varepsilon(s) = \int_0^s H_\varepsilon(r)\,dr.$$

whence, since

$$\int_\Omega (\nabla p_\varepsilon . \nabla(\xi - p_\varepsilon) + H_\varepsilon(p_\varepsilon)(\xi - p_\varepsilon)_{x_n})$$

$$\geq \int_\Gamma B_2^\varepsilon(x, \phi - p_\varepsilon)(\xi - p_\varepsilon)\,d\sigma,$$

for any $\xi \in I\!\!K$ we deduce

$$\int_\Omega (\nabla p . \nabla(\xi - p) + \chi(\xi - p)_{x_n})$$

$$\geq \int_\Gamma \gamma_2 . (\xi - p)\,d\sigma \quad \forall \xi \in I\!\!K.$$

References

[AC] A. Alonso & J. Carrillo *A Unified Formulation of the Boundary Conditions in the Dam Problem.* To appear.

[Al] H.W. Alt *A Free Boundary Problem Associated with the Flow of Groundwater.* Arch. Rat. Mech. Anal. **64**, p.111-126, 1977.

[Ba] C. Baiocchi *Su un problema di frontiera libera connesso a questioni di idraulica.* Ann. Mat. Pura Appl. **92**, p.107-127, 1972.

[Be] J. Bear *Hydraulics of Groundwater.* McGraw-Hill, 1979.

[B1] H. Brézis *Problèmes unilatéraux.* J. Mat. Pures et Appl. **51**, p.1-168, 1972.

[B2] H. Brézis *Opérateurs maximaux monotones et semi-groupes de contractions dans les espaces de Hilbert.* North-Holland, 1973.

[BKS] H. Brézis, D. Kinderlehrer & G. Stampacchia *sur une nouvelle formulation du problème de l'écoulement à travers une digue.* C.R.A.S. de Paris, Série A, **287**, p.711-714, 1978.

[CC] J. Carrillo & M. Chipot *The Dam Problem with Leaky Boundary Conditions.* Appl. Math. Optim. **28**, p.57-85, 1993.

[Li] J.L. Lions *Quelques méthodes de résolution des problèmes aux limites non linéaires.* Dunod Gauthier-Villars, Paris, 1969.

Departamento de Matemática Aplicada
Universidad Complutense de Madrid
Madrid 28040
Spain

M CHIPOT AND A LYAGHFOURI

An existence theorem for an unbounded dam with leaky boundary conditions

1. Introduction

Let Ω be an unbounded locally Lipschitz domain in $\mathbb{R}^n$ ($n \geq 3$). Ω represents an unbounded porous medium. The boundary Γ of Ω is divided into three parts : an impervious part S_1, a part in contact with air S_2, and a part covered by fluid S_3. We denote by $S_{3,i}$, $i \in I$ the different connected components of S_3. Assuming that the flow in Ω has reached a steady state we are concerned with finding the pressure p of the fluid and the part of porous medium where some flow occurs, i.e., the wet subset A of Ω. We suppose that $\Omega \subset \mathbb{R}^{n-1} \times (-\infty, H)$, $H \in \mathbb{R}$.

Let φ be a nonnegative Lipschitz continuous function on $S_2 \cup S_3$. We assume $\varphi = 0$ on S_2, $\varphi \leq H - x_n$ on S_3. When φ is the fluid pressure, φ is given -when renormalized- by :

$$\varphi(x', x_n) = \begin{cases} 0 & \text{if} \quad (x', x_n) \in S_2 \\ h_i - x_n & \text{if} \quad (x', x_n) \in S_{3,i}, \qquad i \in I, \end{cases}$$

where h_i denotes the level of the reservoir covering $S_{3,i}$. We denote by $a(x) = (a_{ij}(x))_{1 \leq i,j \leq n}$ the permeability matrix of the medium at x and we suppose that

$$(1.1) \qquad a_{ij} \in L^\infty(\Omega) \quad \forall i, j = 1, \ldots, n,$$

$$(1.2) \qquad \exists \lambda > 0 \text{ such that } a(x)y \cdot y = \sum_{1 \leq i,j \leq n} a_{ij} y_i y_j \geq \lambda |y|^2 \quad \text{a.e. } x \in \Omega, \forall y \in \mathbb{R}^n.$$

Let $\beta : S_3 \times \mathbb{R} \longrightarrow \mathbb{R}$ be a function satisfying :

$$(1.3) \qquad \beta(., 0) \in L^2_{loc}(S_3),$$

$$(1.4) \qquad x \longmapsto \beta(x, u) \text{ is measurable } \forall u \in \mathbb{R},$$

$$(1.5) \qquad \exists c_0 > 0 : |\beta(x, u_1) - \beta(x, u_2)| \leq c_0 |u_1 - u_2| \quad \text{a.e. } x \text{ in } S_3, \forall u_1, u_2 \in \mathbb{R},$$

$$(1.6) \qquad u \longmapsto \beta(x, u) \text{ is nondecreasing a.e. } x \in S_3,$$

$$(1.7) \qquad \beta(x, u).u \geq 0 \quad \text{a.e. } x \in S_3, \forall u \in \mathbb{R}.$$

($L^2_{loc}(S_3)$ is the L^2_{loc}-space corresponding to the superficial measure on S_3).

Then we consider the following problem :

(P_∞)
$$\begin{cases}
\text{Find } (p, \chi) \in H^1_{loc}(\Omega) \times L^\infty(\Omega) \text{ such that } : \\
p \geq 0, \ 0 \leq \chi \leq 1, \ p(1 - \chi) = 0 \quad \text{a.e. in } \Omega, \\
p = 0 \quad \text{on } S_2, \\
\displaystyle\int_\Omega a(x)(\nabla p + \chi e).\nabla \xi dx \ \leq \ \int_{S_3} \beta(x, \varphi - p).\xi d\sigma(x) \\
\forall \xi \in V_0 \ = \ \left\{ \xi \in H^1(\Omega), \ \xi \geq 0 \text{ on } S_2, \ \text{Supp } \xi \subset \overline{\Omega}_r \text{ for some } r > 0 \right\},
\end{cases}$$

where $\Omega_r = \Omega \cap B(0, r)$, $e = (0, ..., 1)$, $B(0, r)$ is the ball of $\mathbb{R}^n$ of center 0 and radius r, Supp ξ denotes the support of the function ξ.

In [GK] G. Gilardi and D. Kröner consider the problem of an unbounded dam with Dirichlet boundary conditions. They obtain a result of existence of a solution by regularization. Here we impose a leaky boundary conditions. We then obtain a solution of (P_∞) as a monotone limit of a sequence of solutions for bounded subdomains using very simple arguments.

2. Existence of a solution

Let us first introduce an auxillary problem on a truncated domain. For $r > 0$, set
$$\Omega_r = \Omega \cap B(0, r), \quad S_1^r = S_1 \cap \partial\Omega_r, \quad S_2^r = (S_2 \cap \partial\Omega_r) \cup (\partial\Omega_r \cap \Omega), \quad S_3^r = S_3 \cap \partial\Omega_r.$$
We know (see [CC2] or [L]) that the problem

(P_r)
$$\begin{cases}
\text{Find } (p_r, \chi_r) \in H^1(\Omega_r) \times L^\infty(\Omega_r) \text{ such that } : \\
p_r \geq 0, \ 0 \leq \chi_r \leq 1, \ p_r(1 - \chi_r) = 0 \quad \text{a.e. in } \Omega_r, \\
p_r = 0 \quad \text{on } S_2^r, \\
\displaystyle\int_{\Omega_r} a(x)(\nabla p_r + \chi_r e).\nabla \xi dx \ \leq \ \int_{S_3^r} \beta(x, \varphi - p_r).\xi d\sigma(x) \\
\qquad\qquad\qquad\qquad\qquad \forall \xi \in H^1(\Omega_r), \ \xi \geq 0 \text{ on } S_2^r
\end{cases}$$

admits at least one solution (p_r, χ_r) for all $r > 0$. Then we can show :

Theorem 2.1. *We have*
$$\forall r > 0, \qquad 0 \leq p_r(x) \leq H - x_n \qquad a.e. \ x \in \Omega_r.$$

Proof : Let us denote by $(.)^+$ the positive part of a function. Since $\pm(p_r - (H - x_n))^+$ is a test function for (P_r), we have :
$$\int_{\Omega_r} a(x)(\nabla p_r + \chi_r e).\nabla(p_r - (H - x_n))^+ dx =$$
$$\int_{S_3^r} \beta(x, \varphi - p_r).(p_r - (H - x_n))^+ d\sigma(x).$$

65

Since on the set where $p_r > (H - x_n)$, $\chi_r = 1$ we get

$$\int_{\Omega_r} a(x)(\nabla(p_r - (H - x_n))).\nabla(p_r - (H - x_n))^+ dx =$$

$$\int_{S_3^r} \beta(x, \varphi - p_r).(p_r - (H - x_n))^+ d\sigma(x) \leq 0 \qquad \text{(by (1.6) and (1.7)).}$$

Then, see (1.2),

$$\lambda \int_{\Omega_r} |\nabla(p_r - (H - x_n))^+|^2 \leq 0.$$

So, $p_r \leq H - x_n$ a.e. in Ω_r. $\qquad\qquad\qquad\square$

Theorem 2.2. *For all r, $r' > 0$ such that $r < r'$, there exist two pairs (p_r, χ_r) and $(p_{r'}, \chi_{r'})$ of solutions to (P_r) and $(P_{r'})$ such that :*

$$p_r \leq p_{r'}, \qquad \chi_r \leq \chi_{r'}.$$

Proof : Let us consider r and $r' > 0$ such that $r < r'$. For $\varepsilon > 0$ let $p_{\varepsilon,r}$ and $p_{\varepsilon,r'}$ be the solutions corresponding to problem $(P_{\varepsilon,r})$ and $(P_{\varepsilon,r'})$ (see [CC2] and [L]) i.e.:

$$(P_{\varepsilon,r}) \quad \begin{cases} \text{Find } p_{\varepsilon,r}' \in H^1(\Omega_r) \text{ such that } : \\ p_{\varepsilon,r} = 0 \quad \text{on } S_2^r \\ \displaystyle\int_{\Omega_r} a(x)(\nabla p_{\varepsilon,r} + H_\varepsilon(p_{\varepsilon,r})e).\nabla\xi dx = \int_{S_3^r} \beta(x, \varphi - p_{\varepsilon,r}).\xi d\sigma(x) \\ \forall \xi \in H^1(\Omega_r), \quad \xi = 0 \text{ on } S_2^r \end{cases}$$

where $H_\varepsilon(s)$ denotes the fonction $s \longmapsto 1 \wedge \dfrac{s^+}{\varepsilon}$, ($\wedge$ denotes the min of two numbers).
Set

$$V_r = \left\{ \xi \in H^1(\Omega_r) \,/\, \xi = 0 \quad \text{on } S_2^r \right\}$$

$$V_{r'} = \left\{ \xi \in H^1(\Omega_{r'}) \,/\, \xi = 0 \quad \text{on } S_2^{r'} \right\}$$

and for $\delta > 0$, $f_\delta(s) = \left(1 - \dfrac{\delta}{s}\right)^+ \chi([\delta, +\infty))(s)$, where $\chi(A)$ is the characteristic function of the set A.
Assume that $p_{\varepsilon,r}$ has been extended by 0 on $\Omega_{r'} \setminus \Omega_r$ and set $q_\varepsilon = p_{\varepsilon,r} - p_{\varepsilon,r'}$, $\xi = f_\delta(q_\varepsilon)$. Then $\xi \in V_r \cap V_{r'}$, so :

$$(2.1) \qquad \int_{\Omega_r} a(x)(\nabla p_{\varepsilon,r} + H_\varepsilon(p_{\varepsilon,r})e).\nabla\xi dx = \int_{S_3^r} \beta(x, \varphi - p_{\varepsilon,r}).\xi d\sigma(x)$$

$$(2.2) \qquad \int_{\Omega_{r'}} a(x)(\nabla p_{\varepsilon,r'} + H_\varepsilon(p_{\varepsilon,r'})e).\nabla \xi dx = \int_{S_3^{r'}} \beta(x, \varphi - p_{\varepsilon,r'}).\xi d\sigma(x).$$

subtracting (2.2) from (2.1) we get :

$$\int_{\Omega_{r'}} a(x)(\nabla q_\varepsilon).\nabla \xi dx = - \int_{\Omega_{r'}} a(x)((H_\varepsilon(p_{\varepsilon,r}) - H_\varepsilon(p_{\varepsilon,r'}))e).\nabla \xi dx +$$

$$\int_{S_3^{r'}} (\beta(x, \varphi - p_{\varepsilon,r}) - \beta(x, \varphi - p_{\varepsilon,r'})).\xi d\sigma(x).$$

(Note that $\xi = 0$ outside of Ω_r).
Hence, by the monotonicity of β, we have

$$\int_{\Omega_{r'}} a(x)(\nabla q_\varepsilon).\nabla \xi dx \leq - \int_{\Omega_{r'}} a(x)((H_\varepsilon(p_{\varepsilon,r}) - H_\varepsilon(p_{\varepsilon,r'}))e).\nabla \xi dx.$$

Using the coerciveness of a and the Lipschitz continuity of H_ε, we get :

$$\lambda \int_{[q_\varepsilon > \delta] \cap \Omega_{r'}} \frac{|\nabla q_\varepsilon|^2}{q_\varepsilon^2} dx \leq \frac{|a|_\infty}{\varepsilon}. \int_{[q_\varepsilon > \delta] \cap \Omega_{r'}} \frac{|\nabla q_\varepsilon|}{q_\varepsilon} dx,$$

where $[q_\varepsilon > \delta] = \{ x \in \Omega / q_\varepsilon(x) > \delta \}$ and $|a|_\infty$ denotes some constant bounding uniformly the a_{ij} on Ω. Hence, by the Cauchy-Shwarz inequality,

$$\int_{[q_\varepsilon > \delta] \cap \Omega_{r'}} \frac{|\nabla q_\varepsilon|^2}{q_\varepsilon^2} dx \leq \frac{|a|_\infty^2 . |\Omega_{r'}|}{\lambda^2 . \varepsilon^2}.$$

Since

$$\int_{\Omega_{r'}} |\nabla ln(1 + \frac{(q_\varepsilon - \delta)^+}{\delta})|^2 dx = \int_{[q_\varepsilon > \delta] \cap \Omega_{r'}} \frac{|\nabla q_\varepsilon|^2}{q_\varepsilon^2} dx$$

by the Poincaré inequality, we get

$$\int_{\Omega_{r'}} |\nabla ln(1 + \frac{(q_\varepsilon - \delta)^+}{\delta})|^2 dx \leq c = c(\Omega_{r'}, \varepsilon, r', \lambda)$$

and by letting $\delta \to 0$, we deduce $q_\varepsilon \leq 0$ a.e. in $\Omega_{r'}$ that is $p_{\varepsilon,r} \leq p_{\varepsilon,r'}$ and $H_\varepsilon(p_{\varepsilon,r}) \leq H_\varepsilon(p_{\varepsilon,r'})$ a.e. in $\Omega_{r'}$. Passing to the limit in ε (see [CC2]), we get two pairs (p_r, χ_r) and $(p_{r'}, \chi_{r'})$ of solutions to (P_r) and $(P_{r'})$ such that : $p_r \leq p_{r'}$ and $\chi_r \leq \chi_{r'}$ a.e. in $\Omega_{r'}$. $\qquad \square$

In what follows, we denote by (p_r, χ_r) a monotone sequence of solutions to (P_r).

Theorem 2.3. *There exists a pair $(p, \chi) \in L_{loc}^2(\Omega) \times L^\infty(\Omega)$ such that :*

$$p_r \longrightarrow p \quad \text{strongly in } L_{loc}^2(\Omega), \quad p_r \longrightarrow p \quad \text{a.e. in } \Omega$$
$$\chi_r \longrightarrow \chi \quad \text{strongly in } L_{loc}^2(\Omega), \quad \chi_r \longrightarrow \chi \quad \text{a.e. in } \Omega.$$

Proof : Since $(p_r)_r$ and $(\chi_r)_r$ are monotone sequences such that

$$0 \leq \chi_r \leq 1 \qquad 0 \leq p_r \leq H - x_n \quad \text{a.e. in } \Omega,$$

the result is obvious by Beppo-Levi's theorem. $\qquad\qquad\qquad\qquad\square$

We are now able to state our existence result :

Theorem 2.4. *There exists a solution to (P_∞) obtained as a monotone limit of solutions (p_r, χ_r), $r > 0$.*

Proof : Let $\rho > 0$, then for $r >> \rho$, let us first prove that $|p_r|_{1,2,\Omega_\rho} \leq c(\rho)$, where $c(\rho)$ is a constant depending on ρ only, $| \ |_{1,2,\Omega_\rho}$ is the $H^1(\Omega_\rho)$-norm.

Indeed, let $\zeta \in \mathcal{D}(\mathbb{R}^n)$, $(\mathcal{D}(\mathbb{R}^n)$ is the space of C^∞ functions with compact supports), such that :

$$0 \leq \zeta \leq 1, \quad |\nabla\zeta| \leq 2, \quad \zeta = 1 \text{ in } \Omega_\rho, \quad \zeta = 0 \text{ in } \mathbb{R}^n \setminus \Omega_{\rho+1}.$$

Since $\pm\zeta^2 p_r$ is a test function for (P_r), one has

$$\int_{\Omega_{\rho+1}} a(x)(\nabla p_r + \chi_r e).\nabla(\zeta^2 p_r)dx = \int_{S_3^{\rho+1}} \beta(x, \varphi - p_r).(\zeta^2 p_r)d\sigma(x).$$

Moreover, combining this and the coerciveness of a we get

$$\lambda \int_{\Omega_{\rho+1}} |\zeta\nabla p_r|^2 dx = \int_{\Omega_{\rho+1}} \lambda\zeta^2|\nabla p_r|^2 dx \leq \int_{\Omega_{\rho+1}} \zeta^2 a(x)\nabla p_r.\nabla p_r dx$$

$$= -\int_{\Omega_{\rho+1}} 2\zeta p_r a(x)\nabla p_r.\nabla\zeta + \chi_r\zeta^2 a(x)e.\nabla p_r + 2\chi_r\zeta p_r a(x)e.\nabla\zeta dx$$

$$+ \int_{S_3^{\rho+1}} \beta(x, \varphi - p_r).(\zeta^2 p_r)d\sigma(x).$$

Now, using the Cauchy-Schwartz Inequality we derive easily for some constants c

$$\cdot \ \left| \int_{\Omega_{\rho+1}} 2\zeta p_r a(x)\nabla p_r.\nabla\zeta dx \right| \leq 2c \int_{\Omega_{\rho+1}} |\zeta\nabla p_r|.|\nabla\zeta|dx \leq 4c \int_{\Omega_{\rho+1}} |\zeta\nabla p_r|dx$$

$$\leq 4c|\Omega_{\rho+1}|^{1/2} \left(\int_{\Omega_{\rho+1}} |\zeta\nabla p_r|^2 dx \right)^{1/2},$$

$$\cdot \ \left| \int_{\Omega_{\rho+1}} \chi_r\zeta^2 a(x)e.\nabla p_r dx \right| \leq \int_{\Omega_{\rho+1}} c|\zeta\nabla p_r|dx$$

$$\leq c|\Omega_{\rho+1}|^{1/2} \left(\int_{\Omega_{\rho+1}} |\zeta\nabla p_r|^2 dx \right)^{1/2},$$

$$\cdot \ \left| \int_{\Omega_{\rho+1}} 2\chi_r\zeta p_r a(x)e.\nabla\zeta dx \right| \leq \int_{\Omega_{\rho+1}} 4cp_r \leq 4c \int_{\Omega_{\rho+1}} (H - x_n) \leq c|\Omega_{\rho+1}|,$$

$$\cdot \left| \int_{S_3^{\rho+1}} \beta(x, \varphi - p_r).(\zeta^2 p_r) \right| d\sigma(x)$$

$$= \left| \int_{S_3^{\rho+1}} (\beta(x, \varphi - p_r) - \beta(x, 0))(\zeta^2 p_r) d\sigma(x) + \int_{S_3^{\rho+1}} \beta(x, 0)(\zeta^2 p_r) d\sigma(x) \right|$$

$$\leq \int_{S_3^{\rho+1}} c_0 |\varphi - p_r|(\zeta^2 p_r) d\sigma(x) + \int_{S_3^{\rho+1}} |\beta(x, 0)|(\zeta^2 p_r) d\sigma(x) \quad \text{(see (1.5))}$$

$$\leq c(|S_3^{\rho+1}| + |S_3^{\rho+1}|^{1/2}),$$

since the functions ζ, φ, p_r are bounded and $\beta(x, 0)$ belongs to $L^2_{loc}(S_3)$. So, for some constant $c = c_\rho$

$$\int_{\Omega_{\rho+1}} |\zeta \nabla p_r|^2 dx \leq c \left(\int_{\Omega_{\rho+1}} |\zeta \nabla p_r|^2 dx \right)^{1/2} + c.$$

From this follows :

$$\int_{\Omega_{\rho+1}} |\zeta \nabla p_r|^2 dx \leq c,$$

hence

$$\int_{\Omega_\rho} |\zeta \nabla p_r|^2 dx \leq \int_{\Omega_{\rho+1}} |\zeta \nabla p_r|^2 dx \leq c.$$

By Poincaré's Inequality, this implies :

$$|p_r|_{1,2,\Omega_\rho} \leq c' = c'(\rho).$$

Now, from the reflexivity of $H^1(\Omega_\rho)$ and Rellich's theorem, there exists a subsequence (p_{r_k}, χ_{r_k}) such that

$$p_{r_k} \rightharpoonup p^\rho \quad \text{weakly in } H^1(\Omega_\rho), \quad p_{r_k} \longrightarrow p^\rho \quad \text{strongly in } L^2(\Omega_\rho),$$
$$p_{r_k} \longrightarrow p^\rho \quad \text{a.e. in } \Omega_\rho, \quad p_{r_k} \longrightarrow p^\rho \quad \text{strongly in } L^2(S_3^\rho),$$
$$p_{r_k} \longrightarrow p^\rho \quad \text{a.e. in } S_3^\rho,$$
$$\chi_{r_k} \rightharpoonup \chi^\rho \quad \text{weakly in } L^2(\Omega), \quad \text{where } (p^\rho, \chi^\rho) \in H^1(\Omega_\rho) \times L^\infty(\Omega_\rho).$$

From Theorem 2.3 it is clear that : $p^\rho = p$, $\chi^\rho = \chi$ a.e. in Ω_ρ.
So $p \in H^1(\Omega_\rho)$ $\forall \rho$, i.e. $p \in H^1_{loc}(\Omega)$. From (P_{r_k}) we have

$$p_{r_k} \geq 0, \quad p_{r_k} = 0 \text{ on } S_2^\rho, \quad p_{r_k}(1 - \chi_{r_k}) = 0, \quad 0 \leq \chi_{r_k} \leq 1 \text{ a.e. in } \Omega_\rho,$$

and we derive

$$p \geq 0, \quad p = 0 \text{ on } S_2^\rho, \quad p(1 - \chi) = 0, \quad 0 \leq \chi \leq 1 \text{ a.e. in } \Omega_\rho \quad \forall \rho > 0$$

hence
$$p \geq 0, \quad p = 0 \text{ on } S_2, \quad p(1 - \chi) = 0, \quad 0 \leq \chi \leq 1, \text{ a.e. in } \Omega.$$

Let $\xi \in H^1(\Omega)$ such that Supp $\xi \subset \overline{\Omega}_\rho$ and $\xi \geq 0$ on S_2. For r_k large enough we have : $\Omega_\rho \subset \Omega_{r_k}$, so

$$\int_{\Omega_\rho} a(x)(\nabla p_{r_k} + \chi_{r_k} e) . \nabla \xi dx \leq \int_{S_3^\rho} \beta(x, \varphi - p_{r_k}) . \xi d\sigma(x).$$

If $^t a$ detotes the transposed matrix of a since

$$\int_{\Omega_\rho} a(x)(\nabla p_{r_k} + \chi_{r_k} e) . \nabla \xi dx = \int_{\Omega_\rho} \nabla p_{r_k} {}^t a(x) . \nabla \xi + \chi_{r_k} e . {}^t a(x) . \nabla \xi dx,$$

letting $k \to +\infty$, we get :

$$\int_{\Omega_\rho} a(x)(\nabla p + \chi e) . \nabla \xi dx \leq \int_{S_3^\rho} \beta(x, \varphi - p) . \xi d\sigma(x)$$

that is

$$\int_\Omega a(x)(\nabla p + \chi e) . \nabla \xi dx \leq \int_{S_3} \beta(x, \varphi - p) . \xi d\sigma(x)$$

and (p, χ) is a pair of solution to (P_∞). $\qquad \square$

3. Some properties of the solutions

Theorem 3.1. *We suppose Ω, φ, S_2 and S_3 invariant by any translation of direction e_{n-1}, where e_{n-1} denotes by the $(n-1)^{th}$ vector of the canonical basis of $\mathbb{R}^n$. Let (p, χ) be a solution to (P_∞). Then, for any $\lambda \in \mathbb{R}$ the pair $(p_\lambda, \chi_\lambda)$ defined by*

$$p_\lambda(x) = p(x_1, ..., x_{n-2}, x_{n-1} + \lambda, x_n) \quad \text{and} \quad \chi_\lambda(x) = \chi(x_1, ..., x_{n-2}, x_{n-1} + \lambda, x_n),$$

is a solution to (P_∞^λ) where (P_∞^λ) corresponds to $a_\lambda(x) = a(x_1, ..., x_{n-2}, x_{n-1} + \lambda, x_n)$. So if $a(x)$ does not depend on x_{n-1} then $(p_\lambda, \chi_\lambda)$ is a solution of (P_∞) for any $\lambda \in \mathbb{R}$.

Proof : First it is clear that $(p_\lambda, \chi_\lambda) \in H^1_{loc}(\Omega) \times L^\infty(\Omega)$,

$$p_\lambda \geq 0, \quad 0 \leq \chi_\lambda \leq 1, \quad p_\lambda(1 - \chi_\lambda) = 0 \quad \text{a.e. in } \Omega \quad \text{and} \quad p_\lambda = 0 \text{ on } S_2.$$

It remains to verify the inequality in (P_∞). Let $\xi \in H^1(\Omega)$ with Supp $\xi \subset \overline{\Omega}_r$, $\xi \geq 0$ on S_2.

We define $\xi_{-\lambda}$ by $\xi_{-\lambda}(x) = \xi(x_1, ..., x_{n-2}, x_{n-1} - \lambda, x_n)$. Then, $\xi_{-\lambda}$ is a test function for (P_∞) and :

$$\int_\Omega a(x)(\nabla p + \chi e) . \nabla \xi_{-\lambda} dx \leq \int_{S_3} \beta(x, \varphi - p) . \xi_{-\lambda} d\sigma(x).$$

Using the change of variable $(x_1, ..., x_n) \longmapsto (x_1, ..., x_{n-2}, x_{n-1} + \lambda, x_n)$, we obtain

$$\int_\Omega a_\lambda(x)(\nabla p_\lambda + \chi_\lambda e).\nabla \xi dx \leq \int_{S_3} \beta(x, \varphi - p_\lambda).\xi d\sigma(x).$$

$\square$

Remark 3.2. The same results are valid for any direction e_i with $i \in \{1, \ldots, n-1\}$.

Remark 3.3. If $a(x)$ is independent of x_{n-1} and (P_∞) has a unique solution (p, χ) then $p_\lambda = p$ and $\chi_\lambda = \chi$ for all $\lambda \in \mathbb{R}$. Thus p and χ become independent of the variable x_{n-1}.

Proposition 3.4. *Assume that $div(a(x)e) \in L^2_{loc}(\Omega)$. Let (p, χ) be a pair of solution to (P_∞). Then, we have in the distributional sense :*

(3.1)
$$div\big(a(x)(\nabla p + \chi e)\big) = 0$$

(3.2)
$$\frac{\partial \chi}{\partial \nu} \leq 0$$

where $\nu = a(x)(e)$.

Proof : 1- Let $\xi \in \mathcal{D}(\Omega)$, where $\mathcal{D}(\Omega)$ is the space of C^∞ functions with compact supports in Ω. $\pm\xi$ is a test function, so we have :

$$\int_\Omega a(x)(\nabla p + \chi e).\nabla \xi dx = 0$$

which means

$$div\big(a(x)(\nabla p + \chi e)\big) = 0,$$

in the distributional sense.
2- Let $\xi \in \mathcal{D}(\Omega)$, $\xi \geq 0$. For all $\varepsilon > 0$, $\pm H_\varepsilon(p)\xi$ is a test function for (P_∞) and we have :

$$\int_\Omega a(x)(\nabla p + \chi e).\nabla(H_\varepsilon(p)\xi)dx = 0 \iff$$

$$\int_\Omega H_\varepsilon(p)a(x)(\nabla p).\nabla \xi dx + H'_\varepsilon(p)\xi a(x)(\nabla p).\nabla p dx + \int_\Omega a(x)(e).\nabla(H_\varepsilon(p)\xi)dx = 0.$$

This reads also :

$$\int_\Omega H_\varepsilon(p)a(x)(\nabla p).\nabla \xi dx - \int_\Omega div(a(x)(e))H_\varepsilon(p)\xi dx = -\int_\Omega H'_\varepsilon(p)\xi a(x)(\nabla p).\nabla p dx$$

hence

$$\int_\Omega H_\varepsilon(p)a(x)(\nabla p).\nabla \xi dx - \int_\Omega \chi div(a(x)(e))H_\varepsilon(p)\xi dx \leq 0.$$

Letting $\varepsilon \to 0$, we get :

$$\int_\Omega a(x)(\nabla p).\nabla\xi dx - \int_\Omega \chi div(a(x)(e))\xi dx \leq 0,$$

or

$$div(a(x)(\nabla p)) + \chi div(a(x)(e)) \geq 0.$$

Using (3.1), it follows :

$$\chi div(a(x)(e)) - div(\chi a(x)(e)) \geq 0$$

which is (3.2). □

Remark 3.5. We deduce from (3.1) and the usual theorems of regularity (see [GT] or [DL]) that $p \in C_{loc}^{0,\alpha}(\Omega \cup S_2)$. In particular p est continous and $[p > 0]$ is an open set. Moreover if $\forall i, j$, $a_{ij} \in C^\infty(\Omega)$ (resp. is analytic) then $p \in C^\infty([p > 0])$ (resp. is analytic on $[p > 0]$).

References :

[Ad] R.A. Adams : *Sobolev Spaces* .
Academic Press, New-York, (1975).

[Al1] H.W. Alt : A Free boundary Problem Associated With the Flow of Ground Water.
Arch. Rat. Mech. Anal. 64, (1977), 111-126.

[Al2] H.W. Alt : The Fluid Flow through Porous Media. Regularity of the Free Surface.
Manuscripta math 21, (1977), 255 - 272.

[Ba1] C. Baiocchi : Free boundary problems in the theory of fluid flows through porous media.
Proceedings of the international congress of Mathematicians - Vancouver, (1974) 237-243.

[Ba2] C. Baiocchi : Free boundary problems in fluid flow through porous media and variational inequalities.
In : Free Boundary Problems - Proceedings of a Seminar held in Pavia Sept - Oct 1979 Vol 1, (Roma 1980), 175-191.

[BKS] H. Brezis, D. Kinderlehrer and G. Stampacchia : Sur une nouvelle formulation du problème de l'écoulement à travers une digue.
C. R. Acd. Sci Paris Serie A 287, (1978), 711 - 714.

[CC1] J. Carrillo, M. Chipot : On the Dam Problem.
J. Differentiel Equations 45, (1982), 234 - 271.

[CC2] J. Carrillo, M. Chipot : The Dam Problem with Leaky Boundary Conditions.
Applied Mathematics and Optimisation 28, (1993), 57 - 85 .

[CL] M. Chipot, A. Lyaghfouri : To appear.

[DL] R. Dautray, J.L. Lions : *Analyse mathématique et calcul numérique pour les sciences et les techniques. Tome 2. L'opérateur de Laplace.*
Masson. Paris (1987).

[GK] G. Gilardi, D. Kröner : The Dam Problem in Unbounded Domains (preprint).

[GT] D. Gilbarg, N.S. Trudinger : *Elliptic Partial Differential Equations of Second Order.*
Springer - Verlag, New York, (1977).

[L] A. Lyaghfouri : Sur quelques problèmes d'écoulements dans les milieux poreux.
Thesis, University of Metz, June (1994).

Centre d'Analyse Non Linéaire,

Université de Metz, URA-CNRS 399,

Ile du Saulcy 57045 Metz Cedex 01, France.

M CHOULLI[†] AND R DEVILLE[‡]

Local weak solutions for a multivalued evolution equation

We prove that the Cauchy problem $u'(t) \in Bu(t)$ and $u(0) = x$, in a Banach space, has a local weak solution when B is the sum of the generator of a linear and compact C_0 semi-group and a convex w-usco set-valued mapping. We also include an application of this perturbation result to a semi-linear parabolic equation whose nonlinear term is discontinuous.

1. Introduction

Let X be a Banach space, A the generator of a linear C_0 semi-group S and let $F : X \to 2^X$ be a set-valued mapping. We are interested in the existence of a weak solution of the following Cauchy problem:

$$(PC) \qquad \begin{cases} u'(t) \in Au(t) + F(u(t)), \ t \in (0, T), \\ u(0) = x \in X. \end{cases}$$

First, let us specify the definition of our weak solution. We will say that a function u from $C([0, T] : X)$ is a weak solution of the Cauchy problem (PC) if there exists f in $L^1(0, T : X)$ such that

$$f(t) \in F(u(t)) \text{ a.e. } t \text{ and } u(t) = S(t)x + \int_0^t S(t - s)f(s)ds, \ 0 \le t \le T.$$

We recall that a set-valued mapping from a metric space into the nonempty weakly compact and convex subsets of a Banach space Z is said to be convex w-usco provided it is upper semi-continuous when Z is endowed with its weak topology.

Theorem 1. *If A is the generator of a linear and compact C_0 semi-group S and F is convex w-usco, then for each $x \in X$, there exists $T > 0$ for which the Cauchy problem (PC) has at least a weak solution.*

[†] Laboratoire de Mathématiques, URA CNRS 741, Université de Franche-Comté, Route de Gray, 25030 Besançon cedex

[‡] Centre de Recherche en Mathématiques, Université de Bordeaux 1, 351 cours de la Libération, 33405 Talence cedex

A result in this direction, under the assumption that X is a separable Banach space, has been obtained by Papageorgiou [8]. Earlier results by Attouch and Damlamian and others can be found in the book of Vrabie [12].

The method we use in the proof of theorem 1 relies on a compactness argument. This method is not new and it is a useful tool to prove existence results for many nonlinear evolution equations.

2. Proof of theorem 1

We shall need two preliminary lemmas.

Lemma 1. *Let* $F : [0, T] \to 2^X$ *be a convex w-usco set-valued mapping and*

$$S_F^p = \{f \in L^p(0, T : X), \ f(t) \in F(t) \ a.e. \ t\}, \ 1 \le p \le \infty.$$

Then $S_F^p = S_F^\infty \ne \emptyset$ *and* S_F^p *is weakly compact in* $L^p(0, T : X)$ *for all* $p \ge 1$.

Proof. $S_F^p = S_F^\infty \ne \emptyset$ follows from a selection theorem of Jayne and Rogers [7] and the fact that $F([0, T])$ is weakly compact in X. The compactness of S_F^p, $p \ge 1$, is immediate from the results of Diestel, Ruess and Schachermayer [4].

Lemma 2. *Let* $F : X \to 2^X$ *be a convex w-usco set-valued mapping, (u_n) a sequence in* $C([0, T] : X)$ *and* (f_n) *a sequence in* $L^1(0, T : X)$ *such that (u_n) is strongly convergent to u and (f_n) is weakly convergent to f. If $f_n \in S^1_{F(u_n(.))}$ for each n, then* $f \in S^1_{F(u(.))}$.

Proof. For each k, $f \in \overline{co} \bigcup\limits_{n \ge k} f_n$ as a consequence of Mazur's theorem. Hence, for a.e. $t \in (0, T)$,

$$f(t) \in \bigcap_k \overline{co} \bigcup_{n \ge k} f_n(t) \subset \bigcap_k \overline{co} \bigcup_{n \ge k} F(u_n(t)).$$

Let $t \in [0, T]$ and E an open half space containing $F(u(t))$. Since F is upper semi-continuous and $u_n(t)$ is strongly convergent to $u(t)$, E contains $co \bigcup\limits_{n \ge k} F(u_n(t))$ for k large enough. As $F(u(t))$ is the intersection of all closed half spaces which contain it, we have that

$$\bigcap_k \overline{co} \bigcup_{n \ge k} F(u_n(t)) \subset F(u(t)).$$

Remark 1. Lemma 2 tells us that the graph of the set-valued mapping:

$$C([0, T] : X) \to 2^{L^1(0, T : X)} : u \to S^1_{F(u(.))}$$

is closed in $(C([0,T] : X), \text{norm}) \times (L^1(0,T : X), \text{weak topology})$.

Proof of theorem 1. Without loss of generality we assume that S is a semi-group of contractions.

Let $x \in X$. Since F is upper semi-continuous, it is locally bounded. We can then find two positive constants M and r such that

(1) $\qquad \|w\| \leq M$ for each $w \in F(y)$ and $y \in X$, $\|y - x\| \leq r$.

On the other hand, the strong continuity of the semi-group S implies that there exists $T > 0$ such that

(2) $\qquad \|S(t)x - x\| + MT \leq r, \; 0 \leq t \leq T$.

We define the operator $B : L^1(0,T : X) \to C([0,T] : X)$ as follows:

$$Bf = S(t)x + \int_0^t S(t - s)f(s)ds, \; f \in L^1(0,T : X),$$

and on $K = \{u \in C([0,T] : X), \|u(t) - x\| \leq r, \; 0 \leq t \leq T\}$, we consider the set-valued mapping: $H(u) = \{v = Bf, \; f \in S^1_{F(u(.))}\}$, if $u \in K$. From (1) and (2), we deduce that

$$H(K) \subset K \text{ and } H(K) \subset B\{f \in L^\infty(0,T : X), \|f\|_{L^\infty(0,T:X)} \leq M\},$$

and then $H(K)$ is relatively compact in $C([0,T] : X)$ by theorem 1 [1].

We claim that H is closed graph. Indeed, let (u_n) be a sequence in K and $v_n = Bf_n$, where $f_n \in S^1_{F(u_n(.))} = S^\infty_{F(u_n(.))}$ for each n. We suppose that (u_n) and (v_n) are strongly convergent in $C([0,T] : X)$ to u and v respectively. For each $t \in [0,T]$, the set $K_t = F\{u_n(t), n\}$ is relatively weakly compact. This is a consequence of the upper semi-continuity of F. Since $f_n(t) \in K_t$ a.e. t and all n, and (f_n) is bounded in $L^\infty(0,T : X)$, it follows from corollary 1.6 [4] that (f_n) is relatively weakly compact in $L^1(0,T : X)$. Substracting a subsequence if necessary, we assume that f_n converges weakly in $L^1(0,T : X)$ to f. We use lemma 2 to conclued that $f \in S^1_{F(u(.))}$. The operator B is affine and strongly continuous, it is then weakly continuous. Hence $v = Bf$.

We have prove that H is closed graph and its range is relatively compact. Since the values of H are closed convex and $H(K) \subset K$, it follows from the Kakutani and Ky-Fan fixed point theorem (see for instance [11]) that there exists u_* in K such that $u_* \in H(u_*)$. This means that u_* is a weak solution of the Cauchy problem (PC) on $[0,T]$.

Remark 2. In addition of the assumptions of theorem 1, if F satisfies:

$$\sup\{\|w\|, \; w \in F(u)\} \leq C_1\|u\| + C_2, \; u \in X$$

where C_1 et C_2 are some positive constants, then a classical argument shows that (PC) has a global weak solution.

3. Densely defined perturbation

It is possible to consider the case in which F is only densely defined. To this end we introduce the notion a regular semi-group. We say that a linear C_0 semi-group S, generated by A and defined on X, is p-regular ($p>1$) if for each $f \in L^p(0,T:X)$, $\int_0^t S(t-s)f(s)ds$ belongs to $L^p(0,T:D(A))$.

Theorem 2. *Assume that X is reflexive, A is the generator of a linear, p-regular and compact semi-group and F is a locally bounded set-valued mapping from X into the convex and weakly compact subsets of X which is upper semi-continuous on its domain from $(X, norm)$ into $(X, weak\ topology)$. If $D(A)$ is contained in the domain of F, then for each $x \in D(A)$, there exists $T > 0$ for which (PC) has a weak solution.*

We denote by $\mathcal{A}(X)$ the class of linear operators $A : X \to X$ having the following properties:

(i) A is closed and its domain is dense,

(ii) $N(A) = \{0\}$ and $R(A)$ is dense,

(iii) $(-\infty, 0) \subset \rho(A)$ and there exists $M > 0$ such that $\|\lambda(\lambda + A)^{-1}\| \leq M$, $\lambda > 0$,

(iv) $A^{is} \in \mathcal{L}(X)$, $s \in \mathbb{R}$ and there exists $c > 0$, $\theta \in [0, \frac{\pi}{2}[$ such that $\|A^{is}\| \leq ce^{\theta|s|}$, $s \in \mathbb{R}$.

$N(A)$, $R(A)$ and $\rho(A)$ are respectively the kernel, the range and the resolvent set of A.

Let us remark that if A belongs to the class $\mathcal{A}(X)$ then $-A$ generates an holomorphic semi-group.

We recall that a Banach space X is said to be an UMD space if the Hilbert transform is bounded on $L^p(\mathbb{R}, X)$ for some $p \in]1, +\infty[$ and that an UMD space is super-reflexive (see [3] and its references). For instance, Hilbert spaces and L^p ($p \in]1, +\infty[$) spaces are typical examples of UMD spaces.

The corollary bellow is an immediate consequence of theorems 2 and 3.5 [5].

Corollary 1. *Suppose that X is an UMD space, $-A$ belongs to the class $\mathcal{A}(X)$ and F is a locally bounded set-valued mapping from X into the convex and weakly compact subsets of X which is upper semi-continuous on its domain from $(X, norm)$ into $(X, weak\ topology)$. If the semi-group generated by A is compact and the domain of F*

contains $D(A)$, then for each $x \in D(A)$, there exists $T > 0$ for which (PC) has at least a weak solution.

The following lemma will be needed in the proof of theorem 2.

Lemma 3. *let X be a reflexive Banach space, F is a locally bounded set-valued mapping from X into the convex and weakly compact subsets of X. Assume that F is densely defined and it is upper semi-continuous on its domain from $(X, norm)$ into $(X, weak\ topology)$. Then*

$$\overline{F}(x) = \bigcap_{\epsilon > 0} \overline{co} F(B(x, \epsilon) \cap D(F)), \ x \in X$$

is convex w-usco and $\overline{F}_{|D(F)} = F$.

Proof. Since F is locally bounded and densely defined, an easy compactness argument shows that the values of $\overline{F}$ are nonemty convex and weakly compact.

Let $x \in X$ and V be a weakly open subset of X containing $\overline{F}(x)$. Again by a compactness argument, there exists $\epsilon > 0$ such that $\overline{co} F(B(x, \epsilon) \cap D(F)) \subset V$ and then $\overline{F}(B(x, \frac{\epsilon}{2})) \subset V$. Thus $\overline{F}$ is upper semi-continuous from $(X, norm)$ into $(X, weak\ topology)$.

If $x \in D(F)$ and E be an open half space containing $F(x)$, the upper semi-continuity of F at x implies that E contains $co F(B(x, \epsilon) \cap D(F))$ for $\epsilon > 0$ small enough. Thus $\overline{F}(x) = F(x)$ because $F(x)$ is convex.

Proof of theorem 2. From lemma 3, we can find $\overline{F}$ convex w-usco such that $\overline{F}_{|D(F)} = F$. By theorem 1, if $x \in D(A)$ then there exists $T > 0$ for which the Cauchy problem (PC) has a weak solution $u = Bf$ with $f \in S^{\infty}_{\overline{F}(u(.))}$, where B is the operator from the proof of theorem 1. Since S is p-regular, u belongs to $L^p(0, T : D(A))$. This and the fact that $D(A) \subset D(F)$ imply that $f(t) \in F(u(t))$ a.e. t in $[0, T]$.

4. Application

Let Ω be an open bounded subset of $\mathbb{R}^n$ with a smooth boundary, $Q_T = \Omega \times (0, T)$ and let $\partial_P Q_T$ be the parabolic boundary of Q_T.

In this section, we will apply theorem 1 to show the existence of a solution, in a sens to be specified latter, of the following partial differential equation:

$$(E) \qquad \begin{cases} u_t - \Delta u = f(u(.,.)) \text{ in } Q_T, \\ u = 0 \text{ on } \partial_P Q_T, \end{cases}$$

where $f : \mathbb{R} \to \mathbb{R}$ is a measurable function satisfying the growth condition:

$$(3) \qquad |f(s)| \leq a + b|s|, \ s \in \mathbb{R}$$

a and b are given positive constants.
Set

$$\underline{f}(s) = \lim_{r \to 0} \text{ess inf}\{f(t), \ |t - s| \leq r\} \text{ and } \overline{f}(s) = \lim_{r \to 0} \text{ess sup}\{f(t), \ |t - s| \leq r\}.$$

$\underline{f}$ and $\overline{f}$ are respectively lower semi-continuous and upper semi-continuous. Moreover, $\underline{f}(u)$ and $\overline{f}(u)$ are in $L^p(\Omega)$ as soon as u is in $L^p(\Omega)$.

We define the set-valued mapping $E : L^p(\Omega) \to 2^{L^p(\Omega)}$ as follows:

$$E(u) = \{w \in L^p(\Omega), \ \int_\Omega wv \leq \int_\Omega (\overline{f}(u)v^+ - \underline{f}(u)v^-), \ v \in L^{p'}(\Omega)\}. \ (\frac{1}{p'} + \frac{1}{p} = 1)$$

Applying Fatou's lemma and the semi-continuity of $\underline{f}$ and $\overline{f}$ to conclued that the graph of E is sequentially closed in $(L^p(\Omega), \text{norm}) \times (L^p(\overline{\Omega}), \text{weak topology})$.

Lemme 4. $E(u) = \{w \in L^p(\Omega), \ w(x) \in [\underline{f}(u(x)), \overline{f}(u(x))] \ a.e.\}, \ u \in L^p(\Omega)$.

Proof. For $u \in L^p(\Omega)$, we set

$$E'(u) = \{w \in L^p(\Omega), \ w(x) \in [\underline{f}(u(x)), \overline{f}(u(x))] \text{ a.e.}\}.$$

The inclusion $E'(u) \subset E(u)$, for each $u \in L^p(\Omega)$, is easy to see. Conversely, suppose that there exists $u \in L^p(\Omega)$, $w \in E(u)$ and $w \notin E'(u)$. Hence, we can find a measurable subset $C \subset \Omega$, $|C| > 0$ for which $\underline{f}(u(x)) > w(x)$, $x \in C$ or $\overline{f}(u(x)) < w(x)$, $x \in C$. Thus $\int_C \underline{f}(u) > \int_C w$ or $\int_C \overline{f}(u) < \int_C w$. We obtain then the desired contradiction by taking in the definition of $E(u)$ $v = \pm\chi_C$, where χ_C is the characteristic function of the set C.

The case p>1: clearly, E is locally bounded as a consequence of (3) and the Hölder inequality. We deduce from this that E is upper semi-continuous from $(L^p(\Omega), \text{norm})$ into $(L^p(\Omega), \text{weak topology})$ by using the following lemma:

Lemme 5. *Let (Z, d) be a metric space, X be a reflexive Banach space and let F be a locally bounded set-valued mapping from (Z, d) into 2^X whose graph is sequentially closed in $(Z, d) \times (X, \text{weak topology})$. Then F is upper semi-continuous from (Z, d) into $(X, \text{weak topology})$.*

Proof. Suppose that there exists $z \in Z$ such that F is not upper semi-continuous at z. Then we can find V a weakly open subset in X containing $F(z)$, two sequences (z_n) and (x_n) such that

$$d(z_n, z) \to 0, \ x_n \in F(z_n) \text{ and } x_n \notin V, \text{ for each } n.$$

Since F is locally bounded , we can assume that (x_n) is bounded. As X is reflexive, (x_n) has a subsequence (x_m) converging weakly to $x \in X$. Hence, $x \in F(z)$ because the graph of F is sequentially closed in $(Z, d) \times (X$, weak topology), and then $x_m \in V$ if m is large enough. But this contradicts the fact that x_n does not belongs to V for each n.

We Consider the operator $A = \Delta$ with $D(A) = W_0^{1,p}(\Omega) \cap W^{2,p}(\Omega)$. It is well known that A generates an holomorphic and compact semi-group of contractions. We use theorem 1 and L^p-regularity theorems (see for instance [5]) to obtain: there exists $T > 0$, $u \in W^{1,p}(0, T : L^p(\Omega)) \cap L^p(0, T : D(A))$ and $g \in L^p(Q_T)$ such that $g(x, t) \in [\underline{f}(u(x, t))), \overline{f}(u(x, t))]$ a.e. and u is the solution of (E) when $f(u(.,.))$ is replaced by g.

The case p=1: since $\underline{f}$ and $\overline{f}$ satisfy also (3), using the Dunford-Pettis weak compactness criterion, we can easily show that E maps weakly compact subsets in $L^1(\Omega)$ into relatively weakly compact subsets of $L^1(\Omega)$. Hence E is upper semi-continuous by the following lemma:

Lemma 6. *Let (Z, d) be a metric space, X be a Banach space and F a set-valued mapping from Z into 2^X whose graph is sequentially closed in $(Z, d) \times (X$, weak topology). If F maps a compact subsets of Z into the relatively weakly compact subsets of X, then F is upper semi-continuous from (Z, d) into $(X$, weak topology).*

Proof. Identical to the proof of the previous lemma using the fact that the relative weak compactness is the same as the sequential relative weak compactness.

The operator $A = \Delta$ and $D(A) = \{u \in W_0^{1,1}(\Omega), \Delta u \in L^1(\Omega)\}$, defined on $L^1(\Omega)$, generates a compact semi-group (see [1] for more details). As before, theorem 1 shows that there exists $T > 0$, $u \in C([0, T] : L^1(\Omega))$ and $g \in L^1(Q_T)$ such that

$$g(x, t) \in [\underline{f}(u(x, t)), \overline{f}(u(x, t))] \text{ a.e. and } u(., t) = \int_0^t S(t - s)g(., s),$$

where S is the semi-group generated by A.

We notice that in each case, we obtain global weak solutions (see remark 2).

Remark 3. If $f : \mathbb{R} \to \mathbb{R}$ is a measurable function satisfying (3) where $|s|$ is replaced by $|s|^r$ with $r \in [1, \frac{n}{n-p}[$ for $n > p$, then f is contained in a set-valued mapping $F : L^p(\Omega) \to 2^{L^p(\Omega)}$ having a dense domain ($W^{1,p}(\Omega) \subset D(F)$ follows from the inclusion $W^{1,p}(\Omega) \subset L^{rp}(\Omega)$) and whose graph is sequentially closed. But our result does not work in this case because f is not locally bounded. Nevertheless, the conclusion of theorem 1 is true for $p = 2$ (see Frigon and Saccon [6]).

Acknowledgments. We would like to thank Professor Ph. Bénilan for his valuable remarks.

References

[1] Baras P, Hassan J C and Veron L, Compacité de l'opérateur définissant la solution d'une équation d'évolution non homogène, C. R. Acad. Sc., Série A, t. 284 (1977), 799-802.

[2] Bénilan Ph, Crandall M G et Pazy A, Evolutions problems governed by accretive operators, to appear.

[3] Burkholder D L, Martingales and Fourier analysis in Banach spaces, Springer-Verlag, Lectures Notes in Math. 1206 (1986), 61-108.

[4] Diestel J, Ruess W M and Schachermayer W, On weak compacteness in $L^1(\mu, X)$, preprint.

[5] Dore G and Venni A, On the closdeness of the sum of two closed operators, Math. Z 196 (1987), 189-201.

[6] Frigon M and Saccon S, Heat equations with discontinuous nonlinearities on convex and nonconvex constraints, Non. Anal. T M A 17 (10) (1991), 923-946.

[7] Jayne J E and Rogers C A, Borel selectors for upper semi-continuous multi-valued functions, J. Funct. Ana. Vol. 56 (3) (1984), 279-299.

[8] Papageorgiou N S, Boundary value problems for evolutions inclusions, Anna. Pol. Math. L (1990), 251-259.

[9] Pazy A, A class of semi-linear equations of evolution, Israel J. Math. 20 (1) (1975) 23-36.

[10] Pruss J and Sohr H, On operators with bounded imaginary powers in Banach spaces, Math. Z. 203 (1990), 429-452.

[11] D. R. Smart (1974) Fixed point theorems, Cambridge University Press, 66.

[12] Vrabie I I (1987) Compactness methods for nonlinear evolutions, Pitman Monographs and Surveys in Pure and Applied Mathematics 32, Longman Scientific and Technical.

P COLLI AND M GRASSELLI
Nonlinear parabolic problems modelling transition dynamics with memory

1. Introduction

In the framework of a recent joint research we have proposed and compared some phase change models in materials with memory (see [5–10] and references therein) by studying questions like existence, uniqueness, and asymptotic properties for the solutions of the resulting initial and boundary value problems. One of these problems, which could be termed as *parabolic Stefan problem with memory*, consists in finding a couple of functions, the (relative) temperature ϑ and the phase variable χ, solving the integrodifferential equation

$$\partial_t(\alpha_0\vartheta + \beta_0\chi + \alpha * \vartheta + \beta * \chi) - \Delta(k_0\vartheta + k * \vartheta) = f \qquad (1.1)$$

and satisfying the graph relationship

$$\chi = 0 \ \text{if} \ \vartheta < 0, \quad \chi \in [0,1] \ \text{if} \ \vartheta = 0, \quad \chi = 1 \ \text{if} \ \vartheta > 0 \qquad (1.2)$$

in the cylindrical domain $\Omega\times]0,T[$, Ω being a smooth bounded open set in $\mathbf{R}^3$ and $T > 0$ standing for a final time. Regarding (1.1), we should specify that $\partial_t := \partial/\partial t$, the symbol "$*$" denotes the usual convolution product with respect to time over $[0,t]$ $(t \in [0,T])$, and Δ is nothing but the Laplacian acting on the space variable. Moreover, the data α_0, β_0, k_0 are positive constants, α, β, $k :]0,T[\to \mathbf{R}$ represent memory functions, and the right hand side $f : \Omega\times]0,T[\to \mathbf{R}$ accounts for the heat source and for the past histories of temperature and phase proportions up to $t = 0$. The system (1.1–2) is complemented by the initial condition

$$(\alpha_0\vartheta + \beta_0\chi)(\cdot,0) = \eta_0, \qquad (1.3)$$

where $\eta_0 : \Omega \to \mathbf{R}$ provides a sort of initial energy, and by a suitable boundary condition. For instance, letting $\{\Gamma_0, \Gamma_\nu\}$ be a partition of the boundary $\partial\Omega$ into two measurable subsets, one can take

$$\vartheta = 0 \ \ \text{on} \ \Gamma_0\times]0,T[, \quad \partial_\nu(k_0\vartheta + k * \vartheta) = g \ \ \text{on} \ \Gamma_\nu\times]0,T[\qquad (1.4)$$

with $g : \Gamma_\nu \times]0, T[\to \mathbf{R}$ known function depending on the normal heat flux (cf. [5]). Here ∂_ν indicates the outward normal derivative on $\partial\Omega$ and Γ_ν is allowed to coincide with $\partial\Omega$. Thus, the problem (1.1–4) describes a phase transition process influenced by what has already occurred during the evolution. This *memory effect* is governed by the relaxation kernels α, β, k. Note that if $\alpha = \beta = k \equiv 0$, then (1.1–4) reduces to a usual two–phase Stefan problem (see, e.g., [11]).

Let us give a variational formulation of (1.1–4). Set

$$V = \{v \in H^1(\Omega) : v = 0 \text{ a.e. in } \Gamma_0\} \quad \text{and} \quad H = L^2(\Omega).$$

By identifying H with its dual space H', we have $V \subset H \subset V'$ with dense and compact injections. Letting $(\cdot, \cdot)$ denote either the duality pairing between V' and V or the scalar product in H, owing to (1.1) and (1.4) it is easy to derive the equality

$$(\partial_t(\alpha_0 \vartheta + \beta_0 \chi + \alpha * \vartheta + \beta * \chi), v) + \int_\Omega \nabla(k_0 \vartheta + k * \vartheta) \nabla v = (f, v) + \int_{\Gamma_\nu} g v \quad (1.5)$$

for any $v \in V$, in $]0, T[$. Assume that α, β, $k \in W^{1,1}(0, T)$, $f \in L^2(0, T; H)$, $g \in W^{1,1}(0, T; L^2(\Gamma_\nu))$, and that $\eta_0 = \alpha_0 \vartheta_0 + \psi_0 \chi_0$ with $\vartheta_0 \in V$ and $\chi_0 \in H$ fulfilling (1.2). Then the variational problem (1.5), (1.2–3) has one and only one solution $\vartheta \in H^1(0, T; H) \cap L^\infty(0, T; V)$, $\chi \in H^1(0, T; V') \cap L^\infty(0, T; H)$ (cf. [5,6,10]). Now, we aim to rewrite (1.5) in an equivalent form. Integrating (1.5) from 0 to $t \in [0, T]$, with the help of (1.3) we can deduce that

$$\beta_0 \chi + \beta * \chi = \mathcal{R} \quad \text{in } V', \tag{1.6}$$

where

$$(\mathcal{R}, v) := (\eta_0 + 1 * f - \alpha_0 \vartheta - \alpha * \vartheta, v) + \int_{\Gamma_\nu} (1 * g) v - \int_\Omega \nabla(1 * (k_0 \vartheta + k * \vartheta)) \nabla v$$

for any $v \in V$, in $]0, T[$, being obviously $(1 * f)(\cdot, t) = \int_0^t f(\cdot, s) ds$. It is known that (see, e.g., [14, Ch. 2]) there is one function $\gamma \in W^{1,1}(0, T)$, named resolvent of β/β_0 and defined as the solution of the equation $\beta_0 \gamma + \beta * \gamma = \beta$, allowing to solve (1.6) with respect to χ, i.e., to transform (1.6) into the relation $\beta_0 \chi = \mathcal{R} - \gamma * \mathcal{R}$, which holds in $H^1(0, T; V')$. Differentiating with respect to time we easily get $\beta_0 \chi_t = \mathcal{R}_t - \partial(\gamma * \mathcal{R}) = \mathcal{R}_t - \gamma(0)\mathcal{R} - \gamma' * \mathcal{R}$. Hence, recalling that the convolution product is associative and commutative and that $k * \vartheta = k(0)(1 * \vartheta) + k' * 1 * \vartheta$, we finally obtain the equality

$$(\partial_t(\alpha_0 \vartheta + \beta_0 \chi), v) + k_0 \int_\Omega \nabla \vartheta \nabla v = (\mathcal{F} + \mathcal{G}, v), \tag{1.7}$$

where

$$(\mathcal{F}, v) := (-\gamma\eta_0 + f - \gamma * f - (\alpha(0) - \gamma(0)\alpha_0)\vartheta, v)$$
$$- ((\alpha' - \gamma(0)\alpha - \alpha_0\gamma' - \gamma' * \alpha) * \vartheta, v),$$

$$(\mathcal{G}, v) := \int_{\Gamma_\nu} (g - \gamma * g)v - (k(0) - \gamma(0)k_0) \int_\Omega \nabla(1 * \vartheta)\nabla v$$
$$- \int_\Omega \nabla((k' - \gamma(0)k - k_0\gamma' - \gamma' * k) * (1 * \vartheta))\nabla v,$$

for any $v \in V$, a.e. in $]0, T[$. Observe that $\mathcal{F}$ and $\mathcal{G}$ depend on ϑ and $1 * \vartheta$, respectively, in a non local way and that the kernels $(\alpha' - \gamma(0)\alpha - \alpha_0\gamma' - \gamma' * \alpha)$ and $(k' - \gamma(0)k - k_0\gamma' - \gamma' * k)$ belong to $L^1(0, T)$. On the other hand, since (cf., e.g., [14]) $\|\nu * z\|_{L^p(0,T;H)} \leq \|\nu\|_{L^1(0,T)}\|z\|_{L^p(0,T;H)}$ for any $\nu \in L^1(0, T)$ and any $z \in L^p(0, T; H)$ with $p \in [1, \infty]$, it turns out that $\mathcal{F} \in L^2(0, T; H)$ and $\mathcal{G} \in W^{1,1}(0, T; V')$.

The above problem (in the formulation characterized by (1.7)) provides an example of the abstract Cauchy problem we are going to investigate in this paper. We write it in terms of the so-called freezing index (see [13]) $u = 1 * \vartheta$. Letting $F : u' \mapsto F[u']$ and $G : u \mapsto G[u]$ define two nonlinear mappings such that

$$F : L^2(0, t; H) \to L^2(0, t; H), \quad G : W^{1,1}(0, t; V) \to W^{1,1}(0, t; V') \quad \forall\, t \in]0, T],$$

we deal with the evolution equation

$$d_t\,(u'(t) + Au'(t)) + Lu'(t) \ni F\,[u']\,(t) + G\,[u]\,(t) \quad \text{for a.e. } t \in]0, T[, \qquad (1.8)$$

where $d_t := d/dt$, A is a maximal monotone operator from V to V', and $L : V \to V'$ is linear, continuous, and selfadjoint. Moreover, denoting by I the identity (or injection) operator, the existence of a constant $\ell > 0$ making the operator $\ell I + L$ strongly monotone is supposed, thus requiring some coerciveness for L. Suitable boundedness and continuity properties are assumed for F and G (cf. the later (2.4-7)) including however the simple (linear) case given by the expressions of $\mathcal{F}$ and $\mathcal{G}$. In this concern, let us point out that one could allow the right hand side f of (1.1) and the boundary datum g in (1.4) to depend on ϑ and $1 * \vartheta$, respectively. A further remark regards the general conditions postulated for the operator A, whereas the graph in (1.2) yields a subdifferential mapping bounded in H. Thus, our setting may cover a larger class of applications to partial differential equations, especially to systems.

By supplying (1.8) with the initial conditions

$$u(0) = u_0, \quad (u' + Au')\,(0) \ni e_0, \qquad (1.9)$$

an existence and uniqueness theorem is established for the related problem. First we examine the explicit version of (1.8), where F and G are independent of the solution, and deduce preliminary results by exploiting standard techniques of maximal monotone operators (see [3,1]). In respect of the explicit problem, let us quote, for instance, the works [4,12,2] and their references as well, since there similar abstract equations are dealt with. Then, to show that the actual problem (1.8–9) is well posed we make use of an appropriate fixed point argument. The Lipschitz continuous dependence of the solution on the initial data is also proved. In the last part of the paper we address the analysis to the situation of A subdifferential. In this case an existence and uniqueness result is obtained under weaker assumptions on F, G, and e_0, even if the operator A does not satisfy any boundedness property.

2. Main result

We start by recalling some notation and stating precise hypotheses on the data. Let $T > 0$ and let V and H two Hilbert spaces such that $V \subset H \equiv H' \subset V'$ with dense and continuous injections. Both the scalar product in H and the duality pairing between V' and V are indicated by $(\cdot, \cdot)$. For the sake of simplicity, we denote by $\|\cdot\|, |\cdot|, \|\cdot\|_*$ the norms in V, H, V', respectively. Then, we require that

$$A \text{ is a maximal monotone operator from } V \text{ to } V' \tag{2.1}$$

with domain $D(A) \subseteq V$. Moreover,

$$L : V \to V' \text{ is linear continuous symmetric} \tag{2.2}$$

and weakly coercive in the sense that

$$\exists \, \ell > 0 \quad : \quad \ell |v|^2 + (Lv, v) \geq c \, \|v\|^2 \quad \forall \, v \in V, \text{ for some } c > 0. \tag{2.3}$$

The assumptions on the mappings $F : L^2(0, T; H) \to L^2(0, T; H)$, $G : W^{1,1}(0, T; V) \to W^{1,1}(0, T; V')$ read as follows. There are two positive constants C_F and C_G such that

$$\|F[v]\|_{L^2(0,t;H)}^2 \leq C_F \left(1 + \|v\|_{L^2(0,t;H)}^2 \right), \tag{2.4}$$

$$\|F[v_1] - F[v_2]\|_{L^1(0,t;H)}^2 \leq C_F \|v_1 - v_2\|_{L^1(0,t;H)}^2 \tag{2.5}$$
$$\forall \, t \in \,]0, T], \quad \forall \, v, v_1, v_2 \in L^2(0, T; H),$$

$$\|G[v](0)\|_*^2 + \|d_t G[v]\|_{L^1(0,t;V')}^2 \leq C_G \left(1 + \|v(0)\|^2 + \|v'\|_{L^1(0,t;V)}^2 \right), \tag{2.6}$$

$$\|G[v_1] - G[v_2]\|_{L^2(0,t;V')}^2 \leq C_G \|v_1 - v_2\|_{L^2(0,t;V)}^2 \tag{2.7}$$
$$\forall \, t \in \,]0, T], \quad \forall \, v, v_1, v_2 \in W^{1,1}(0, T; V).$$

The initial datum u_0 is just supposed to satisfy

$$u_0 \in V, \tag{2.8}$$

while $e_0 \in V'$ has to belong to the range of the operator $I + A$, that is,

$$e_0 = v_0 + w_0 \quad \text{for some } v_0 \in D(A) \text{ and } w_0 \in Av_0. \tag{2.9}$$

We can finally present the main result of this note.

Theorem 1. *Assume that (2.1–9) hold. Then there exists one and only one pair*

$$u \in H^2(0,T;H) \cap W^{1,\infty}(0,T;V), \quad w \in H^1(0,T;V') \tag{2.10}$$

fulfilling

$$(u' + w)'(t) + Lu'(t) = F\left[u'\right](t) + G\left[u\right](t), \tag{2.11}$$

$$w(t) \in Au'(t) \tag{2.12}$$

for a.e. $t \in \,]0,T[$ and

$$u(0) = u_0, \quad (u' + w)(0) = e_0. \tag{2.13}$$

Remark 1. In view of (2.10), one could wonder about the values $u'(0)$ and $w(0)$, which do not follow directly from (2.13). Nonetheless, it will become clear in the sequel that necessarily $u'(0) = v_0$ and $w(0) = w_0$.

Remark 2. The question arises of what kind of nonlinearities can be considered in the above framework. Referring, for instance, to the case $V = H^1(\Omega)$ and $H = L^2(\Omega)$, as an example one could take

$$(F\left[u'\right](t), v) = \int_\Omega \int_0^t \alpha_1(\cdot, t - \tau, u_t(\cdot, \tau))d\tau\, v + \int_\Omega \alpha_2(\cdot, t, u_t(\cdot, t))v,$$

$$(G\left[u\right](t), v) = \int_\Omega \int_0^t k_1(\cdot, t - \tau, u(\cdot, \tau), \nabla u(\cdot, \tau))d\tau\, \nabla v + \int_{\partial\Omega} k_2(\cdot, t, u(\cdot, t))v,$$

for $v \in V$, with suitable requirements on the functions $\alpha_1, \alpha_2, k_2 : \Omega \times]0, T[\times \mathbf{R} \to \mathbf{R}$ and $k_1 : \Omega \times]0, T[\times \mathbf{R} \times \mathbf{R}^3 \to \mathbf{R}^3$ in order to satisfy (2.4–7).

The proof of Theorem 1 is detailed in the next sections. Note that it is not restrictive working under the condition

$$(Lv, v) \geq c\,\|v\|^2 \quad \forall\, v \in V \tag{2.14}$$

in place of (2.3). Indeed, if (2.14) does not hold, it suffices to add the term $\ell u'(t)$ to both sides of equation (2.11) and incorporate it into $Lu'(t)$ and $F\left[u'\right](t)$, respectively. Thus, (2.2) and (2.13) are fulfilled by the new L, as well as inequalities like (2.4–5) remain true for the modified F.

3. The explicit problem

In this section we prepare an auxiliary result concerning the problem (2.11–13) with a given right hand side in (2.11). Henceforth, we denote by C a constant such that

$$\|v\|_* \le C\,|v| \quad \forall\, v \in H, \quad |v| \le C\,\|v\| \quad \text{and} \quad \|Lv\|_* \le C\,\|v\| \quad \forall\, v \in V. \tag{3.1}$$

Also, for $z \in H^1(0,T;H) \cap L^\infty(0,T;V)$ we define the auxiliary quantity

$$\|z\|_t^2 := \|z'\|_{L^2(0,t;H)}^2 + c\,\|z\|_{L^\infty(0,t;V)}^2 \quad \forall\, t \in\,]0,T].$$

Then the following statement summarizes properties of the explicit problem.

Lemma 1. *Given two functions*

$$f \in L^2(0,T;H), \quad g \in W^{1,1}(0,T;V'), \tag{3.2}$$

under the assumptions (2.1–2), (2.14), and (2.8–9) there exists a unique pair (u,w) satisfying (2.10), (2.13), and

$$(u' + w)'(t) + Lu'(t) = f(t) + g(t), \tag{3.3}$$
$$w(t) \in Au'(t) \tag{3.4}$$

for a.e. $t \in\,]0,T[$. Moreover, there is a constant C_1, depending on C, $\|v_0\|$, and c, such that for any $t \in\,]0,T]$ one has

$$\|u'\|_t^2 \le C_1\left(1 + \|f\|_{L^2(0,t;H)}^2 + \|g(0)\|_*^2 + \|g'\|_{L^1(0,t;V')}^2\right). \tag{3.5}$$

Furthermore, letting $\{u_{0i}, e_{0i}, f_i, g_i\}$, $i = 1,2$, be two sets of data and letting (u_i, w_i), $i = 1,2$, represent the corresponding solutions of the problem (3.3–4), (2.13), the estimate

$$\|u_1 - u_2\|_t^2 \le C_2\Big(\|u_{01} - u_{02}\|^2 + \|e_{01} - e_{02}\|_*^2$$
$$+ \int_0^t \|f_1 - f_2\|_{L^1(0,\tau;H)}^2\, d\tau + \|g_1 - g_2\|_{L^2(0,t;V')}^2\Big) \tag{3.6}$$

holds for any $t \in\,]0,T]$, where the constant C_2 depends only on c and T.

Proof. Let us first prove (3.6). Thanks to (2.13), an integration of (3.3) yields

$$u_1'(\tau) - u_2'(\tau) + w_1(\tau) - w_2(\tau) + L(u_1 - u_2)(\tau) - L(u_{01} - u_{02})$$
$$= e_{01} - e_{02} + \int_0^\tau (f_1 - f_2)(s)\,ds + \int_0^\tau (g_1 - g_2)(s)\,ds \quad \forall\, \tau \in [0,T]. \tag{3.7}$$

Next, we multiply (3.7) by $u_1'(\tau) - u_2'(\tau)$ (cf. (2.10)) and integrate from 0 to $t \in]0, T]$. As $(w_1 - w_2, u_1' - u_2') \geq 0$ a.e. in $]0, T[$ because of (2.12) and (2.1), recalling (2.2) it is easy to check that $\tilde{u} := u_1 - u_2$ satisfies

$$\|\tilde{u}'\|_{L^2(0,t;H)}^2 + \frac{1}{2}(L(\tilde{u}(t) - \tilde{u}_0), \tilde{u}(t) - \tilde{u}_0) \leq (\tilde{e}_0, \tilde{u}(t) - \tilde{u}_0) + \int_0^t (\int_0^\tau \tilde{f}(s)ds, \tilde{u}'(\tau))d\tau$$

$$+ (\int_0^t \tilde{g}(\tau)d\tau, \tilde{u}(t)) - \int_0^t (\tilde{g}(\tau), \tilde{u}(\tau))d\tau \quad \forall t \in]0, T], \tag{3.8}$$

where $\tilde{u}_0 := u_{01} - u_{02}$, $\tilde{e}_0 := e_{01} - e_{02}$, $\tilde{f} := f_1 - f_2$, and $\tilde{g} := g_1 - g_2$. Making use of (2.14) and of the elementary inequality

$$ab \leq (\varepsilon/2)a^2 + (2\varepsilon)^{-1}b^2 \quad \forall\, a, b \in \mathbf{R}, \ \forall \varepsilon > 0, \tag{3.9}$$

from (3.8) one can deduce that

$$\frac{1}{2}\|\tilde{u}'\|_{L^2(0,t;H)}^2 + \frac{c}{4}\|\tilde{u}(t) - \tilde{u}_0\|^2 \leq \frac{1}{c}\|\tilde{e}_0\|_*^2 + \frac{1}{2}\int_0^t \left|\int_0^\tau \tilde{f}(s)ds\right|^2 d\tau$$

$$+ 2\|\tilde{u}\|_{C^0([0,t];V)}\int_0^t \|\tilde{g}(\tau)\|_* d\tau \quad \forall t \in]0, T]. \tag{3.10}$$

Since $\|\tilde{u}(t)\|^2 \leq 2\|\tilde{u}_0\|^2 + 2\|\tilde{u}(t) - \tilde{u}_0\|^2$, by handling (3.10) it is not difficult to conclude that

$$\frac{1}{8}\|\tilde{u}\|_t^2 \leq \frac{c}{2}\|\tilde{u}_0\|^2 + \frac{2}{c}\|\tilde{e}_0\|_*^2 + \int_0^t \|\tilde{f}\|_{L^1(0,\tau;H)}^2 d\tau + \frac{4}{\sqrt{c}}\|\tilde{u}\|_t\|\tilde{g}\|_{L^1(0,t;V')}$$

for any $t \in]0, T]$. Now it suffices to apply (3.9) to the last term and the Hölder inequality in order to obtain (3.6). Note that, in particular, (3.6) entails uniqueness (cf. (3.7) as well) for the problem (3.3–4), (2.13). In order to show the existence of a solution, we approximate (3.3–4) by

$$(u_\varepsilon' + \varepsilon L u_\varepsilon' + w_\varepsilon)'(t) + L u_\varepsilon'(t) = f_\varepsilon(t) + g_\varepsilon(t), \tag{3.11}$$

$$w_\varepsilon(t) \in A u_\varepsilon'(t) \tag{3.12}$$

and take the initial conditions

$$u_\varepsilon(0) = u_0, \quad (u_\varepsilon' + \varepsilon L u_\varepsilon' + w_\varepsilon)(0) = e_{0\varepsilon}, \tag{3.13}$$

where $\varepsilon > 0$, $f_\varepsilon \in C^0([0,T];H)$, $g_\varepsilon \in C^0([0,T];V')$, and $e_{0\varepsilon} \in V'$. Also, the sequences $\{f_\varepsilon\}$, $\{g_\varepsilon\}$, $\{e_{0\varepsilon}\}$ are intended to approach f, g, e_0, respectively, as $\varepsilon \searrow$

88

0. Due to (2.1–2) and (2.14), it results that (see, e.g., [1, Cor. 1.3, p. 48]) the operator $I + \varepsilon L + A$ is maximal monotone and strongly monotone from V to V', with domain $D(A)$ and range V'. Hence the inverse operator $(I + \varepsilon L + A)^{-1}$ is single–valued and Lipschitz continuous from V' to V. This property, coupled with (2.2) and (2.14) ($L : V \to V'$ may be viewed as a Riesz isomorphism), ensures that there exists one and only one solution $y_\varepsilon \in C^1([0,T]; V')$ of the following Cauchy problem

$$y'_\varepsilon(t) + L(I + \varepsilon L + A)^{-1} y_\varepsilon(t) = f_\varepsilon(t) + g_\varepsilon(t) \quad \forall \, t \in [0,T], \tag{3.14}$$

$$y_\varepsilon(0) = e_{0\varepsilon}. \tag{3.15}$$

Indeed, observe that (3.14) is just an ordinary differential equation in V'. Setting $u_\varepsilon(t) = u_0 + \int_0^t (I + \varepsilon L + A)^{-1} y_\varepsilon(\tau)d\tau$ for any $t \in [0,T]$, it is a standard matter to verify that u_ε and $w_\varepsilon = y_\varepsilon - u'_\varepsilon - \varepsilon L u'_\varepsilon$ fulfil (3.11–12) for any $t \in [0,T]$ and (3.13). Conversely, if $(u_\varepsilon, w_\varepsilon)$ yields a solution of (3.11–13), then one easily sees that $y_\varepsilon = u'_\varepsilon + \varepsilon L u'_\varepsilon + w_\varepsilon$ solves (3.15–16). This equivalence argument allows us to infer that there is a unique pair $u_\varepsilon \in C^2([0,T]; V)$, $w_\varepsilon \in C^1([0,T]; V')$ satisfying (3.13) and (3.11–12) for any $t \in [0,T]$. Moreover, choosing (cf. (2.13) and (2.9)) $e_{0\varepsilon} = e_0 + \varepsilon L v_0$ it turns out that

$$u'_\varepsilon(0) = v_0, \quad w_\varepsilon(0) = w_0 \tag{3.16}$$

for any $\varepsilon > 0$. Now, letting (cf. (3.2)) $f_\varepsilon \to f$ strongly in $L^2(0,T; H)$ as $\varepsilon \searrow 0$ and simply $g_\varepsilon = g$, we derive uniform estimates for $(u_\varepsilon, w_\varepsilon)$. Testing (3.11) by $u''_\varepsilon(t)$ and integrating, with the help of (2.14), (2.2), (3.16), and (3.1) we get

$$\|u''_\varepsilon\|^2_{L^2(0,\tau;H)} + \varepsilon c \|u''_\varepsilon\|^2_{L^2(0,\tau;V)} + \int_0^\tau (w'_\varepsilon(t), u''_\varepsilon(t))dt + \frac{c}{2} \|u'_\varepsilon(\tau)\|^2$$

$$\leq \frac{C}{2} \|v_0\|^2 + \int_0^\tau (f_\varepsilon(t), u''_\varepsilon(t))dt + \int_0^\tau (g(t), u''_\varepsilon(t))dt \quad \forall \, \tau \in \,]0,T]. \tag{3.17}$$

Thanks to (3.12) and to the monotonicity of A, we have that $\int_0^\tau (w'_\varepsilon(t), u''_\varepsilon(t))dt \geq 0$. In addition, on account of (3.9) we deduce that

$$\int_0^\tau (f_\varepsilon(t), u''_\varepsilon(t))dt \leq \frac{1}{2} \|f_\varepsilon\|^2_{L^2(0,\tau;H)} + \frac{1}{2} \|u''_\varepsilon\|^2_{L^2(0,\tau;H)}$$

and, via integrations by parts,

$$\int_0^\tau (g(t), u''_\varepsilon(t))dt = (g(0), u'_\varepsilon(\tau) - v_0) + \int_0^\tau (g'(t), u'_\varepsilon(\tau) - u'_\varepsilon(t))dt$$

$$\leq 2 \|u'_\varepsilon\|_{C^0([0,\tau];V)} \left(\|g(0)\|_* + \|g'\|_{L^1(0,\tau;V')} \right).$$

89

Hence, from (3.17) it follows that

$$\|u_\varepsilon''\|^2_{L^2(0,\tau;H)} + 2\varepsilon c \|u_\varepsilon''\|^2_{L^2(0,\tau;V)} + c \|u_\varepsilon'(\tau)\|^2 \le C \|v_0\|^2 + \|f_\varepsilon\|^2_{L^2(0,t;H)}$$
$$+ 4 \|u_\varepsilon'\|_{C^0([0,t];V)} \left(\|g(0)\|_* + \|g'\|_{L^1(0,t;V')} \right) \quad \text{for } 0 < \tau \le t \le T.$$

Taking the maximum of the left hand side with respect to τ, it is straightforward to check that

$$\|u_\varepsilon'\|^2_t + 2\varepsilon c \|u_\varepsilon''\|^2_{L^2(0,t;V)} \le 2C \|v_0\|^2 + 2 \|f_\varepsilon\|^2_{L^2(0,t;H)}$$
$$+ \frac{8}{\sqrt{c}} \|u_\varepsilon'\|_t \left(\|g(0)\|_* + \|g'\|_{L^1(0,t;V')} \right).$$

At this point, a further application of (3.9) enables us to determine a constant C_1 (the one playing in (3.5)) such that

$$\|u_\varepsilon'\|^2_t + \varepsilon \|u_\varepsilon''\|^2_{L^2(0,t;V)} \le C_1 \left(1 + \|f_\varepsilon\|^2_{L^2(0,t;H)} + \|g(0)\|^2_* + \|g'\|^2_{L^1(0,t;V')} \right) \quad (3.18)$$

for any $t \in \,]0,T]$. Then, recalling (3.13), (3.16) and letting $0 < \varepsilon \le 1$, we conclude that $\|u_\varepsilon\|_{H^2(0,T;H) \cap W^{1,\infty}(0,T;V)}$ and $\sqrt{\varepsilon} \|u_\varepsilon\|_{H^2(0,T;V)}$ are bounded independently of ε. Further, a comparison in (3.11) yields the boundedness of $\|w_\varepsilon\|_{H^1(0,T;V')}$. Therefore there exist two functions u and w such that, in principle for subsequences, $u_\varepsilon \to u$ weakly star in $H^2(0,T;H) \cap W^{1,\infty}(0,T;V)$ and $w_\varepsilon \to w$ weakly in $H^1(0,T;V')$ as ε tends to 0. To verify that the pair (u,w) gives the wanted solution, we first note that u and w must fulfil both (3.16) and (2.13) (see Remark 1). On the other hand, $\varepsilon u_\varepsilon \to 0$ strongly in $H^2(0,T;V)$ so that, passing to the limit in (3.11), in virtue of (2.2) we establish (3.3). It remains to prove that (3.4) holds for a.e. $t \in \,]0,T[$. Owing to (3.12) it suffices to infer the inequality (cf., e.g., [3, pp. 25–27])

$$\limsup_{\varepsilon \searrow 0} \int_0^T (w_\varepsilon(t), u_\varepsilon'(t))dt \le \int_0^T (w(t), u'(t))dt. \quad (3.19)$$

By using (3.11), (3.13), (3.3), (2.13) and arguing as in the deduction of (3.8), it is not difficult to obtain

$$\int_0^T (w_\varepsilon(t), u_\varepsilon'(t))dt = \int_0^T (e_0 + \varepsilon L v_0 + \int_0^t (f_\varepsilon + g)(\tau)d\tau, u_\varepsilon'(t))dt - \|u_\varepsilon'\|^2_{L^2(0,T;H)}$$
$$- \int_0^T (\varepsilon L u_\varepsilon'(t), u_\varepsilon'(t))dt - \frac{1}{2}(L(u_\varepsilon(T) - u_0), u_\varepsilon(T) - u_0)$$

and the analogous identity for $\int_0^T (w(t), u'(t))dt$. Since $u_\varepsilon(T) \to u(T)$ weakly in V as $\varepsilon \searrow 0$, then (3.19) comes as a consequence of (2.9), (2.14), and of the weak lower semicontinuity of norms. Finally, taking the lower limit in (3.18) we easily recover (3.5) and thus achieve the proof of the lemma.

4. Local existence and uniqueness

In order to show that the problem (2.11–13) admits a unique local solution, we apply a fixed point argument and exploit the Contraction Mapping Principle. Consider a function z belonging to the following space

$$X := \{v \in H^2(0,\tau;H) \cap W^{1,\infty}(0,\tau;V) : \ v(0) = u_0, \ \|v'\|_\tau^2 \leq M\}, \qquad (4.1)$$

with $\tau \in]0,T]$ and $M > 0$ to be specified. Let now $u \in H^2(0,\tau;H) \cap W^{1,\infty}(0,\tau;V)$ and $w \in H^1(0,\tau;V')$ satisfy (2.13) and

$$u''(t) + w'(t) + Lu'(t) = F[z'](t) + G[z](t), \qquad (4.2)$$

(2.12) for a.e. $t \in]0,\tau[$. Due to the assumptions on F and G and to Lemma 1, the solution (u,w) of (4.2), (2.12–13) is uniquely determined and we can set $u = B(z)$ defining thus the operator B. We claim that B maps X into X provided M is suitably chosen and τ is small enough. Indeed, $u(0) = u_0$ and (3.5), (2.4), (2.6), and (3.1) imply that

$$
\begin{aligned}
\|u'\|_\tau^2 &\leq C_1 \left(1 + \|F[z']\|_{L^2(0,\tau;H)}^2 + \|G[z](0)\|_*^2 + \|d_t G[z]\|_{L^1(0,\tau;V')}^2 \right) \\
&\leq C_1 + C_1 C_F \left(1 + C^2 \|z'\|_{L^2(0,\tau;V)}^2 \right) + C_1 C_G \left(1 + \|u_0\|^2 + \|z'\|_{L^1(0,\tau;V)}^2 \right) \\
&\leq C_3 + C_4 \tau \|z'\|_\tau^2,
\end{aligned}
\qquad (4.3)
$$

where $C_3 = C_1 + C_1 C_F + C_1 C_G + C_1 C_G \|u_0\|^2$ and $C_4 = C_1 C_F C^2 + C_1 C_G T$. Hence, taking $M = 2C_3$ and $\tau \leq 1/(2C_4)$ it is ensured that $B : X \to X$.

Next, on account of the weak star lower semicontinuity of norms it is not difficult to see that the space X, endowed with the distance $d(z_1,z_2) := \|z_1 - z_2\|_\tau$ ($z_1, z_2 \in X$), is a complete metric space. We are looking for conditions on τ allowing B to result a contraction mappping. If $z_1, z_2 \in X$ and $u_1 = B(z_1)$, $u_2 = B(z_2)$, then we have that

$$
\begin{aligned}
\|u_1 - u_2\|_\tau^2 &\leq C_2 \left(\int_0^\tau \|F[z_1'] - F[z_2']\|_{L^1(0,t;H)}^2 \, dt + \|G[z_1] - G[z_2]\|_{L^2(0,\tau;V')}^2 \right) \\
&\leq C_2 \int_0^\tau \left(C_F T \|z_1 - z_2\|_{L^2(0,t;H)}^2 + C_G \|(z_1 - z_2)(t)\|^2 \right) dt \\
&\leq C_2 \max\{C_F T, C_G\} \tau \|z_1 - z_2\|_\tau^2
\end{aligned}
\qquad (4.4)
$$

because of (3.6), (2.13), (2.5), the Hölder inequality, and (2.7). Therefore, setting $C_5 = C_2 \max\{C_F T, C_G\}$, it is sufficient to choose $\tau = \min\{T, 1/(2C_4), 1/(2C_5)\}$ to conclude that the operator B has a unique fixed point $u \in X$. Obviously, such u and the corresponding w give the solution to the problem (2.11–13) on the interval $[0,\tau]$.

5. A priori estimate and continuous dependence

Here the proof of Theorem 1 is going to be completed by showing global existence. To this aim it suffices to derive a global estimate for the solution (u, w) on the whole interval $[0, T]$. We also prove a Lipschitz continuous dependence of the solution upon the initial data. As in the previous section, the basic tools are the hypotheses (2.4-7) and the statement of Lemma 1.

First, recalling (4.3) and using the Hölder inequality we infer that

$$\|u'\|_\tau^2 \le C_3 + C_4 \int_0^\tau \|u'\|_s^2 \, ds, \tag{5.1}$$

where τ is now an arbitrary value in $[0, T]$. By integrating (5.1) from 0 to t and then applying the Gronwall lemma we deduce that $\int_0^t \|u'\|_\tau^2 \, d\tau \le C_3 T \exp(C_4 T)$ for any $t \in [0, T]$. Hence, thanks to (5.1), (2.13), and (3.1) it is straightforward to find a constant C_6 such that $\|u\|_{H^2(0,\tau;H) \cap W^{1,\infty}(0,\tau;V)} \le C_6$ for any $\tau \in \,]0, T]$. At this point, one can provide an analogous bound for $\|w\|_{H^1(0,T;V')}$ in virtue of the relationship (see (2.11) and (2.13))

$$w(t) = \int_0^t (F[u'] + G[u])(\tau)d\tau + e_0 - u'(t) + L(u_0 - u(t)) \quad \forall\, t \in [0, T]. \tag{5.2}$$

Then, consider two pairs of initial data (u_{0i}, e_{0i}), $i = 1, 2$, and denote by (u_1, w_1) and (u_2, w_2) the solutions of the problem (2.11-13) with u_0, e_0 replaced by u_{01}, e_{01} and u_{02}, e_{02}, respectively. Owing to (3.6) and (4.4) it turns out that

$$\|u_1 - u_2\|_t^2 \le C_7 \left(\|u_{01} - u_{02}\|^2 + \|e_{01} - e_{02}\|_*^2 + \int_0^t \|u_1 - u_2\|_\tau^2 \, d\tau \right) \quad \forall\, t \in \,]0, T],$$

where, for instance, $C_7 = C_2 \max\{1, C_F T, C_G\}$. Since the function $t \mapsto \|u_1 - u_2\|_t^2$ is continuous (cf. (2.10)), a direct application of the Gronwall lemma yields

$$\|u_1 - u_2\|_t^2 \le C_7 e^{C_7 t} \left(\|u_{01} - u_{02}\|^2 + \|e_{01} - e_{02}\|_*^2 \right) \quad \forall\, t \in \,]0, T]. \tag{5.3}$$

On the other hand, writing (5.2) for w_1, w_2 and taking the difference, by (2.5), (2.7), and (3.1) it is easy to calculate a constant C_8 fulfilling

$$\|w_1 - w_2\|_{L^2(0,t;V')} \le C_8 \left(\|u_1 - u_2\|_t + \|e_{01} - e_{02}\|_* + \|u_{01} - u_{02}\| \right) \tag{5.4}$$

for any $t \in \,]0, T]$. The two estimates (5.3) and (5.4) enable us to state the following result.

Theorem 2. *Assume that (2.1-7) hold. For $i = 1, 2$ let u_{0i}, e_{0i} fulfil (2.8-9) and let (u_i, w_i) be the corresponding solution of (2.11-13). Then there exists a constant C_9, depending only on c, T, C_F, C_G, and C, such that*

$$\|u_1 - u_2\|_{H^1(0,T;H) \cap C^0([0,T];V)} + \|w_1 - w_2\|_{L^2(0,T;V')}$$
$$\le C_9 \left(\|u_{01} - u_{02}\| + \|e_{01} - e_{02}\|_* \right). \tag{5.5}$$

6. The case of A subdifferential

This section is devoted to the investigation of the problem (1.8–9) when the operator A is the subdifferential of a convex and lower semicontinuous function ϕ. Thus, in addition to (2.1) we suppose that

$$A = \partial\phi \quad \text{with} \quad \phi : V \to]-\infty, +\infty] \quad \text{proper convex lower semicontinuous.} \quad (6.1)$$

As usual, we denote by $\phi^*(z) := \sup\{(z, v) - \phi(v), \; v \in V\}$, $z \in V'$, the convex conjugate of ϕ. We also introduce the auxiliary function $\psi(v) := (1/2)|v|^2 + \phi(v)$, $v \in V$, which is still convex and lower semicontinuous. Observe that $\partial\psi = I + A$ and $\partial\psi^* = (I + A)^{-1}$ (see, e.g., [3] or [1]). Concerning the data, we replace the assumptions (2.4), (2.6) by the weaker ones

$$\|F[v]\|^2_{L^1(0,t;H)} \leq C_F \left(1 + \|v\|^2_{L^1(0,t;H)}\right) \quad \forall\, t \in]0, T], \;\; \forall\, v \in L^1(0, T; H), (6.2)$$

$$\|G[v]\|^2_{L^2(0,t;V')} \leq C_G \left(1 + \|v\|^2_{L^2(0,t;V)}\right) \quad \forall\, t \in]0, T], \;\; \forall\, v \in L^2(0, T; V), (6.3)$$

and let $e_0 \in D(\psi^*)$, that is,

$$\psi^*(e_0) = \sup\{(e_0, v) - \psi(v), \; v \in V\} < +\infty, \quad (6.4)$$

instead of requiring $e_0 \in D(\partial\psi^*)$ as in (2.9). In view of (2.5) and (2.7), F and G act now on $L^1(0, T; H)$ and $L^2(0, T; V)$, respectively, and the conditions (6.2–3) are ensured provided, for instance, $F[0] \in L^1(0, T; H)$, $G[0] \in L^2(0, T; V')$. Since, in general, $D(\partial\psi^*)$ is a proper subset of $D(\psi^*)$, the hypothesis (6.4) turns out to be an extension of (2.9).

Remark 3. In the example considered in the Introduction, where (cf. (1.2)) $\phi(v) = \int_\Omega \max\{v, 0\}$ makes sense for $v \in L^1(\Omega)$, the reduction of (2.9) yields that

the function $\min\{e_0, \max\{0, e_0 - 1\}\}$ belongs to $H^1(\Omega)$,

while the related (6.4) is certainly satisfied if e_0 lies just in $L^2(\Omega)$.

Within the above framework, we have the following result.

Theorem 3. *Assume that (2.1–3), (2.5), (2.7–8), and (6.1–4) hold. Then there exists one and only one pair*

$$u \in W^{1,\infty}(0, T; H) \cap H^1(0, T; V), \quad w \in L^\infty(0, T; V') \quad (6.5)$$

fulfilling

$$(u' + w) \in W^{1,1}(0, T; V') \tag{6.6}$$

and solving the problem (2.11–13). Moreover, the solution depends continuously on the initial data as in Theorem 2.

Proof. The method already used in the paper applies also to this case. In particular, we point out that the proofs of (3.6) and (5.5) are exactly the same, due to (2.5) and (2.7). Hence it will suffice to state an analogue of Lemma 1 for a pair of given functions $f \in L^1(0, T; H)$ and $g \in L^2(0, T; V')$. More precisely, we show that the problem (3.3–4), (2.13) admits a unique solution (u, w) and we verify the estimate

$$\|u'\|^2_{L^\infty(0,t;H) \cap L^2(0,t;V)} \leq C_{10} \left(1 + \|f\|^2_{L^1(0,t;H)} + \|g\|^2_{L^2(0,t;V')}\right) \quad \forall\, t \in {]0, T]}, \tag{6.7}$$

where the constant C_{10} depends only on $\psi^*(e_0)$, $\|e_0\|$, T, C, and c. Consider then the regularized problem (3.11–13), take $e_{0\varepsilon} = e_0$ for any $\varepsilon > 0$, and let

$$f_\varepsilon \to f \quad \text{strongly in } L^1(0, T; H), \qquad g_\varepsilon \to g \quad \text{strongly in } L^2(0, T; V') \tag{6.8}$$

as $\varepsilon \searrow 0$. The needed uniform bounds for $(u_\varepsilon, w_\varepsilon)$ are derived first multiplying (3.11) by $u'_\varepsilon(t)$ and integrating with respect to t. Since $u'_\varepsilon \in A^{-1} w_\varepsilon$ a.e. in ${]0, T[}$ because of (3.12) and $A^{-1} = \partial \phi^*$, it turns out that (cf., e.g., [3, p. 73]) $\int_0^t (w'_\varepsilon, u'_\varepsilon) = \phi^*(w_\varepsilon(t)) - \phi^*(w_\varepsilon(0))$. Therefore, with the help of (2.2), (2.14), and (3.9) we deduce that

$$\frac{1}{2} |u'_\varepsilon(t)|^2 + \frac{\varepsilon c}{2} \|u'_\varepsilon(t)\|^2 + \frac{c}{2} \int_0^t \|u'_\varepsilon(\tau)\|^2 \, d\tau \leq \frac{1}{2} |u'_\varepsilon(0)|^2 + \frac{\varepsilon}{2}(L u'_\varepsilon(0), u'_\varepsilon(0))$$

$$+ \phi^*(w_\varepsilon(0)) - \phi^*(w_\varepsilon(t)) + \|u'_\varepsilon\|_{L^\infty(0,t;H)} \int_0^t |f_\varepsilon(\tau)| \, d\tau + \frac{1}{2c} \int_0^t \|g_\varepsilon(\tau)\|^2_* \, d\tau \tag{6.9}$$

for any $t \in {]0, T]}$. Owing to the inclusion $w_\varepsilon(0) \in \partial \phi(u'_\varepsilon(0))$, it results that (see, e.g., [1, p. 54]) $\phi^*(w_\varepsilon(0)) + \phi(u'_\varepsilon(0)) = (w_\varepsilon(0), u'_\varepsilon(0))$. Hence, by (3.13), (2.14), and (6.4) we infer that

$$\frac{1}{2} |u'_\varepsilon(0)|^2 + \frac{\varepsilon}{2}(L u'_\varepsilon(0), u'_\varepsilon(0)) + \phi^*(w_\varepsilon(0))$$

$$= (e_0, u'_\varepsilon(0)) - \psi(u'_\varepsilon(0)) - \frac{\varepsilon}{2}(L u'_\varepsilon(0), u'_\varepsilon(0)) \leq \psi^*(e_0). \tag{6.10}$$

Next, fix some element $\bar{v} \in D(\phi)$ and note that $\phi^*(w_\varepsilon(t)) \geq (w_\varepsilon(t), \bar{v}) - \phi(\bar{v})$. As

$$w_\varepsilon(t) = \int_0^t (f_\varepsilon + g_\varepsilon)(\tau) d\tau + e_0 - u'_\varepsilon(t) - \varepsilon L u'_\varepsilon(t) - \int_0^t L u'_\varepsilon(\tau) d\tau \tag{6.11}$$

in virtue of (3.11) and (3.13), using (3.1) and (3.9) it is straightforward to check that

$$-\phi^*(w_\varepsilon(t)) \leq \phi(\bar{v}) + |\bar{v}| \, \|f_\varepsilon\|_{L^1(0,t;H)} + \frac{1}{2} \|g_\varepsilon\|^2_{L^2(0,t;V')} + \frac{T}{2} \|\bar{v}\|^2 + \|e_0\| \, \|\bar{v}\|$$

$$+ \frac{1}{4} |u'_\varepsilon(t)|^2 + |\bar{v}|^2 + \frac{\varepsilon c}{4} \|u'_\varepsilon(t)\|^2 + \frac{c}{4} \|u'_\varepsilon\|^2_{L^2(0,T;V)} + \frac{\varepsilon+T}{c} C^2 \|\bar{v}\|^2 . \quad (6.12)$$

Then, letting $0 < \varepsilon \leq 1$, thanks to (6.9–10) and (6.12) one can easily determine a constant C_{11}, having the same dependences as C_{10}, such that

$$\|u'_\varepsilon\|^2_{L^\infty(0,t;H)} + \varepsilon c \, \|u'_\varepsilon\|^2_{L^\infty(0,t;V)} + c \, \|u'_\varepsilon\|^2_{L^2(0,t;V)}$$

$$\leq C_{11} \left(1 + (1 + \|u'_\varepsilon\|_{L^\infty(0,t;H)}) \, \|f_\varepsilon\|_{L^1(0,t;H)} + \|g_\varepsilon\|^2_{L^2(0,t;V')} \right)$$

Finally, a further application of (3.9) allows us to get the needed estimate (cf. (6.7))

$$\|u'_\varepsilon\|^2_{L^\infty(0,t;H)\cap L^2(0,t;V)} + \varepsilon \, \|u'_\varepsilon\|^2_{L^\infty(0,t;V)}$$

$$\leq C_{10} \left(1 + \|f_\varepsilon\|^2_{L^1(0,t;H)} + \|g_\varepsilon\|^2_{L^2(0,t;V')} \right) \quad (6.13)$$

for any $t \in \,]0,T]$. Taking (6.8) and (3.13) into account, from (6.13) and (6.11) it follows that $\|u_\varepsilon\|_{W^{1,\infty}(0,T;H)\cap H^1(0,T;V)}$, $\sqrt{\varepsilon}\|u_\varepsilon\|_{W^{1,\infty}(0,T;V)}$, and $\|w_\varepsilon\|_{L^\infty(0,T;V')}$ are bounded independently of ε. Moreover, by comparison in (3.11) we recover the same property for $\|(u'_\varepsilon + \varepsilon L u'_\varepsilon + w_\varepsilon)' - f_\varepsilon\|_{L^2(0,T;V')}$. Therefore there exist u, w, ξ such that, possibly extracting subsequences,

$$u_\varepsilon \to u \quad \text{weakly star in} \quad W^{1,\infty}(0,T;H)\cap H^1(0,T;V), \quad (6.14)$$

$$w_\varepsilon \to w \quad \text{weakly star in} \quad L^\infty(0,T;V'), \quad (6.15)$$

$$(u'_\varepsilon + \varepsilon L u'_\varepsilon + w_\varepsilon)' - f_\varepsilon \to \xi \quad \text{weakly in} \quad L^2(0,T;V') \quad (6.16)$$

as $\varepsilon \searrow 0$. Since $\varepsilon L u'_\varepsilon \to 0$ strongly in $L^\infty(0,T;V')$, (6.14–15) and (6.8) enable us to identify $\xi = (u' + w)' - f$ in (6.16) and to pass to the limit in (3.11) obtaining (3.3). Now, (6.6) is a direct consequence of (3.11) and the initial conditions (2.13) can be easily proved by exploiting (3.13) and (6.16). To show (3.4) it suffices to proceed as in the proof of Lemma 1. Then (cf. (3.6)) the pair (u, w) yields the unique solution to the problem (3.3–4), (2.13) (whence the convergences (6.14–16) hold for the whole sequences) and, in view of (6.13–14) and (6.8), u satisfies (6.7). This inequality and the assumptions (6.2–3) have to replace (3.5), (2.4), (2.6) in the fixed point argument of Section 4 and in the *a priori* estimate of Section 5. As the necessary modifications are quite obvious, the details are left to the reader.

95

Remark 4. Let us point out that the existence and uniqueness result stated by Theorem 3 provides functions u and w not so regular as in Theorem 1 (cf. (6.5–6) and (2.10)). Indeed, A being a subdifferential, we can find a weaker solution under weaker hypotheses on F, G, e_0 by avoiding the test of (2.11) with u''. However, let us point out that, in the present case, it is straightforward to deduce the regularity (2.10) from Lemma 1, provided that e_0 fulfils (2.9) and simply $F : L^\infty(0,T;H) \to L^2(0,T;H)$, $G : H^1(0,T;V) \to W^{1,1}(0,T;V')$.

REFERENCES

[1] V. BARBU, *Nonlinear semigroups and differential equations in Banach spaces*, Noordhoff International Publishing, Leyden 1976.

[2] V. BARBU AND A. FAVINI, *Existence for implicit differential equations in Banach spaces*, Atti Accad. Naz. Lincei Cl. Sci. Fis. Mat. Natur. Rend. (9) Mat. Appl. **3** (1992), 203-215.

[3] H. BRÉZIS, *Opérateurs maximaux monotones et semi-groupes de contractions dans les espaces de Hilbert*, North–Holland, Amsterdam 1973.

[4] R. W. CARROLL AND R. E. SHOWALTER, *Singular and degenerate Cauchy problems*, Math. Sci. Engrg. **27**, Academic Press, New York 1976.

[5] P. COLLI AND M. GRASSELLI, *Phase transition problems in materials with memory*, J. Integral Equations Appl. **5** (1993), 1–22.

[6] P. COLLI AND M. GRASSELLI, *Phase transitions in materials with memory*, in Progress in partial differential equations: calculus of variations, applications (C. Bandle, J. Bemelmans, M. Chipot, M. Grüter, and J. Saint Jean Paulin eds.), Pitman Res. Notes Math. Ser. **267**, Longman Scientific & Technical 1992, pp. 173–186.

[7] P. COLLI AND M. GRASSELLI, *An existence result for a hyperbolic phase transition problem with memory*, Appl. Math. Lett. **5** (1992), 99–102.

[8] P. COLLI AND M. GRASSELLI, *Hyperbolic phase change problems in heat conduction with memory*, Proc. Roy. Soc. Edinburgh Sect. A **123** (1993), 571–592.

[9] P. COLLI AND M. GRASSELLI, *Justification of a hyperbolic approach to phase changes in materials with memory*, Asymptotic Anal. (1994), to appear.

[10] P. COLLI AND M. GRASSELLI, *Convergence of parabolic to hyperbolic phase change models with memory*, Preprint IAN–CNR **917**, Pavia 1994, 1–30.

[11] A. DAMLAMIAN, *Some results on the multi-phase Stefan problem*, Comm. Partial Differential Equations **2** (1977), 1017–1044.

[12] E. DI BENEDETTO AND R. E. SHOWALTER, *Implicit degenerate evolution equations and applications*, SIAM J. Math. Anal. **12** (1981), 731–751.

[13] G. Duvaut, *Résolution d'un problème de Stéfan*, C. R. Acad. Sci. Paris Sér. A **276** (1973), 1461–1463.

[14] G. Gripenberg, S-O. Londen, and O. Staffans, *Volterra integral and functional equations*, Encyclopedia Math. Appl. **34**, Cambridge University Press, Cambridge 1990.

Pierluigi Colli
Dipartimento di Matematica, Università di Pavia
Via Abbiategrasso 209, 27100 Pavia, Italy

Maurizio Grasselli
Dipartimento di Matematica, Politecnico di Milano
Via Bonardi 9, 20133 Milano, Italy

J ESCHER AND G SIMONETT
On a multi-dimensional moving boundary problem

This paper aims to present some recent analytic results for a class of moving boundary problems in several space dimensions. Typically, these equations arise in laminar flow of fluids through porous media. More precisely, we consider the following situation: Let Γ_0 denote a fixed, impermeable layer in a homogeneous and isotropic porous medium. We assume that some portion of the solid matrix above Γ_0 is occupied by an incompressible Newtonian fluid. In addition, we suppose that there is a sharp interface Γ_f, separating the wet region Ω_f enclosed by Γ_0 and Γ_f, respectively, from the dry part, i.e. we consider a saturated fluid-air flow. In order to illustrate the above situation, let us introduce the following class of admissible interfaces:

$$\mathfrak{A}_0 := \{ f \in BC^2(\mathbb{R}^n, \mathbb{R}) \,;\, \inf_{x \in \mathbb{R}^n} f(x) > 0 \},$$

where $n \in \mathbb{N}$, $n \geq 1$, is fixed. Given $f \in \mathfrak{A}_0$, let

$$\Omega_f := \{ (x, y) \in \mathbb{R}^n \times (0, \infty) \,;\, 0 < y < f(x) \}.$$

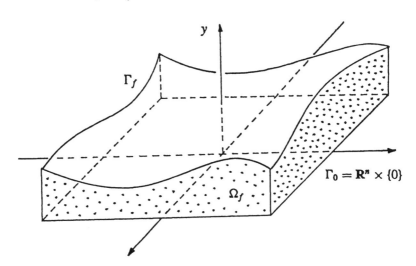

The boundaries of Ω_f are given by

$$\Gamma_0 := \mathbb{R}^n \times \{0\},$$
$$\Gamma_f := \text{graph}(f) := \{ (x, y) \in \mathbb{R}^n \times (0, \infty) \,;\, y = f(x) \}.$$

In order to describe the evolution of the free interface f, it is convenient to introduce a so-called velocity potential or piezometric head u, see [4, p. 175], [5, p. 32], or [8, p. 24], by setting

$$u(x, y) := \frac{p(x, y)}{g \cdot \rho(x, y)} + y \quad \text{for} \quad (x, y) \in \Omega_f. \tag{1}$$

Here, p denotes the pressure in the fluid, ρ the fluid density, and g stands for the gravity acceleration. We now assume that the flow is governed by Darcy's law, i.e., the velocity v of the fluid is given by

$$v(x, y) = -K(x, y)\nabla_{n+1}u(x, y) \quad \text{for} \quad (x, y) \in \Omega_f, \tag{2}$$

where K, the hydraulic conductivity, is a symmetric, uniformly positive definite $(n + 1) \times (n + 1)$−matrix, i.e. $K(x, y) \in \mathcal{L}_{sym}^+(\mathbb{R}^{n+1})$, uniformly in (x, y). Introducing the viscosity μ of the fluid and the permeability $k(x, y) \in \mathcal{L}_{sym}^+(\mathbb{R}^{n+1})$, $(x, y) \in \Omega_f$, of the porous medium, we have that

$$K(x, y) = k(x, y)\frac{g \cdot \rho(x, y)}{\mu(x, y)}.$$

This shows that the hydraulic conductivity, in general, depends on properties of both the fluid phase and the solid matrix of the porous medium. However, in a isotropic and homogeneous medium we have $k = k_0 \cdot id_{\mathbb{R}^{n+1}}$, for some positive constant k_0 ([4], p.296) and for an incompressible fluid ρ/μ is constant too ([7, p.4]). Consequently, we take the hydraulic conductivity to be

$$K(x, y) = K_0 \cdot id_{\mathbb{R}^{n+1}} \quad \text{for} \quad (x, y) \in \Omega_f, \tag{3}$$

for some positive constant K_0.

Next we note that, due to the incompressibiliy of the fluid phase, the equation of continuity $\nabla_{n+1} \cdot v = 0$ is satisfied in Ω_f. Hence we infer from (2) and (3) that the velocity potential u has to be harmonic in Ω_f, i.e.

$$\Delta_{n+1}u = 0 \quad \text{in} \quad \Omega_f. \tag{P_1}$$

In a further step we have to discuss the boundary conditions for u. As already mentioned, the boundary Γ_0 models an impermeable layer in the porous medium. Hence, we have the no-flux condition

$$\partial_y u = 0 \quad \text{on} \quad \Gamma_0. \tag{P_2}$$

It is no essential restriction to normalize the pressure in the air to be identically zero. Moreover, it is reasonable to assume that the pressure p is continuous. In absence of surface tension, this is equivalent to the conservation of momentum, see [8, p.18]. Hence we have that $p = 0$ on the free boundary Γ_f. Together with (1) we conclude that

$$u(x, y) = y \quad \text{for} \quad (x, y) \in \Gamma_f,$$

which we can also write in the form

$$u = f \quad \text{on} \quad \Gamma_f. \tag{P_3}$$

As usual for free boundary problems, there is a second condition for Γ_f, describing the motion of the interface. Thus suppose there is a time dependent family $\{\Gamma_{f(t)} ; \ t \geq 0\}$ of admissible interfaces, i.e.,

$$\{f(t) ; \ t \geq 0\} \subset \mathfrak{A}_0.$$

Introducing

$$F(t, x, y) := y - f(t, x) \quad \text{for} \quad (x, y) \in \Omega_{f(t)}, \ t \geq 0, \tag{4}$$

it is obvious that $\Gamma_{f(t)} = \{(x, y) \in \mathbb{R}^{n+1} ; \ F(t, x, y) = 0\}$, $t \geq 0$. Using the continuity of fluid flux across the interface, we obtain

$$\frac{\partial F}{\partial t} + \left(v | \nabla_{n+1} F\right) = 0 \quad \text{on} \quad \Gamma_{f(t)},$$

see p. 320 in [4]. Invoking (4) and again (2), (3), we conclude that

$$\partial_t f + K_0\left(\nabla_{n+1} u | (-\nabla_n f(t, x), 1)\right) = 0 \quad \text{on} \quad \Gamma_{f(t)}. \tag{5}$$

Introducing the normalized outer unit vector v on $\Gamma_{f(t)}$,

$$v(t, x) := \frac{(-\nabla_n f(t, x), 1)}{\sqrt{1 + |\nabla_n f(t, x)|^2}} \quad \text{for} \quad x \in \mathbb{R}^n, \ t \geq 0,$$

and the derivative along v, i.e.

$$\partial_v u := \left(\nabla_{n+1} u | v\right),$$

we rewrite (5) as

$$\partial_t f + K_0\sqrt{1 + |\nabla_n f|^2}\, \partial_v u = 0 \quad \text{on} \quad \Gamma_f(t). \tag{P_4}$$

Of course, for $t = 0$ the initial shape of the interface is prescribed, i.e. we also assume that the initial condition

$$f(0) = f_0 \quad \text{on} \quad \mathbb{R}^n \tag{P_5}$$

is satisfied for some $f_0 \in \mathfrak{A}_0$. Moreover, we suppose that there is a positive constant d such that

$$f_0(x) \to d \quad \text{as} \quad |x| \to \infty. \tag{P_6}$$

For the remainder of this note we assume that K_0 and d are fixed positive constants. Let us now summarize the above considerations by addressing the following free boundary problem in $n + 1$ space dimensions:

Given $f_0 \in \mathfrak{A}_0$, try to find a pair of functions (u, f) possessing the regularity

$$f \in C([0, \infty), \mathfrak{A}_0) \cap C^1([0, \infty), BC^1(\mathbb{R}^n)),$$
$$u(t, \cdot) \in BC^2(\overline{\Omega}_{f(t)}, \mathbb{R}), \ t \geq 0, \tag{6}$$

and satisfying pointwise the following set of equations

$$\begin{aligned}
\Delta_{n+1} u(t, \cdot) &= 0 && \text{in } \Omega_{f(t)}, \ t \geq 0, \\
\partial_{n+1} u(t, \cdot) &= 0 && \text{on } \Gamma_0, \ t \geq 0, \\
u(t, \cdot) &= f(t) && \text{on } \Gamma_{f(t)}, \ t \geq 0, \\
\lim_{|z| \to \infty} u(t, z) &= d && \text{for } t \geq 0, \\
\partial_t f + K_0\sqrt{1 + |\nabla_n f|^2}\, \partial_v u &= 0 && \text{on } \Gamma_{f(t)}, \ t > 0, \\
f(0) &= f_0 && \text{on } \mathbb{R}^n.
\end{aligned} \tag{$P)_{f_0}$}$$

Let us add the following

100

Remarks: a) Observe that the first four equations in $(P)_{f_0}$ represent just an elliptic boundary value problem for the velocity potential u, in which the time t appears as a free parameter. Particularly, there is no smoothing effect for u with respect to time. Let us also remark that the above elliptic boundary value problem has to be solved on an apriori unknown domain. Indeed, the free boundary of the domain is not known it advance and is to be determined as a part of the problem. However, assume that $f \in \mathfrak{A}_0 \cap h_d^{2+\alpha}(\mathbb{R}^n)$ is given, where $h_d^{2+\alpha}(\mathbb{R}^n)$ denotes an appropriate closed subspace of the Hölder space $C^{2+\alpha}$, see definition (9) below. Then it can be shown (see [9, Theorem 3.5]) that there exists a unique classical Hölder solution u_f of

$$\Delta_{n+1}u = 0 \quad \text{in} \quad \Omega_f, \quad \partial_{n+1}u = 0 \quad \text{on} \quad \Gamma_0, \quad u = f \quad \text{on} \quad \Gamma_f, \quad \lim_{\substack{|z| \to \infty \\ z \in \Omega_f}} u(z) = d. \quad (7)$$

b) Using the above notation, we introduce the following operator, acting on the free boundary f:

$$\Phi(f) := K_0 \sqrt{1 + |\nabla_n f|^2}\, \partial_\nu u_f \quad \text{for} \quad f \in \mathfrak{A}_0 \cap h_d^{2+\alpha}(\mathbb{R}^n).$$

We now rewrite the last two equations in $(P)_{f_0}$ as

$$\partial_t f + \Phi(f) = 0, \quad f(0) = f_0. \quad (8)$$

It can be shown that Φ is a nonlinear, nonlocal pseudo-differential operator of first order. Hence problem (8) is a nonlinear evolution equation for the interface f. Moreover, there is no smoothing effect for Φ. Thus (8) has to be considered as a fully nonlinear problem. However, it turns out that on a appropriate phase space V, which we shall describe below, the evolution equation (8) can be solved.

c) Observe that $u \equiv d$, $f \equiv d$ is a solution of $(P)_{f_0}$ to the initial datum $f_0 \equiv d$. Consequently, $f \equiv d$ is an equilibrium of the evolution equation (8).

d) It is instructive to precisely write down the third equation in $(P)_{f_0}$:

$$u\big(t, (x, f(t, x))\big) = f(t, x), \quad x \in \mathbb{R}^n, \ t \geq 0. \quad \square$$

We are now going to state our results for problem $(P)_{f_0}$. In order to do this, we need some preparation. Let $S(\mathbb{R}^n)$ denote the Schwartz space, i.e., the Fréchet space of all rapidly decreasing smooth functions on $\mathbb{R}^n$ and define

$$h^s(\mathbb{R}^n) := \text{closure of } S(R^n) \text{ in } BUC^s(\mathbb{R}^n), \ s \geq 0, \quad (9)$$

the so-called little Hölder spaces of order s. Here $BUC^k(\mathbb{R}^n)$, $k \in \mathbb{N}$, stands for the Banach space of all functions on $\mathbb{R}^n$ having bounded and uniformly continuous derivatives up to order k and $BUC^{k+\beta}(\mathbb{R}^n)$, $k \in \mathbb{N}$, $\beta \in (0, 1)$, denotes the Banach space of all functions in $BUC^k(\mathbb{R}^n)$ having uniformly β–Hölder continuous derivatives of order k. It is known, see [13], that $S(\mathbb{R}^n)$ is not dense in $BUC^s(\mathbb{R}^n)$. Hence $h^s(\mathbb{R}^n)$ is a proper subspace of $BUC^s(\mathbb{R}^n)$. In addition, we define the affine spaces $h_d^s(\mathbb{R}^n) := \{d + g ; g \in h^s(\mathbb{R}^n)\}$ and we restrict our class $\mathfrak{A}_0$ of admissible interfaces to be

$$\mathfrak{A} := \mathfrak{A}_0 \cap h_d^{2+\alpha}(\mathbb{R}^n) = \left\{ f \in BUC^{2+\alpha}(\mathbb{R}^n) ; \ f - d \in h^{2+\alpha}(\mathbb{R}^n) \text{ and } \inf_{x \in \mathbb{R}^n} f(x) > 0 \right\}.$$

101

Recall that for each $f \in \mathfrak{A}$ the solution of the elliptic problem (8) is denoted by u_f. Moreover, set

$$\kappa_f := \frac{f^2}{(1 + f^2 + |\nabla_n f|^2)(1 + |\nabla_n f|^2)} \quad \text{for} \quad f \in \mathfrak{A}$$

and define

$$V := \left\{ f \in \mathfrak{A} ; \ \partial_{n+1} u_f(x, f(x)) < \kappa_f(x), \ x \in \mathbb{R}^n \right\}.$$

Observe that $\inf_{x \in \mathbb{R}^n} \kappa_f(x) > 0$ for each $f \in \mathfrak{A}$. Hence Remark c) implies that $f \equiv d$ belongs to V. In fact, it can be shown that V is an open neighborhood of d in $\mathfrak{A}$ and that

$$\text{diam}_{2+\alpha}(V) := \sup_{f, g \in V} \|f - g\|_{BUC^{2+\alpha}} = \infty.$$

We refer to Lemma 5.10 in [9] for a proof of these properties of V. The space V plays an important role in an analytical treatment of the free boundary problem $(P)_{f_0}$. This is reflected in the following

Theorem 1: *Let $f_0 \in V$ be given. Then there exists a unique classical Hölder solution (u, f) of the free boundary problem $(P)_{f_0}$ on a maximal interval of existence $J := [0, t^+(f_0))$ with*

$$f \in C(J, V) \cap C^1(J, h_d^{1+\alpha}(\mathbb{R}^n))$$

$$u(t, \cdot) \in h_d^{2+\alpha}(\overline{\Omega}_{f(t)}), \ t \in J.$$

The map $[(t, f_0) \mapsto f]$ defines a smooth semiflow on V.
In addition, if $t^+ := t^+(f_0) < \infty$ and if $f : [0, t^+) \to V$ is uniformly continuous then either

$$\lim_{t \to t^+} \|f(t)\|_{BUC^{2+\alpha}} = \infty \quad \text{or} \quad \lim_{t \to t^+} \inf_{g \in \partial V} \|f(t) - g\|_{BUC^{2+\alpha}} = 0. \tag{10}$$

Before pointing out some steps in the proof of Theorem 1, let us remark that the above results are generalizations of those obtained in [9], where the case $n = 1$ is considered. Also in the one-dimensional case an existence result of a local solution for rather smooth initial conditions is given in [12], using the Nash-Moser theorem to solve the abstract evolution equation for the free boundary. It should be mentioned that there is no uniqueness result in [12] and that the solution constructed in [12] is less regular than the initial value. Hence it is impossible to continue this local solution to a maximal solution and to get dynamical statements comparable to (10). There is a different approach to problems of type $(P)_{f_0}$ on bounded domains, based on variational inequalities, cf. [1], [5], [8], [10], and the references therein. Of course, in this variational setting, one only gets weak solutions. But as already mentioned, there is no smoothing effect for the nonlinear evolution equation (8). Hence it is impossible to use elliptic theory to improve the regularity of weak solutions. Besides these analytical results there a several numerical investigations of problem $(P)_{f_0}$, see [5], [8], [11], [14].

Sketch of the proof: a) It follows from the considerations in Remarks a) and b) that it suffices to prove existence and uniqueness of a solution of the abstract nonlinear evolution equation (8). Thus let $f \in \mathfrak{A}$ be given and define

$$\varphi_f(x, y) := \left(x, 1 - \frac{y}{f(x)}\right) \quad \text{for} \quad (x, y) \in \overline{\Omega}_f.$$

Then it is not difficult to verify that φ_f is a diffeomorphism of class $C^{2+\alpha}$ mapping Ω_f onto the strip $\Omega := \mathbb{R}^n \times (0, 1)$. Let φ_*^f and φ_f^* denote the corresponding push forward and pull back operators induced by φ_f, i.e., $\varphi_*^f u := u \circ \varphi_f^{-1}$ for $u \in C(\Omega_f)$ and $\varphi_f^* v := v \circ \varphi_f$ for $v \in C(\Omega)$, respectively. With these notation we now introduce the transformed differential operators

$$\mathcal{A}(f)v := \varphi_*^f \Delta_{n+1}(\varphi_f^* v), \qquad \mathcal{B}(f) := \varphi_*^f \big(\nabla_{n+1}(\varphi_f^* v)|(-\nabla_n f, 1)\big) \quad \text{for} \quad v \in C^2(\overline{\Omega}). \quad (11)$$

Let $a_{jk}(f)$, $a_j(f)$, and $b_j(f)$, $1 \le j, k \le n+1$, denote the coefficients of $\mathcal{A}(f)$ and $\mathcal{B}(f)$, respectively, i.e.,

$$\mathcal{A}(f)v = \sum_{j,k=1}^{n+1} a_{jk}(f)\partial_j\partial_k v + \sum_{j=1}^{n+1} a_j(f)\partial_j v, \qquad \mathcal{B}(f)v = \sum_{j=1}^{n+1} b_j(f)\partial_j v, \quad v \in C^2(\overline{\Omega}).$$

Moreover, let v_f denote the unique solution of the following transformed elliptic boundary value problem

$$\mathcal{A}(f)v = 0 \text{ in } \Omega, \quad v = f - d \text{ on } \mathbb{R}^n \times \{0\}, \quad \partial_{n+1} v = 0 \text{ on } \mathbb{R}^n \times \{1\}, \quad \lim_{\substack{|z| \to \infty \\ z \in \Omega}} v(z) = 0. \quad (12)$$

b) We also transform the abstract evolution equation (8). Hence let $W \subset h^{2+\alpha}(\mathbb{R}^n)$ be given by $W := V - d := \{g \in h^{2+\alpha}(\mathbb{R}^n) ; g + d \in V\}$ and define

$$\Phi_0(g) := \mathcal{B}(g + d)v_{g+d} \quad \text{for} \quad g \in W.$$

Clearly, W is open in $h^{2+\alpha}(\mathbb{R}^n)$ since V is open in $h_d^{2+\alpha}(\mathbb{R}^n)$. Moreover, it is not too difficult to see that the evolution equation according to Φ_0, i.e.,

$$\partial_t g + \Phi_0(g) = 0, \quad g(0) = g_0 := f_0 - d, \quad (14)$$

is equivalent to (8). We also remark that Φ_0 maps W smoothly into $h^{1+\alpha}(\mathbb{R}^n)$, i.e., $\Phi_0 \in C^\infty(W, h^{1+\alpha}(\mathbb{R}^n))$, see Lemma 4.3 in [9]. Consequently, its Fréchet derivative $\partial\Phi_0(g)$, $g \in W$, belongs to $\mathcal{L}(h^{2+\alpha}(\mathbb{R}^n), h^{1+\alpha}(\mathbb{R}^n))$. In the following we consider $\partial\Phi_0(g)$ as an unbounded linear pseudo-differential operator in $h^{1+\alpha}(\mathbb{R}^n)$. Our goal is to prove that $-\partial\Phi_0(g)$ generates a strongly continuous analytic semigroup on $h^{1+\alpha}(\mathbb{R}^n)$, i.e.,

$$\partial\Phi_0(g) \in \mathcal{H}(h^{1+\alpha}(\mathbb{R}^n)). \quad (15)$$

Indeed, this result enables us to apply the theory of maximal regularity due to Da Prato and Grisvard [6], see also [3], [13], and [15]. In order to verify (15), we fix $g \in W$ and $x \in \mathbb{R}^n$ and we try to associate to $\partial\Phi_0(g(x))$ an appropriate Fourier operator $A_{g,x}$, i.e. a pseudo-differential operator with constant coefficients, such that $A_{g,x} \in \mathcal{H}(h^{1+\alpha}(\mathbb{R}^n))$ and such that $\partial\Phi_0(g(x)) - A_{g,x}$ can be controlled in $\mathcal{L}(h^{2+\alpha}(\mathbb{R}^n), h^{1+\alpha}(\mathbb{R}^n))$. In fact it follows from the results in Sections 5 and 6 in [9] that such a procedure is possible. More precisely, let $\mathcal{F}_n$ denote the Fourier transform in $S(\mathbb{R}^n)$ and define

$$A_{g,x}h := \mathcal{F}_n^{-1}\big(a_{g,x}(\cdot)(\mathcal{F}_n h)(\cdot)\big) \quad \text{for} \quad h \in S(\mathbb{R}^n),$$

where the symbol $a_{g,x}$ is given by

$$a_{g,x}(\xi) := i\big((\vec{b} + \vec{c}\,|\xi)\big) - b\lambda(\xi)\Big\{1 - \frac{w}{a}|\xi|^2 \frac{d(\xi)+1}{d(\xi)|\lambda(\xi)+1|^2}\Big\}, \quad \xi \in \mathbb{R}^n. \tag{16}$$

Using the coefficients defined in (11), we have

$$\vec{b} := \vec{b}(g,x) := \big(b_1(g+d)(x), \ldots, b_n(g+d)(x)\big),$$

$$\vec{c} := \vec{c}(g,x) := -\nabla_n v_{g+d}(x,0) - \frac{2}{g(x)+d}\partial_{n+1}v_{g+d}(x,0)\nabla_n g(x),$$

$$b := b_{n+1}(g+d)(x), \quad a := a_{n+1,n+1}(g+d)(x,0),$$

$$w := w(g,x) := -\frac{\partial_{n+1}v_{v+g}(x,0)}{g(x)+d}.$$

To explain the meaning of $\lambda(\xi)$ and $d(\xi)$ appearing in (16), let

$$\vec{a} := \vec{a}(g,x) := \big(a_{1,n+1}(g+d)(x,0), \ldots, a_{n,n+1}(g+d)(x,0)\big),$$

$$a_0(\xi) := a_0(g,x,\xi) := \sum_{j,k=1}^n a_{jk}(g+d)(x)\xi^j\xi^k, \quad \xi \in \mathbb{R}^n,$$

and define the parameter dependent quadratic polynomial

$$q_\xi(z) := 1 + a_0(\xi) + 2i(\vec{a}\,|\xi)z - az^2 \quad \text{for} \quad z \in \mathbb{C}.$$

Then it can be shown that, given $\xi \in \mathbb{R}^n$, there is exactly one root of $q_\xi(\cdot)$ with positive real part. Let $\lambda(\xi) \in \mathbb{C}$ denote this root and $d(\xi)$ its real part. In addition, similar to the proof of Lemma 2.2 and Remark B.1c) in [9], the following estimate for $d(\xi)$ can be established:

$$d^2(\xi) \geq \frac{1+|\xi|^2}{a^2\big(1 + (g(x)+d)^2 + |\nabla_n g(x)|^2\big)} \quad \text{for} \quad \xi \in \mathbb{R}^n. \tag{17}$$

c) Once worked out the symbol $a_{g,x}$ of the Fourier operator $A_{g,x}$, the tools of harmonic analysis are accessible. More precisely, it follows from a general result due to Amann (see [2]), based on Mihlin-Hörmander's multiplier theorem and parameter dependent norms in Besov spaces, that $A_{g,x}$ belongs to $\mathcal{H}(h^{1+\alpha}(\mathbb{R}^n))$, provided there is a positive constant α such that the following ellipticity condition holds:

$$\operatorname{Re} a_{g,x}(\xi) \geq \alpha\sqrt{1+|\xi|^2} \quad \text{for all} \quad \xi \in \mathbb{R}^n. \tag{18}$$

It is not difficult to verify that b is negative and that

$$a = a_{n+1,n+1}(g+d)(x,0) = \frac{1+|\nabla_n g(x)|^2}{(g(x)+d)^2}. \tag{19}$$

Hence it follows from $b < 0$, (17), and (19) that estimate (18) is true, provided

$$w < \frac{(g(x)+d)^2}{(1+(g(x)+d)^2 + |\nabla_n g(x)|^2)(1+|\nabla_n g(x)|^2)}. \tag{20}$$

Finally, set $f := g + d$ and observe that

$$\partial_{n+1}u_f(x,f(x)) = -\frac{\partial_{n+1}v_{g+d}(x,0)}{g(x)+d} = w.$$

This shows that (20) is satisfied for each $f \in V$ and completes our argumentation. $\quad\square$

REFERENCES

[1] **J. M. Aitchinson, C. M. Elliott & J. R. Ockendon:** Percolation in gently sloping beaches. IMA J. Appl. Math. **30** (1983), 269-287.

[2] **H. Amann:** "Linear and Quasilinear Parabolic Problems". 1994. Book in preparation.

[3] **S. B. Angenent:** Nonlinear analytic semiflows. Proc. Roy. Soc. Edinburgh **115** A (1990), 91-107.

[4] **J. Bear & Y. Bachmat:** "Introduction to Modeling of Transport Phenomena in Porous Media". Kluwer Academic Publisher, Boston, 1990.

[5] **J. Crank:** "Free and Moving Boundary Problems". Clarendon Press, Oxford, 1984.

[6] **G. Da Prato & P. Grisvard:** Equations d'évolution abstraites nonlinéaires de type parabolique. Ann. Mat. Pura Appl., (4) **120** (1979), 329-396.

[7] **R. Dautray & J. L. Lions:** "Mathematical Analysis and Numerical Methods for Science and Technology". Springer, Berlin, 1992.

[8] **C. M. Elliott & J. R. Ockendon:** "Weak and Variational Methods for Moving Boundary Problems". Pitman, Boston, 1982.

[9] **J. Escher & G. Simonett:** Maximal regularity for a free boundary problem. Submitted.

[10] **P. Forsyth & H. Rasmussen:** J. Inst. Math. Appl. **24** (1979), 411-424.

[11] **B. Hunt:** Vertical recharge of unconfined aquifer. J. Hydr. Divn., **97** (1971), 1017-1030.

[12] **H. Kawarada & H. Koshigoe:** Unsteady flow in porous media with a free surface. Japan J. Indust. Appl. Math., **8** (1991), 41-82.

[13] **A. Lunardi:** "Analytic Semigroups and Optimal Regularity in Parabolic Equations". Birkhäuser, Basel, 1994.

[14] **H. Rasmussen & D. Salhani:** Unsteady porous flow with a free surface. IMA J. Appl. Math. **27** (1981), 307-318.

[15] **G. Simonett:** Zentrumsmannigfaltigkeiten für quasilineare parabolische Gleichungen. Institut für angewandte Analysis und Stochstik, Report Nr. 2, Berlin, 1992.

Institute of Mathematics
University of Basel
Rheinsprung 21
CH-4051 Basel, Switzerland

Department of Mathematics
Vanderbilt University
Nashville, TN 37240
current address
Department of Mathematics
University of California
Los Angeles, CA 90024

A L GLADKOV

Cauchy problem for equations of nonlinear heat conductibility with convection in classes of growing functions

We consider the Cauchy problem for the porous medium equation with nonlinear convection. The emphasis is on the existence and uniqueness of nonnegative solutions in a classes of growing at an infinity functions. The convection term in the equation introduces an asymmetry, a different conditions at a minus and plus infinity for the initial data are needed. A behaviour of growing solutions for large time is studied too. A constructed examples show a certain exactness of the obtaining results.

1. Introduction

In this paper we consider nonnegative solutions of the Cauchy problem

$$(I) \begin{cases} u_t = (u^m)_{xx} + c(u^n)_x & \text{in} \quad S_T = R \times [0, T] & (1.1) \\ u(x, 0) = u_0(x) & \text{on} \quad R & (1.2) \end{cases}$$

where $m > 1$, $m \geq n \geq 1$, $c > 0$, $u_0(x)$ is a nonnegative continuous function which can be unbounded on R.

The equation (1.1) arises as a model for a number of different physical phenomena ([1],[2]). Like the porous medium equation, (1.1) is a degenerate parabolic equation. We cannot expect that Problem (I) has a classical solution for any smooth initial data. In defining generalized solution to this problem, we follow [3],[4].

Definition 1. A function $u(x, t)$ is said to be a *solution* of equation (1.1) in S_T if (*i*) u is defined, nonnegative and continuous in S_T; and (*ii*) satisfies the integral equality

$$\int_{t_1}^{t_2}\int_{x_1}^{x_2}\{uf_t + u^m f_{xx} - cu^n f_x\}\,dx\,dt = \int_{x_1}^{x_2} uf\,\Big|_{t_1}^{t_2}\,dx + \int_{t_1}^{t_2} u^m f_x\,\Big|_{x_1}^{x_2}\,dt \qquad (1.3)$$

for all bounded rectangles $P \equiv [x_1, x_2] \times [t_1, t_2] \subseteq S_T$ and nonnegative functions $f \in C^{2,1}(P)$ such that $f(x_1, t) = f(x_2, t) = 0$ for all $t \in [t_1, t_2]$. If (1.3) is satisfied with $\geq$ instead of $=$ we say that u is a *subsolution*, if with $\leq$ *supersolution*.

Definition 2. A function $u(x, t)$ is said to be a *solution* of Problem (I) in S_T if it is a solution of equation (1.1) in S_T and satisfies (1.2).

There are a large number of papers devoted to the equation (1.1). In particular, a various theorems of existence and uniqueness for bounded solutions of the Cauchy problem have been proved for more general equations in [3]-[5]. The existence and uniqueness classes of the Cauchy problem for the porous medium equation and the porous medium equation with an absorption have been studied. For example here it is possible to mention the articles [6] -[13].

The main goal of this work is to get some existence and uniqueness results for Problem (I) in a classes of growing at an infinity functions. There are examples which show that a growing at initial time solutions of our Cauchy problem may decrease to zero in any point as $t \to \infty$. Therefore we investigate a qualitative nature of behaviour of the solutions for large time also.

It is easy to show the correctness of the next results ([10]).

Theorem 1. *Suppose that $\varphi(x,t) \geq 0$ is a supersolution of the equation (1.1) in S_T, and let u_0 be bounded above by $\varphi(x,0)$. Then there exists a minimal solution $u(x,t)$ of Problem (I) such that*

$$u(x,t) \leq \varphi(x,t) \ \text{in} \ S_T. \tag{1.4}$$

Corollary 1. *Let u and v be two minimal solutions of Problem (I) with initial data u_0 and v_0 such that $u_0 \leq v_0$. Then for every $t > 0$ we have*

$$u(t) \leq v(t). \tag{1.5}$$

In order to describe the existence and uniqueness classes of Problem (I) we must distinguish the following four cases: $1 \leq n < (m+1)/2$, $n = (m+1)/2$, $(m+1)/2 < n < m$, $n = m$.

In Section 2 we consider the case $1 \leq n < (m+1)/2$. It corresponds to a diffusion dominating a convection for a growing solutions. The existence and uniqueness results are of the same nature as for the equation (1.1) without convection. Denote M_i and α_i are a positive and nonnegative constants respectively, $S = R \times [0, \infty)$.

Theorem 2. *Assume that $1 \leq n < (m+1)/2$ and*

$$0 \leq u_0(x) \leq M_1(\alpha_1 + |x|)^p \ \text{on} \ R. \tag{1.6}$$

If $p < 2/(m-1)$ then Problem (I) admits a solution in S. When $p = 2/(m-1)$ Problem (I) admits a solution in S_T and

$$u(x,t) \leq M_2(\alpha_2 + |x|)^{2/(m-1)}. \tag{1.7}$$

Solutions of Problem (I) satisfying (1.7) are uniquely determined in S_T by their initial data.

It is shown a certain exactness of Theorem 2. In particular, if u_0 grows at an infinity more quickly than a right hand side of the inequality (1.7) then our problem does not admit a solution in any strip S_T.

Let $v_b(x)$ $(b \geq 0)$ be a stationary solution of the equation (1.1). It is easy to check that v_b satisfies to the ordinary differential equation

$$(v_b^m)' + cv_b^n = b. \tag{1.8}$$

This solution is written out in explicit form for $b = 0$. Denote it $w_a(x)$. We have

$$w_a(x) = \{c_0[a - x]_+\}^{1/(m-n)}, a \in R. \tag{1.9}$$

Here $c_0 = c(m - n)/m$, $[h(x)]_+ = \max\{0, h(x)\}$. The next theorem gives information on the behaviour of growing minimal solution of our Cauchy problem as $t \to \infty$.

Theorem 3. Let $1 \leq n < m$.

1). *Suppose that*

$$u_0 \leq v_b \tag{1.10}$$

for some $b \geq 0$. Then $u(x.t)$ is bounded for any $x \in R$.

2). *Assume that $1 \leq n < (m + 1)/2$, $u(x.t)$ exists in S and*

$$\liminf_{x \to -\infty} u_0(x)/w_a(x) > 1 \tag{1.11}$$

for some $a \in R$. Then

$$\lim_{t \to +\infty} u(x.t) = +\infty \quad \text{for any} \quad x \in R. \tag{1.12}$$

3). *Suppose that both inequalities*

$$\limsup_{x \to -\infty} u_0(x)/w_a(x) < 1 \quad \text{and} \quad \limsup_{x \to +\infty} u_0(x) \leq \alpha_3 \tag{1.13}$$

are true for some $a \in R$. Then

$$\limsup_{t \to +\infty} u(x.t) \leq \alpha_3 \quad \text{for any} \quad x \in R. \tag{1.14}$$

If an addition

$$\lim_{x \to +\infty} u_0(x) = \alpha_3, \tag{1.15}$$

then

$$\lim_{t \to +\infty} u(x.t) = \alpha_3 \quad \text{for any} \quad x \in R. \tag{1.16}$$

4). *Assume that $\lim_{x \to +\infty} u_0(x) = +\infty$ and $u(x.t)$ exists in S. Then (1.12) is valid. If $\liminf_{x \to +\infty} u_0(x) > 0$ then $u(x.t) > 0$ for any $x \in R$ by $t > t(x)$.*

A point 2 of Theorem 2 has not got a sense for $n \geq (m + 1)/2$ because a solution of Problem (I) does not exist in S when inequality (1.11) is true.

108

An analysis of the stationary solutions of the equation (1.1) shows that for the case $n = (m+1)/2$ a global existence theorem must be another than in a previous case. In Section 3 we prove more strong global existence result for our problem.

Theorem 4. *Assume that $n = (m+1)/2$ and*

$$u_0(x) \leq w_a(x), \quad x \leq 0, \tag{1.17}$$

$$u_0(x) \leq M_3(\alpha_4 + |x|)^p, \quad x \geq o \tag{1.18}$$

with $p < 2/(m-1)$. Then Problem (I) admits a solution in S.

A constructed examples show the exactness of Theorem 4. The uniqueness and local existence theorem is the same as in a previous case.

Denote $c_* = c(m-n)n/[m(2m-n)]$. The following existence and uniqueness theorem is proved in Section 4.

Theorem 5. *Assume that $(m+1)/2 < n < m$ and $u_0(x)$ satisfies both inequalities (1.17) and (1.18) with $p = 1/(n-1)$. Then minimal solution $u(x,t)$ of Problem (I) exists in S_T for some $T > 0$ and*

$$u(x,t) \leq [M_4(\alpha_5 + |x|)]^{1/(m-n)}, x \leq 0, \tag{1.19}$$

$$u(x,t) \leq M_5(\alpha_6 + |x|)^{1/(n-1)}, x \geq 0, \tag{1.20}$$

where $M_4 = c_0$. Solutions Problem (I) satisfying (1.19), (1.20) with $M_4 < c_*$ are uniquely determined in S_T by their initial data.

Theorem 5 is exact in some sense. In particular, if u_0 grows at a plus infinity more quickly than a right hand side of (1.20) then our Cauchy problem has not got, a solution in any strip S_T. It is shown that the uniqueness result of Problem (I) in a class of functions satisfying (1.19), (1.20) with $M_4 = c_0$ is not valid generally speaking.

In Section 5 we consider the case $n = m$. Denote $\varepsilon(x)$ a nonnegative function for which $\lim_{x \to -\infty} \varepsilon(x) = 0$.

Theorem 6. *Assume that $n = m$ and u_0 satisfies both inequalities (1.18) and*

$$u_0(x) \leq M_6 \exp(-cx/m) + \alpha_7, \quad x \leq 0. \tag{1.21}$$

If $p < 1/(n-1)$ in (1.18) then Problem (I) admits a solution in S. When $p = 1/(n-1)$ a minimal solution $u(x,t)$ exists in S_T for some $T > 0$. Solutions of Problem (I) satisfying both (1.18) with $p = 1/(n-1)$ and

$$u(x,t) \leq \varepsilon(x) \exp(-cx/m), \quad x \leq 0 \tag{1.22}$$

are uniquely determined in S_T by their initial data.

An examples show that it is impossible to replace the function $\varepsilon(x)$ in (1.22) on any positive constant without loss of the uniqueness. If u_0 grows at a plus infinity more quickly than a right hand side of the inequality (1.18) with $p = 1/(n-1)$ then Problem (I) has not got a solution in any strip S_T. When the growth of u_0 is the same as one the solution of Problem (I) blows up in finite time. The next theorem gives information on the behaviour of growing minimal solution of Problem (I) as $t \to \infty$.

Theorem 7. *Let $n = m$.*

1). *Assume that u_0 satisfies both (1.21) and second inequality in (1.13). Then (1.14) is true for minimal solution of Problem (I). If an addition u_0 satisfies to (1.15) then (1.16) is valid.*

2). *The same as 4) in Theorem 3.*

2. The case $1 \le n < (m+1)/2$.

Proof of Theorem 2. The existence part of Theorem 2 follows from Theorem 1 with using of the auxiliary functions $\varphi_1(x,t) = M_1(M_7 + x^2)^{p/2} \exp t$ and $\varphi_2(x,t) = M_1[(1 - M_8 t)^{-1}(\alpha_8 + x^2)]^{1/(m-1)}$. In order to prove the uniqueness part of Theorem 2 we next state a basic technical result ([10]).

Lemma 1. *Let $\underline{u}$ is a subsolution and $\overline{u}$ is a supersolution of (1.1) with initial data $\underline{u}_0$ and $\overline{u}_0$. Assume moreover that both $\underline{u}$ and $\overline{u}$ satisfying (1.7) and one of them is positive in S_T. If $(\underline{u}_0 - \overline{u}_0)_+$ is integrable then*

$$\int\limits_R (\underline{u}(t) - \overline{u}(t))_+ \, dx \le \int\limits_R (\underline{u}_0 - \overline{u}_0))_+ \, dx \qquad (2.1)$$

for every $0 < t \le T$.

Proof. Because many arguments are the same as in [10]-[14], we shall omit a some details of the proof.

From Definition 1 with $x_1 = -l$, $x_2 = l$, $t_1 = 0$ and $t_2 = \tau$ we have

$$\int\limits_0^\tau \int\limits_{-l}^l (\underline{u} - \overline{u})[f_t + b(af_{xx} - cf_x)] \, dx \, dt - \int\limits_0^\tau (\underline{u}^m - \overline{u}^m)f_x \Big|_{-l}^l \, dt + \int\limits_{-l}^l (\underline{u}_0 - \overline{u}_0)f(0) \, dx$$

$$\ge \int\limits_{-l}^l [\underline{u}(\tau) - \overline{u}(\tau)]f(\tau) \, dx, \qquad (2.2)$$

where

$$\begin{cases} b = \dfrac{\underline{u}^n - \overline{u}^n}{\underline{u} - \overline{u}}, & a = \dfrac{\underline{u}^m - \overline{u}^m}{\underline{u}^n - \overline{u}^n} \quad \text{if} \quad \underline{u} \ne \overline{u} \\[2mm] b = n\underline{u}^{n-1}, & a = (m/n)\underline{u}^{m-n} \quad \text{if} \quad \underline{u} = \overline{u}. \end{cases}$$

Denote $Q_{l,\tau} = [-l, l] \times [0, \tau]$. Owing to an assumptions of Theorem 2 the following inequalities are true in $Q_{l,\tau}$:

$$r_l \le b(x, t) \le M_9(\alpha_9 + x^2)^{(n-1)/(m-1)}, \qquad (2.3)$$

$$r_l \le a(x, t) \le M_{10}(\alpha_{10} + x^2)^{(m-n)/(n-1)} \qquad (2.4)$$

for some $r_l > 0$. In $Q_{l,\tau}$ we construct two sequences of smooth positive functions $b_{j,l}(x,t)$ and $a_{j,l}(x,t)$ with the properties:

1^0. $\{b_{j,l}\}$ and $\{a_{j,l}\}$ are monotonically decreasing and uniformly tend to b and a respectively when $j \to \infty$;

2^0. $b_{j,l}$ and $a_{j,l}$ satisfy to (2.3) and (2.4) respectively.

Let $\omega(x)$ be an infinitely differentiable function with compact support, such that Supp $\omega(x) \subset (-l_0, l_0)$, $0 \leq \omega(x) \leq 1$. By classical linear parabolic theory the next initial-boundary value problem

$$\begin{cases} f_t + b_{jl}(a_{jl}f_{xx} - cf_x) = 0 & \text{in} \quad Q_{l,\tau} & (2.5) \\ f(x,\tau) = \omega(x), \quad f(\pm l, t) = 0 & & (2.6) \end{cases}$$

has got unique solution $f(x,t)$ for $l \geq l_0$. As a consequence of the maximum principle we have

$$0 \leq f(x,t) \leq \max_{|x| \leq l} \omega(x) \leq 1 \quad \text{in} \quad Q_{l,\tau}.$$

In the next lemma we shall give some another properties of the function $f(x,t)$.

Lemma 2. *There exist positive constants $M_{11}(\gamma, \tau)$, $M_{12}(\gamma, \tau)$ and $M_{13}(\gamma, \tau, l)$ such that*

(a) $\quad f(x,t) \leq M_{11}(\alpha_{11} + x^2)^{-\gamma} \quad \text{in} \quad S_\tau,$

(b) $\quad |f_x(\pm l, t)| \leq M_{12}(\alpha_{11} + l^2)^{-\gamma - 1/2} \quad \text{for} \quad 0 \leq t \leq \tau,$

(c) $\quad \int_0^\tau \int_{-l}^l b_{jl} a_{jl}(f_{xx})^2 \, dx \, dt \leq M_{13}.$

Proof. Point (a) is proved by comparison with the auxiliary function

$$z_1(x,t) = M_{14} \exp[d(\tau - t)]\{(\alpha_{11} + x^2)^{-\gamma} - (\alpha_{11} + l^2)^{-\gamma}\}. \quad (2.7)$$

Part (b) is a consequence of the relations

$$f(x,t) \leq z_1(x,t), \quad f(\pm l, t) = z_1(\pm l, t).$$

For the proof of (c) we can multiply (2.5) by f_{xx} and integrate the obtaining identity over $Q_{l,\tau}$. $\quad \Diamond$

Rewrite (2.2) in the form

$$\int_{-l}^l [\underline{u}(\tau) - \overline{u}(\tau)]\omega(x) \, dx \leq \int_{-l}^l [\underline{u}_0 - \overline{u}_0]_+ \, dx - \int_0^\tau (\underline{u}^m - \overline{u}^m)f_x \Big|_{-l}^l \, dt$$

111

$$+ \int_0^\tau \int_{-l}^l (\underline{u} - \overline{u})\{(ba - b_{jl}a_{jl})f_{xx} - c(b - b_{jl})f_x\}\, dx\, dt. \tag{2.8}$$

Due to Lemma 2 the second integral of the right hand side of (2.8) tends to 0 as $l \to \infty$ and the last integrand tends to 0 as $j \to \infty$ when l is kept fixed. Thus if we let $j \to \infty$ and then $l \to \infty$ from (2.8) one obtains (2.1). $\Diamond$

Remark 1. As it is easy to see Lemma 1 takes place if we replace $(\underline{u} - \overline{u})_+$ by $(\overline{u} - \underline{u})_+$. Therefore we get that

$$\|\overline{u}(t) - \underline{u}(t)\|_1 \le \|\overline{u}_0 - \underline{u}_0\|_1$$

for every $t > 0$.

In order to complete the proof of Theorem we can note that the conclusion follows from Remark 1 if one of solutions is positive. For another case we refer to [10]. $\Diamond$

As consequence of Theorem 2 together with Corollary 1 the following comparison result is true.

Corollary 2.   *Let u be a subsolution satisfying (1.7) and v an arbitrary supersolution of (1.1) in S_T corresponding to initial values u_0 and v_0. Then*

$$u_0 \le v_0 \quad \Rightarrow \quad u \le v \quad \text{in} \quad S_T. \tag{2.9}$$

Remark 2. We now show the exactness of the existence part of Theorem 2. Suppose that u_0 satisfies (1.6) with $p = 2/(m - 1)$ and

$$\lim_{x \to -\infty} u_0/|x|^{2/(m-1)} = M_{15}.$$

Then we consider the comparison functions

$$v_{\mp}(x,t) = [d_1 - d_2 t]^{-1/(m-1)} \left\{ d_3 + d_4(d_1 - d_2 t)^{(m-n)/(m-1)} \mp x \right\}_+^{2/(m-1)}.$$

Here d_i $(i = \overline{1,4})$ are a some constants. A direct check shows that $v_-(x,t)$ is a subsolution of (1.1) satisfying both inequalities (1.7) and $v_-(x,0) \le u_0(x)$ by $d_1 > M_{15}^{1-m}$, $0 < d_2 < 2m(m + 1)/(m - 1)$, if d_3 and d_4 are large enough. Because of Corollary 2 we have

$$v_- \le u \quad \text{in} \quad S_T$$

with $T < d_1/d_2$ and therefore u may exist only a finite time. Moreover, if $M_{15} = \infty$ due to comparison with v_- we can conclude that Problem (I) does not admit a solution in any strip S_T. In the case

$$\lim_{x \to +\infty} u_0(x)/|x|^{2/(m-1)} = M_{16}$$

112

we get an analogous results from a comparison of solution u with v_+.

For the proof of Theorem 3 we need in a comparison principle for the Cauchy problem of the ordinary differential equations. Denote $D(z) \equiv z' - f(x, z)$ and $B(z) \equiv z'' - g(x, z, z')$, where $f(x, z)$ and $g(x, z, z')$ are a continuous functions monotonically nondecreasing by z.

Lemma 3. *Let* a). $D(w) > D(z)$ *and* $w(x_0) \geq z(x_0)$ *or* b). $D(w) \geq D(z)$ *and* $w(x_0) > z(x_0)$. *Then* $w(x) > z(x)$ *by* $x > x_0$.

Lemma 4. *Let* $B(w) > B(z)$ *by* $x \geq x_0$ *and* $w(x_0) \geq z(x_0)$, $w'(x_0) \geq z'(x_0)$. *Then* $w(x) > z(x)$ *by* $x > x_0$.

A proofs of Lemma 3 and Lemma 4 are equally easy.

Proof of Theorem 3. Part 1 follows from Theorem 1 with $\varphi(x, t) = u_b(x)$.

2). To begin with, we introduce the travelling wave solution of (1.1) of the form

$$\omega_k(x) = \nu_k(\xi), \quad \xi = [kt + \overline{x} - x]_+, \quad k > 0 \tag{2.10}$$

with condition $\nu_k(0) = 0$. It is not difficult to verify that $z(\xi) = \nu_k^{m-1}(\xi)$ satisfies to the problem

$$z' - c(m-1)z^{(n-1)/(m-1)}/m - k(m-1)/m = 0, \tag{2.11}$$

$$z(0) = 0. \tag{2.12}$$

Using Lemma 3 we can get that

$$(c_0\xi)^{(m-1)/(m-n)} \leq z(\xi) \leq (c_0\xi)^{(m-1)/(m-n)} + p\xi, \tag{2.13}$$

where $p > k(m-1)(m-n)/[m(m-2n+1)]$. In order to compare u with ω_k, we notice that

$$u_0(x) \geq \{[c_0 + \varepsilon][a - x]_+\}^{1/(m-n)} \geq \{z([\overline{x} - x]_+\}^{1/(m-1)} = \omega_k(x, 0) \tag{2.14}$$

with a, $\overline{x}$ and ε suitably chosen. From (2.14) and Corollary 2 (the comparison principle will be proved later in the case $(m+1)/2 < n < m$) it follows that

$$\{c_0[kt + \overline{x} - x]_+\}^{1/(m-n)} \leq \omega_k(x, t) \leq u(x, t)$$

and hence $u(x, t) \to +\infty$ as $t \to +\infty$ in any point $x \in R$.

3). Suppose that (1.13) is fulfilled. Due to the comparison u with the travelling wave solution of the form $\rho_k(x, t) = \mu_k(\overline{x} - kt - x)$, $k > 0$ we can get (1.14) by analogy with a previous case.

If an addition (1.16) is true then the desired result can be proved by comparing of a solution of Problem (I) with $\zeta_k(x, t) = \gamma_k([kt + \overline{x} + x]_+)$, $\gamma_k(0) = 0$ $(k > 0)$.

4). This part follows from the comparison u with $\zeta_k(x, t)$. $\diamond$

3. The case $n = (m+1)/2$.

Proof of Theorem 4. We shall construct a supersolution of Problem (I) of the following form

$$u_A(x,t) = \begin{cases} \theta_1(\sigma)\exp(\beta t), & x \geq \overline{x} \\ \theta_2(\sigma)\exp(\beta t), & x < \overline{x}, \end{cases}$$

where $\sigma = |x - \overline{x}|\exp[-\beta(m-n)t]$, β is some positive number, $\overline{x} \in R$. It is easy to verify that $v_i(\sigma) = \theta_i^m(\sigma)$ $(i = 1,2)$ must satisfy to the differential inequalities

$$L_i v_i \equiv v_i'' + c_i(v_i^{n/m})' + \beta(m-n)\sigma(v_i^{1/m})' - \beta v_i^{1/m} \leq 0 \tag{3.1}$$

with $c_1 = c$ and $c_2 = -c$. Let

$$L_1 v_1 = 0, \tag{3.2}$$

$$v_i(0) = M_{17}, \quad v_i'(0) = 0, \quad i = 1,2. \tag{3.3}$$

The constant M_{17} will be chosen below. Applying Lemma 4 we obtain for solution v_1 of problem (3.2), (3.3) the following estimates

$$M_{18}(\sigma^2 + M_{19})^{p/2} \leq v_1(\sigma) \leq M_{20}(\sigma + M_{21})^{m/(m-n)}. \tag{3.4}$$

Here $p < m/(m-n)$, β, M_{18} and M_{20} are any positive constants, M_{19} and M_{21} are sufficiently large ones. Moreover, $M_{18}M_{19}^{p/2} \leq M_{17} \leq M_{20}M_{21}^{m/(m-n)}$. An immediate check shows that

$$v_2(\sigma) = \left\{(c_0\sigma)^2 + M_{17}^{2(m-n)/m}\right\}^{m/[2(m-n)]} \tag{3.5}$$

satisfies (3.1), (3.3) for $\beta = \beta(c,m,n)$ large enough. Taking account (3.4), (3.5) we get

$$u_0(x) \leq u_A(x,0)$$

if M_i and $\overline{x}$ are suitably chosen. To end the proof we now use Theorem 1. $\diamond$

Remark 3. The assumptions of Theorem 4 are optimal. Indeed, let u_0 satisfies (1.18) and

$$\lim_{x \to -\infty} u_0(x)/|x|^{1/(m-n)} = M_{22}, \tag{3.6}$$

where $M_{22} > c_0^{1/(m-n)}$. We shall show that Problem (I) has not got a solution in S. Note that $1/(m-n) = 2/(m-1)$ in the case $n = (m+1)/2$. As it easy to see equation (1.1) admits a solutions with a separable variables

$$u_r(x,t) = [\phi(t)(q - lx)]_+^{1/(m-n)}, \tag{3.7}$$

where $l > 0$, $q \in R$ and $\phi(t)$ satisfies to the ordinary differential equation

$$\phi' = \lambda\phi^3 - \rho\phi^2 \tag{3.8}$$

with $\lambda = l^2 m(m+1)/(m-1)$, $\rho = lc(m+1)/2$. Let

$$\phi(0) = 1. \tag{3.9}$$

Deciding problem (3.8), (3.9) we obtain

$$\lambda \ln\{(\rho - \lambda\phi)/[(\rho - \lambda)\phi]\}/\rho^2 + (1 - \phi)/(\rho\phi) = t, \quad \rho \neq \lambda, \tag{3.10}$$

$$\phi(t) \equiv 1, \quad \rho = \lambda.$$

In particular, from (3.10) it follows that

$$\phi(t) = 1/(\rho t) + o(1/t) \quad \text{as} \quad t \to \infty$$

while $\rho > \lambda$ and $\phi(t)$ blows up in finite time

$$T_s = \lambda \ln[\lambda/(\lambda-\rho)]/\rho^2 - 1/\rho = 4m \ln\{2lm/[2lm - c(m-1)]\}/[c^2(m^2-1)] - 2/[cl(m+1)]$$

while $\rho < \lambda$. Obviously, $u_r(x, t)$ exists only finite time T_s when $l > c_0$. Moreover,

$$T_s \to 0 \quad \text{as} \quad l \to \infty. \tag{3.11}$$

Note that by (3.6)-(3.8) we have

$$u_r(x, 0) \leq u_0(x)$$

for some suitable q and $l > c_0$. We now apply Corollary 2 and conclude that

$$u_r(x, t) \leq u(x, t) \tag{3.12}$$

and hence a solution of Problem (I) blows up in finite time.

If we let $M_{22} = \infty$ in (3.6) then due to (3.11), (3.12) Problem (I) does not admit a solution in any strip S_T. The exactness of the condition (1.18) has been shown in Remark 2.

4. The case $(m+1)/2 < n < m$.

Proof of Theorem 5. At first we shall establish existence part of Theorem 5. A direct computation shows that

$$u_L(x, t) = [\tau - \gamma t]^{-1/(2n-m-1)} \theta(\sigma)$$

is a supersolution of Problem (I). Here

$$\theta(\sigma) = \begin{cases} M_{23}(\sigma^2 + M_{24})^{1/[2(n-1)]}, & x \geq 0 \\ [(c_0\sigma + M_{25}^{(m-n)/n})^{m/(m-n)} - cM_{25} \arctan \sigma]^{1/m}, & x < 0, \end{cases}$$

115

$\sigma = |x|[\tau - \gamma t]^{(m-n)/(2n-m-1)}$ and constants M_{23}, M_{24}, M_{25}, γ and τ are suitably chosen. A desired existence result now follows from Theorem 1. The proof of uniqueness part of Theorem 5 is in all similar to the proof of one in Theorem 2. For estimate of solution f of the problem (2.5), (2.6) we can use the comparison functions $z_1(x,t)$ defined in (2.7) and

$$z_2(x,t) = M_{26}[(M_{27} - x)^{-p} - (M_{27} + l)^{-p}], \quad p > m/(m-n),$$

respectively in $Q_{l,\tau}^+ = Q_{l,\tau} \cap \{x \geq 0\}$ and $Q_{l,\tau}^- = Q_{l,\tau} \cap \{x \leq 0\}$. $\diamond$

From Theorem 5 and Corollary 1 it follows useful comparison principle.

Corollary 3. *Let u be a subsolution satisfying (1.19), (1.20) with $M_4 < c_*$ and v an arbitrary supersolution of (1.1) in S_T corresponding to initial values u_0 and v_0. Then*

$$u_0 \leq v_0 \quad \Rightarrow \quad u \leq v \quad in \quad S_T. \tag{4.1}$$

Remark 4. We show that no solution exists in any strip S_T if

$$\lim_{x \to +\infty} u_0(x)/|x|^{1/(n-1)} = +\infty$$

and hence the condition (1.20) of Theorem 5 is exact. Indeed, a direct computation shows that

$$h(x,t) = [(\mu + x)_+/(M_{28} - cnt)]^{1/(n-1)}$$

is a subsolution of Problem (I) for any M_{28} and suitably chosen μ. Moreover, $h(x,t)$ satisfies to (1.19), (1.20) with $M_4 < c_*$. Using Corollary 3 we deduce that

$$h(x,t) \leq u(x,t).$$

But $h(x,t)$ blows up in finite time $T = M_{28}/(cn)$.

Remark 5. As we shall see below uniqueness of Problem (I) does not take place in class of functions satisfying (1.19), (1.20) with $M_4 = c_0$. To demonstrate this we consider solution $\omega_k(x,t)$ of (1.1) defined in (2.10). As in the proof of Theorem 3 $z(\xi)$ satisfies to (2.11), (2.12). Applying Lemma 3 we have

$$(c_0\xi)^{(m-1)/(m-n)} \leq z(\xi) \leq [c_0(\xi + d)]^{(m-1)/(m-n)} - p(\xi + d), \tag{4.2}$$

where c_0 and p were defined in (1.9) and (2.13) respectively, d is sufficiently large. From (1.9) and (4.2), and if $a = \overline{x} + d$ and $x \leq \overline{x}$, we get

$$w_a(x) = \{c_0[\overline{x} + d - x]_+\}^{1/(m-n)} = \{c_0([\overline{x} - x]_+ + d)\}^{1/(m-n)}$$

$$\geq \left\{[c_0([\overline{x} - x]_+ + d)]^{(m-1)/(m-n)} - p([\overline{x} - x]_+ + d)\right\}^{1/(m-1)} \geq \omega_k(x,0). \tag{4.3}$$

For $x \geq \bar{x}$ we have $\omega_k(x,0) = z(0) = 0$ and again

$$\omega_k(x,0) \leq w_a(x). \tag{4.4}$$

We now consider Problem (I) with initial data $u_0(x) = \omega_k(x,0)$. By Theorem 1, and from (4.3) and (4.4) it follows that there exists solution $u(x,t)$ of Problem (I) bounded in any point $x \in R$. On the other hand, the same problem has got a solution $\omega_k(x,t)$ which tends to infinity as $t \to \infty$ in any point $x \in R$.

5. The case $n = m$.

Proof of Theorem 6. To establish existence part we shall construct a supersolutions of Problem (I) as in a previous cases. At first let T be any positive constant and (1.18), (1.21) hold with $p < 1/(n-1)$. A direct check shows that

$$V(x,t) = \begin{cases} M_{29}(\alpha_{12} + x^2)^q \exp(\alpha t), & x \geq 0 \\ [M_{30}\exp(-cx) + cM_{30}\arctan x + \alpha_{13}]^{1/n}\exp(\alpha t), & x < 0 \end{cases}$$

is a supersolution of Problem (I) if $q < 1/[2(n-1)]$, $\alpha > 0$, M_i and α_i are suitably chosen. Using Theorem 1 with $\varphi(x,t) = V(x,t)$ we obtain a global existence result for Problem (I).

In order to prove a local existence of solution for Problem (I) we can consider a solution of (1.1) with a separable variables

$$V_s(x,t) = [M_{31} - (n-1)st]^{-1/(n-1)}g(x), \tag{5.1}$$

where $s > 0$, $g(x) = g_1(|x|)$ for $x \geq 0$ and $g(x) = g_2(|x|)$ for $x \leq 0$. Moreover, $h_i(\xi) = g_i^n(\xi)$ $(i = 1, 2, \xi = |x|)$ satisfy to the initial conditions (3.3) and the following differential equations

$$h_i'' + c_i h_i' - s h_i^{1/n} = 0,$$

where c_i were defined in (3.1). Applying Lemma 4 we deduce that

$$M_{32}(\alpha_{14} + \xi^2)^{n/[2(n-1)]} \leq h_1(\xi), \quad M_{33}\exp(c\xi) - cM_{33}\arctan\xi + \alpha_{15} \leq h_2(\xi) \tag{5.2}$$

if s and M_{17} are large enough. From (5.1) and (5.2) it follows that

$$u_0(x) \leq V_s(x,0)$$

if M_{31} is small enough and therefore according to Theorem 1 there exists a solution of Problem (I) in S_T for $T < M_{31}/[(n-1)s]$.

The proof of uniqueness part of Theorem 6 is in all similar to the proof of one in Theorem 2. It should be noted only that $a_{jl} \equiv 1$ in (2.5) and for estimate of solution

117

f of the problem (2.5), (2.6) we can use the comparison functions $z_1(x,t)$ defined in (2.7) and

$$z_3(x,t) = M_{34}[\exp(cx) - \exp(-cl)]$$

respectively in $Q_{l,r}^+$ and $Q_{l,r}^-$.   ◇

Remark 6. The optimality of condition (1.18) in the existence part of Theorem 6 has been demonstrated in Remark 4. We now show that it is impossible to replace $\varepsilon(x)$ in (1.22) on any positive constant without loss of the uniqueness. Indeed, the equation (1.1) has a solutions

$$u_s(x) = [M_{35}\exp(-cx) + \alpha_{16}]^{1/n},$$

$$\lambda_k(x,t) = \{k[\exp(c(n-1)[kt - x + \overline{x}]_+/n) - 1]/c\}^{1/(n-1)}.$$

It is easy to see that

$$\lambda_k(x,0) \leq u_s(x)$$

for any M_{35} and α_{16}, if $\overline{x}$ and $k > 0$ are suitably chosen. By virtue of Theorem 1 this implies that Problem (I) with $u_0(x) = \lambda_k(x,0)$ admits unbounded and bounded by $u_s(x)$ in any point $x \in R$ solutions.

Proof of Theorem 7. Part 1 is proved by comparing u either with a solutions of equation (1.1)

$$y_k(x,t) = \{k + (M_{36} - k)\exp[c(n-1)(\overline{x} - x - kct)/n]\}^{1/(n-1)}$$

and $\gamma_k([kt + \overline{x} + x]_+)$, where $k > 0$, $\overline{x} \in R$, or with ones

$$p(x,t) = \{d\exp[\gamma t - c(n-1)x/n] - \gamma n/[c^2(n-1)]\}_+^{1/(n-1)},$$

where constants d and γ may have any sign. Part 2 is established by the same way as in Theorem 3.   ◇

Remark 7. The results of this paper can be easily extended to a number of related models. For instance, we may consider the equations of the following form

$$u_t = \Delta\varphi(u) + \sum_{i=1}^{N}[c_i(x)\psi(u)]_{x_i},$$

where $\varphi(p)$ and $\psi(p)$ behave like an exponential functions as $p \to \infty$.

REFERENCES

1. J. Bear, *Dynamics of Fluids in Porous Media*, American Elsevier Publishing Company, New York, 1972.

2. J.R. Phillip, *Evaporation and moisture and heat fields in the soil*, J. of Meteorology **14** (1957), 354-366.

3. J.I. Diaz, R. Kersner, *On a nonlinear degenerate parabolic equation in infiltration or evaporation through a porous medium*, J. Diff. Equat. **69** (1987), 368-403.

4. B.H. Gilding, *Improved theory for a nonlinear degenerate parabolic equation*, Annali Scu. Norm. Sup. Pisa. Cl. sci. **16** (1989), 165-224.

5. A.S. Kalashnikov, *On the character of the propagation of perturbation in processed described by quasilinear degenerate parabolic equations*, in Proceedings of Seminars Dedicated to I.G. Petrovskii 1 (1975), 135-144.

6. A.S. Kalashnikov, *The Cauchy problem in the class of increasing functions for equations of the nonstationary seepage type*, Vestnik Moskov. Univ. Ser. 1 Math. Mech. **6** (1963), 17-27.

7. D.G. Aronson, L.A. Caffarelli, *The initial trace of a solution of the porous media equation*, Trans. Amer. Math. Soc. **280** (1983), 351-366.

8. P. Benilan, M. Crandall, M. Pierre, *Solutions of the porous medium equation in R under optimal conditions on initial values*, Indiana Univ. Math. J. **33** (1984), 51-87.

9. B.E.G. Dahlberg, C.E. Kenig, *Non-negative solutions of the porous medium equation*, Comm. P. D. E. **9** (1984), 409-437.

10. S. Kamin, L.A. Peletier, J.L. Vazquez. *A nonlinear diffusion-absorption equation with unbounded initial data*, in Nonlinear Diffusion Equations and Their Equilibrium States **3** (1992), 243-263.

11. J.B. McLeod, L.A. Peletier, J.L. Vazquez, *Solutions of a nonlinear ODE appearing in the theory of diffusion with absorption* , Differ. and Integral Equat. **4** (1991), 1-14.

12. A.L. Gladkov, *The Cauchy problem for some quasilinear degenerate parabolic equations with absorption*, Sibirskii matematicheskii zhurnal **34** (1993), 47-64.

13. J.L. Vazquez, M. Walias, *Existence and uniqueness of solutions of diffusion-absorption equation with general data*, Differ. and Integral Equat. **7** (1994), 15 - 36.

14. R. Kersner, *Degenerate parabolic equations with general nonlinearities*, Nonlin. Anal., TMA 4 (1980), 1043-1062.

Department of Mathematics
Vitebsk Pedagogical Institute
210036 Vitebsk, Belarus

A ITO, N KENMOCHI AND M NIEZGÓDKA

Large-time behaviour of non-isothermal models for phase separation

This paper is concerned with a non-isothermal model of phase separation in a binary mixture. The model is described as a couple of two nonlinear parabolic equations subject to appropriate initial-boundary conditions. In this paper we mainly treat the one-dimensional problem and discuss the large-time behavior of the solution.

1. Introduction

Let us consider the following system of two parabolic Pads as a non-isothermal model for phase separation in a binary system:

$$[\rho(u) + \lambda(w)]_t - \Delta u = f(t,x) \quad \text{in } Q := (0, +\infty) \times \Omega, \tag{1.1}$$

$$w_t - \Delta\{-\kappa\Delta w + \xi + g(w) - \lambda'(w)u\} = 0 \quad \text{in } Q, \tag{1.2}$$

$$\xi \in \beta(w) \quad \text{in } Q, \tag{1.3}$$

$$\frac{\partial u}{\partial n} + n_0 u = h(t,x) \quad \text{on } \Sigma := (0, +\infty) \times \Gamma, \tag{1.4}$$

$$\frac{\partial w}{\partial n} = \frac{\partial}{\partial n}\{-\kappa\Delta w + \xi + g(w) - \lambda'(w)u\} = 0 \quad \text{on } \Sigma, \tag{1.5}$$

$$u(0,x) = u_0(x), \quad w(0,x) = w_0(x) \quad \text{in } \Omega. \tag{1.6}$$

We denote the system (1.1)-(1.6) by P, and in this system $u = u(t,x)$ is a function related to the temperature field $\theta(t,x)$ $(= \rho(u(t,x)))$, and $w = w(t,x)$ is an order parameter (local concentration of one component of the mixture). Here Ω is a bounded domain in $R^N, 1 \leq N \leq 3$, with smooth boundary Γ; $\rho(\cdot)$ is an increasing function with open domain $D(\rho)$ and range $R(\rho)$ in R, locally bi-Lipschitz; $\lambda(\cdot)$, $g(\cdot)$ are smooth functions on R, $\lambda'(\cdot)$ the derivative of λ; $\beta(\cdot)$ is a maximal monotone graph in $R \times R$, with bounded domain $D(\beta)$ in R; $n_0 > 0$, $\kappa > 0$ are constants; f, h, u_0, w_0 are prescribed data.

We refer to [1, 2, 5, 7, 8] for physical interpretation of problem P and to [9] for solvability of the one-dimensional problem without non-smooth term ξ in (1.2).

Typical examples of ρ, β, λ, g are respectively given by

$$\rho(u) = -\frac{1}{u}, \quad -\infty < u < 0, \tag{1.7}$$

$$\beta = \partial I_{[\sigma_*,\sigma^*]} \quad \text{(subdifferential of the indicator function } I_{[\sigma_*,\sigma^*]}),$$

$$\lambda(w) = \frac{1}{2}a_0 w^2 + a_1 w + a_2 \quad \text{with constants } a_i, \ i = 0,1,2,$$

and

$$g(w) = c_0 w^3 - c_1 w \quad \text{with positive constants } c_i, \ i = 0,1.$$

To handle system P which involves a class of singular functions ρ such as (1.7), we apply the viscosity approach, namely, approximate P by problem $P_\mu, 0 < \mu < 1$, consisting of (1.1)-(1.6) with (1.2) and (1.5) replaced respectively by

$$w_t - \Delta\{\mu w_t - \kappa\Delta w + \xi + g(w) - \lambda'(w)u\} = 0 \quad \text{in } Q \tag{1.2}'$$

and

$$\frac{\partial w}{\partial n} = \frac{\partial}{\partial n}\{\mu w_t - \kappa\Delta w + \xi + g(w) - \lambda'(w)u\} = 0 \quad \text{on } \Sigma. \tag{1.5}'$$

Problem P is sometimes denoted by P_0.

2. Existence and uniqueness for $P_\mu, 0 \leq \mu \leq 1$

We first specify assumptions on the class of data, particularly on the function ρ.

(ρ) ρ is a maximal monotone graph in $R \times R$ whose domain $D(\rho)$ and range $R(\rho)$ are open in R, and is locally bi-Lipschitz continuous as a function from $D(\rho)$ onto $R(\rho)$, and furthermore there are constants $A_0 > 0$ and α with $1 \leq \alpha < 2$ such that

$$|\rho(r_1) - \rho(r_2)| \geq \frac{A_0|r_1 - r_2|}{|r_1 r_2|^\alpha + 1} \quad \text{for } \forall r_1, r_2 \in D(\rho). \tag{2.1}$$

(β) β is a maximal monotone graph in $R \times R$ such that $\overline{D(\beta)} = [\sigma_*, \sigma^*]$ for constants σ_*, σ^* with $-\infty < \sigma_* < \sigma^* < +\infty$.

(λ) λ is a C^2-function from R into itself.

(g) g is a C^1-function from R into itself.

Furthermore, suppose that n_0, κ are positive constants and f, h, u_0, w_0 satisfy the following hypotheses.

(H1) $f \in W_{loc}^{1,2}(R_+; L^2(\Omega))$.

(H2) $h \in W_{loc}^{1,2}(R_+; L^2(\Gamma)) \cap L^\infty(R_+; L^\infty(\Gamma))$ such that

$$n_0 \sup D(\rho) \geq h(t,x) \geq n_0 \inf D(\rho) \quad \text{for a.e. } (t,x) \in \Sigma$$

and there are positive constants A_1 and A_1' such that

$$\rho(r)(n_0 r - h(t,x)) \geq -A_1|r| - a_1' \quad \text{for all } r \in D(\rho) \text{ and a.e. } (t,x) \in \Sigma.$$

(H3) $u_0 \in H^1(\Omega)$ such that $\rho(u_0) \in L^2(\Omega)$.

(H4) $w_0 \in H^2(\Omega)$ such that $\dfrac{\partial w_0}{\partial n} = 0$ a.e. on Γ and there is $\xi_0 \in L^2(\Omega)$ satisfying

$$\xi_0 \in \beta(w_0) \quad \text{a.e. in } \Omega, \quad -\kappa \Delta w_0 + \xi_0 \in H^1(\Omega).$$

Now we introduce the variational formulation of P and P_μ.

Definition 2.1. For $0 \leq \mu \leq 1$ and $0 < T < +\infty$ a couple $\{u,w\}$ of functions $u : [0,T] \to H^1(\Omega)$ and $w : [0,T] \to H^2(\Omega)$ is called a (weak) solution of P_μ on $[0,T]$, if:

(w1) $u \in L^2(0,T; H^1(\Omega))$, $\rho(u) \in C_w([0,T]; L^2(\Omega))(=$ space of all weakly continuous functions from $[0,T]$ into $L^2(\Omega))$, $\rho(u)'(= \dfrac{d}{dt}\rho(u)) \in L^1(0,T; H^1(\Omega)^*)$, $w \in L^2(0,T; H^2(\Omega)) \cap L^\infty(0,T; H^1(\Omega))$, $w' \in L^2(0,T; H^1(\Omega)^*)$, $\lambda(w)' \in L^1(0,T; H^1(\Omega)^*)$, where $H^1(\Omega)^*$ stands for the dual space of $H^1(\Omega)$.

(w2) $\rho(u)(0) = \rho(u_0)$ (hence $u(0) = u_0$) and $w(0) = w_0$.

(w3) For a.e. $t \in [0,T]$ and all $z \in H^1(\Omega)$

$$\frac{d}{dt}(\rho(u(t)) + \lambda(w(t)), z) + a(u(t), z) + (n_0 u(t) - h(t), z)_\Gamma = (f(t), z),$$

where $a(u,z) = \displaystyle\int_\Omega \nabla u \cdot \nabla z \, dx$ and $(\cdot, \cdot)$ (resp. $(\cdot, \cdot)_\Gamma$) denotes the standard inner product in $L^2(\Omega)$ (resp. $L^2(\Gamma)$).

(w4) For a.e. $t \in [0, T]$,

$$\frac{\partial}{\partial n} w(t) = 0 \quad \text{a.e. on } \Gamma,$$

and there is a function $\xi \in L^2(0, T; L^2(\Omega))$ such that

$$\xi \in \beta(w) \quad \text{a.e. in } (0, T) \times \Omega$$

and

$$\frac{d}{dt}(w(t), \eta - \mu\Delta\eta) + \kappa(\Delta w(t), \Delta\eta) - (g(w(t)) + \xi(t) - \lambda'(w(t))u(t), \Delta\eta) = 0$$

for all $\eta \in H^2(\Omega)$ with $\dfrac{\partial\eta}{\partial n} = 0$ a.e. on Γ and a.e. $t \in [0, T]$.

A couple $\{u, w\}$ of functions $u : R_+ \to H^1(\Omega)$ and $w : R_+ \to H^2(\Omega)$ is called a (weak) solution of P_μ on R_+, if it is a solution of P_μ on $[0, T]$ for every finite $T > 0$.

Theorem 2.1. *(Uniqueness, $0 \leq \mu \leq 1$) Assume that (ρ), (β), (λ), (g), (H1)-(H4) hold. Let $0 < T < +\infty$, $0 \leq \mu \leq 1$, and $\{u, w\}$ be a solution of P_μ on $[0, T]$. Then the solution $\{u, w\}$ is unique in the class of $u \in L^\infty(0, T; H^1(\Omega))$.*

In proving Theorem 2.1, the key is to take advantage of inequality (2.1) of condition (ρ) together with an interpolation inequality of the form

$$|z|_{L^q(\Omega)} \leq \delta|z|_{H^1(\Omega)} + C_\delta|z|_{H^1(\Omega)}. \quad \text{for all } z \in H^1(\Omega) \text{ and } 1 \leq q < 6,$$

where δ is an arbitrary positive number and C_δ is a certain positive constant dependent only on δ.

Theorem 2.2. *(Existence, $0 < \mu \leq 1$) Assume (ρ), (β), (λ), (g) and (H1)-(H4) hold as well as*

$$\sigma_* < \frac{1}{|\Omega|} \int_\Omega w_0(x)dx =: m_0 < \sigma^*. \tag{2.2}$$

Then for every finite $T > 0$ and $0 < \mu \leq 1$, problem P_μ admits a solution $\{u, w\}$ on $[0, T]$ such that

$$\begin{cases} u \in L^\infty(0, T; H^1(\Omega)), \\ w \in L^\infty(0, T; H^2(\Omega)), \quad w' \in L^\infty(0, T; L^2(\Omega)) \cap L^2(0, T; H^1(\Omega)), \\ \xi \in L^\infty(0, T; L^2(\Omega)), \end{cases} \tag{2.3}$$

where ξ is as in (w4) of Definition 2.1.

In the case of $0 < \mu \leq 1$, we can reformulate P_μ as a system of parabolic evolution equations involving time-dependent subdifferentials in Hillbert space $L^2(\Omega)$, and construct a solution $\{u, w\}$ of P_μ, having regularity properties (2.3), in such a framework. By Theorem 2.1, the solution of P_μ due to Theorem 2.2 is unique.

Theorem 2.3. *(Existence, $\mu = 0$) Assume that (ρ), (β), (λ), (g), (H1)-(H4) and (2.2) hold. Then, for every finite $T > 0$, problem P admits at least one solution $\{u, w\}$ on $[0, T]$.*

A solution of P is obtained as a limit of the viscosity solutions $\{u_\mu, w_\mu\}$ given by Theorem 2.2, as $\mu \to 0$. But, in general, regularity properties (2.3) may be lost, when $\mu \to 0$.

In the case of space dimension $N = 1$, the viscosity approximation yields also a regular (hence unique) solution of P on every finite time interval $[0, T]$.

Theorem 2.4. *Assume that $N = 1$ and (ρ), (β), (λ), (g),(H1)-(H4) and (2.2) hold. Then, for every finite $T > 0$ problem P admits one and only one solution $\{u, w\}$ on $[0, T]$ which has regularity properties*

$$u \in L^2(0, T; H^2(\Omega)), \quad u' \in L^2(0, T; L^2(\Omega))$$

in addition to (2.3).

In the case of $N = 2$ or 3, the existence of a solution of P satisfying (2.3) remains an open question. We have so far noticed the only one paper [9] that proved the existence and uniqueness for the unconstrained problem (the problem P without term ξ) in one-dimensional space.

We refer to [6] for detailed proofs of Theorems 2.1-2.4.

3. Large time behaviour of the solution of 1D problem

In this section we suppose that $N = 1$ and $\Omega = (-L, L)$ with a positive number L. To discuss the one-dimensional problem P we further suppose that

$$f \in L^2(R_+; L^2(-L, L)), \quad \sup_{t \geq 0} |f|_{W^{1,2}(t,t+1; L^2(-L,L))} < +\infty \tag{3.1}$$

$$\begin{cases} \sup_{t \geq 0} \{|h_+|_{W^{1,2}(t,t+1)} + |h_-|_{W^{1,2}(t,t+1)}\} < +\infty, \\ h_\pm - h^\infty \in L^2(R_+) \quad \text{for a constant } h^\infty, \end{cases} \tag{3.2}$$

where $h_+(t) := h(t, L)$ and $h_-(t) := h(t, -L)$, and

$$\beta = \partial I_{[\sigma_*, \sigma^*]}. \tag{3.3}$$

124

Theorem 3.1. *(Global estimates) Assume that $N = 1$, (ρ), (β), (λ), (g), $(H1)$-$(H4)$, (2.2), (3.1)-(3.3) hold. Let $u^\infty := \dfrac{h^\infty}{n_0}$ and $\{u, w\}$ be the solution of P on R_+. Then*

$$u - u^\infty \in L^2(R_+; H^1(-L, L)), \quad u \in L^\infty(R_+; H^1(-L, L)), \tag{3.4}$$

$$\sup_{t \geq 0} |u'|_{L^2(t, t+1; L^2(-L, L))} < +\infty, \tag{3.5}$$

$$w \in L^\infty(R_+; H^2(-L, L)), \quad w' \in L^\infty(R_+; H^1(-L, L)^*) \cap L^2(R_+; H^1(-L, L)^*) \tag{3.6}$$

and

$$\sup_{t \geq 0} |w'|_{L^2(t, t+1; H^1(-L, L))} < +\infty. \tag{3.7}$$

The global estimates (3.4)-(3.7) can be inferred from several energy inequalities which are obtained by multiplying (formally) (1.1) and (1.2) by $u - u^\infty$ and w', respectively, and by multiplying $\dfrac{\partial}{\partial t}(1.2)$ by w', etc. See [3, 4] for a detailed proof of Theorem 3.1.

As an easy consequence of Theorem 3.1 we see that $u(t)$ converges to the constant $u^\infty := \dfrac{h^\infty}{n_0}$ as $t \to +\infty$ and the ω-limit set $\omega(u_0, w_0)$ of the order parameter $w(t)$, given by

$$\omega(u_0, w_0) := \{v \in H^1(-L, L); w(t_n) \to v \text{ in } H^1(-L, L) \text{ as } n \to +\infty$$

$$\text{for some } t_n \uparrow +\infty\},$$

is non-empty. More precisely, we have:

Theorem 3.2. *Under the same assumptions as in Theorem 3.1, the following statements hold:*

(a) $u(t) \to u^\infty$ *weakly in* $H^1(-L, L)$ *as* $t \to +\infty$.

(b) $\omega(u_0, w_0)$ *is compact and connected in* $H^1(-L, L)$, *and bounded in* $H^2(-L, L)$. *Furthermore, any ω-limit point $v \in \omega(u_0, w_0)$ is a solution of the steady-state* *problem, denoted by* P^∞,

$$P^\infty \begin{cases} -\kappa v_{xx} + \gamma + g(v) - \lambda'(v)u^\infty = \nu & a.e. \ in \ (-L, L), \\[2mm] \gamma \in L^2(-L, L), \quad \gamma \in \partial I_{[\sigma_*, \sigma^*]}(v) & a.e. \ in \ (-L, L), \\[2mm] v_x(-L) = v_x(L) = 0, \\[2mm] \nu = \dfrac{1}{2L} \displaystyle\int_{-L}^{L} \{\gamma + g(v) - \lambda'(v)u^\infty\}dx, \\[4mm] \dfrac{1}{2L} \displaystyle\int_{-L}^{L} v dx = m_0. \end{cases}$$

(c) $\displaystyle\lim_{t \to +\infty} \{\frac{\kappa}{2}|w_x(t)|^2_{L^2(-L,L)} + \int_{-L}^{L}(\hat{g}(w(t)) - \lambda(w(t))u^\infty)dx\}$ *exists, where* $\hat{g}$ *is any prim-itive of* g.

When we need to indicate explicitly the data σ_*, σ^*, $u^\infty(= \dfrac{h^\infty}{n_0})$ and m_0, etc., we denote P^∞ by $P^\infty(\sigma_*, \sigma^*; u^\infty, m_0)$.

In the remainder of this section, under additional restrictions on u^∞, g and λ we consider the steady state problem P^∞ with $m_0 = 0$, namely $P^\infty(\sigma_*, \sigma^*; u^\infty, 0)$ which is very restricted, but still interesting from some physical points of view.

For simplicity $P^\infty(\sigma_*, \sigma^*; u^\infty, 0)$ denoted by P_0^∞, and $q(v)$ is defined by

$$q(v) := g(v) - \lambda'(v)u^\infty, \quad v \in R.$$

We suppose that q satisfies the following conditions (q1)-(q3):

(q1) (oddness) $q(v) = -q(-v)$ for all $v \in R$.

(q2) (N-shape) There are points ζ_{0-}, ζ_M, ζ_m and ζ_{0+} such that $\zeta_{0-} < \zeta_M < 0 < \zeta_m < \zeta_{0+}$,

$$q(\zeta_{0-}) = q(\zeta_{0+}) = 0,$$

$$q'(= \frac{d}{dv}q) > 0 \quad on \ (-\infty, \zeta_M) \cup (\zeta_m, +\infty),$$

$$q' < 0 \quad on \ (\zeta_M, \zeta_m),$$

and

$$q'(\zeta_M) = q(\zeta_m) = 0.$$

(q3) (convexity-concavity) $q'' \geq 0$ on $(0, +\infty)$, $q''(0) = 0$ and $q'' \leq 0$ on $(-\infty, 0)$.

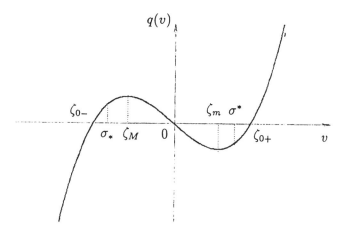

Also, the solution set of P_0^∞ is denoted by S and its structure is to be investigated in the following decomposition

$$S = S_c + S_0 + S_1, \tag{3.8}$$

where $S_c := \{v; v \text{ is a constant solution of } P_0^\infty\}$, $S_0 := \{v; v \text{ is a non-constant solution of } P_0^\infty \text{ such that } \sigma_* < v < \sigma^* \text{ on } [-L, L]\}$ and $S_1 := \{v; v \text{ is a non-constant solution of } P_0^\infty \text{ such that } v(x) = \sigma_* \text{ or } \sigma^* \text{ for some } x \in [-L, L]\}$.

In the decomposition (3.8), it is easy to see that $S_c = \{0\}$, and it follows from [10] that S_0 is a finite set. But, in general, the structure of S_1 is very complicated.

Concerning the relation between σ_*, σ^*, ζ_{0-} and ζ_{0+} we have the following two cases:

$$\sigma_* \leq \zeta_{0-}, \quad \sigma^* \geq \zeta_{0+}. \tag{3.9}$$

$$\zeta_{0-} < \sigma_* < 0 \quad \text{or} \quad 0 < \sigma^* < \zeta_{0+}. \tag{3.10}$$

Theorem 3.3. *In addition to all the hypotheses of Theorem 3.1, assume that $m_0 = 0$ and (q1)-(q3) are satisfied. Then:*

(1) *If (3.9) holds, then $S_1 = \emptyset$ and hence $\omega(u_0, w_0)$ is a singleton in $\{0\} + S_0$. In this case $w(t)$ converges in $H^1(-L, L)$ to a solution of P_0^∞ as $t \to +\infty$.*

(2) *If (3.10) holds, then one of the following two cases (i) and (ii) occurs:*

(i) *$\omega(u_0, w_0)$ is a singleton. In this case, $w(t)$ converges in $H^1(-L, L)$ to a solution of P_0^∞ as $t \to +\infty$.*

127

(ii) $w(u_0, w_0)$ *contains a continuum of solutions of* P_0^∞. *In this case, after a certain large time, the number of 0-points of* $w(t, \cdot)$ *(i.e. the number of* $x \in [-L, L]$ *with* $w(t, x) = 0$*) is finite and constant.*
Moreover, $\lim_{t \to +\infty} \kappa |w_x(t)|^2_{L^2(-L,L)}$ *exists.*

We can prove Theorem 3.3 by taking advantage of concrete expressions of solutions to P_0^∞. In fact, any solution v of P_0^∞ behaves as the figure shows below; in general, $v(x)$ consists of a finite number of non-constant parts ($\sigma_* < v < \sigma^*$) and flat parts ($v = \sigma_*$ or σ^*).

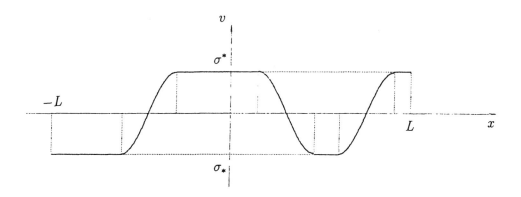

In the terminology of phase separation, Theorem 3.3 (2) (ii) says that the number of the interfaces is invariant for large t, though their locations move very slowly in time as $t \to +\infty$.

For more detailed expressions of solutions to P_0^∞ and a proof of Theorem 3.3, we refer to [3]. Also, in [3] there is given an example in which the ω-limit set $\omega(u_0, w_0)$ contains a continuum of the corresponding steady-state solutions.

References

1. H. W. Alt and I. Pawlow, Existence of solutions for non-isothermal phase separation, Adv. Math. Sci. Appl. **1**(1992), 319-409.

2. J. F. Blowey and C. M. Elliott, The Cahn-Hilliard gradient theory for phase separation with non-smooth free energy, Part I: Mathematical analysis, European J. Appl. Math. **2**(1991), 233-280.

3. A. Ito and N. Kenmochi, Asymptotic behaviour of solutions to non-isothermal phase separation model with constraints in one-dimensional space, Tech. Rep. Math. Chiba Univ., Vol. 9 No. 12, 1993.

4. N. Kenmochi and M. Niezgódka, Large time behaviour of a nonlinear system for phase separation, pp. 12-22, in "Progress in partial differential equations: the Metz surveys 2", Pittman Research Notes Math. Ser. Vol. 290, 1993.

5. N. Kenmochi and M. Niezgódka, Nonlinear system for non-isothermal diffusive phase separation, to appear in J. Math. Anal. Appl.

6. N. Kenmochi and M. Niezgódka, Viscosity approach to modelling non-isothermal diffusive phase separation, Tech. Rep. Math. Sci. Chiba Univ. Vol. 9, No. 8, 1993.

7. N. Kenmochi, M. Niezgódka and I. Pawlow, Subdifferential operator approach to the Cahn-Hilliard equation with constraint, to appear in J. Differential Equations.

8. O. Penrose and P. C. Fife, Thermodynamically consistent models of phase-field type for the kinetics of phase transitions, Physica D **43** (1990), 44-62.

9. W. Shen and S. Zheng, On the coupled Cahn-Hilliard equations, Common.P.D.E. **18**(1993), 711-727.

10. S. Zheng, Asymptotic behaviour of the solution to the Cahn-Hilliard equation, Applicable Anal. **23**(1986), 165-184.

A. Ito: Department of Mathematics Graduate School of Sciense and Technology
Chiba University, Chiba, 263 Japan

N. Kenmochi: Department of Mathematics Faculty of Education
Chiba University, Chiba, 263 Japan

M. Niezgódka: Interdisciplinary Centre for Mathematical and Computational
Modelling, Warsaw University, Banacha 2, 02-097 Warsaw, Poland

V L KAMYNIN

On convergence of the solutions of inverse problems for parabolic equations with weakly converging coefficients

We consider passage to the limit in the sequence of inverse problems

$$
u_t - \sum_{i,j=1}^{n} \frac{\partial}{\partial x_j}\left(a_{ij}^m(t,x)\frac{\partial u}{\partial x_i}\right) + \sum_{i=1}^{n} a_i^m(t,x)u_{x_i} +
$$

$$
+ b^m(t,x)u = f(t)g^m(t,x) + \sum_{j=1}^{n} \frac{\partial h_j^m(t,x)}{\partial x_j} , \quad (1^m)
$$

$$
u\Big|_{t=0} = u_0^m(x), \tag{2^m}
$$

$$
u\Big|_{x\in\partial\Omega} = 0, \tag{3}
$$

$$
\int_{\Omega} u(t,x)w^m(t,x)dx = \phi^m(t), \tag{4^m}
$$

$m=1,2,\ldots,\infty$; here $(t,x)\in Q \equiv \left\{(t,x):0<t\leqslant T, \ x\in\bar\Omega\right\}$, Ω - is

bounded domain in $\mathbb{R}^n$ with smooth boundary.

In these problems the functions $u(t,x)$, $f(t)$ are unknown, and functions a_{ij}^m, a_i^m, b^m, g^m, h_j^m, u_0^m, w^m, ϕ^m, $m=1,2,\ldots,\infty$, are given. We assume the weak L_2-convergence as $m \to \infty$ of input data of the problems $(1^m)-(4^m)$ with $m\neq\infty$ to the corresponding input data of the limit problem $(1^\infty)-(4^\infty)$.

The problem under consideration associated with mathematical simulation of physical processes in strongly inhomogeneous media in the case when part of the properties of these media are unknown but it is possible to obtain additional information of the process described by condition (4^m).

The problems of the type $(1^m)-(4^m)$ can also be considered as control problems where we have to find the unknown control function $f(x)$ ensuring the solution $u(t,x)$ satisfying condition (4^m).

The problem considered in present paper connected with the problem of G-convergence and our aim is to find the additional conditions under which the weak convergence of the coefficients of the equation (1^m) to the coefficients of the equation (1^∞) implies G-convergence of corresponding operators of inverse problems.

Such kind of problems are widely investigated in the case of direct problems for parabolic and elliptic equations [1,2,8,13,etc.]. On the contrary, in the case of inverse problems the are only few papers on that subject. We indicate the papers [6,7] and also the papers [3,4] where some interesting results were obtained in the case of passage to the limit in optimal control problems.

It should be noted that setting of the inverse problems in the form $(1^m)-(4^m)$ considered in [5,10,11] where the questions of unique solvability of such problems were studied. We consider the problems $(1^m)-(4^m)$ under weaker restrictions on the input data·than in papers mentioned, so in the first part of present paper we prove the theorem of unique solvability of the problems $(1^m)-(4^m)$ which does not follow from the previous results. This theorem also permits us to obtain uniform with respect to m estimates of the solutions being used below in passing to the limit in the sequence of problems $(1^m)-(4^m)$.

Let us offer certain notations used in the paper. We

define $\quad Q_{t_2}^{t_1} \equiv \left\{(t,x): t_1 \leqslant t \leqslant t_2, x \in \bar{\Omega}\right\}, \quad Q_t^0 \equiv Q_t, \quad Q_T \equiv Q,$

$$G^m(t) = \int_\Omega g^m(t,x) w^m(t,x) dx, \quad m=1,2,\ldots,\infty.$$

The Banach spaces $L_q(Q)$, $L_p(t_1,t_2;L_q(\Omega))$, $\overset{O}{W}{}_2^1(\Omega)$, $C^{0,\alpha}(Q)$, $C(t_1,t_2;L_q(\Omega))$ with corresponding norms are understood in usual sense. We also introduce the following spaces: $W_2^{-1}(\Omega)$ is a space dual to $\overset{O}{W}{}_2^1(\Omega)$;

$$W(Q) = \left\{ w(t,x): w \in L_\infty(0,T; \overset{O}{W}{}_2^1(\Omega)), w_t \in L_\infty(0,T;L_2(\Omega)) \right\},$$

$$|w|_{W(Q)} = |w|_{L_\infty(0,T; \overset{O}{W}{}_2^1(\Omega))} + |w_t|_{L_\infty(0,T;L_2(\Omega))};$$

$$V(Q) = \left\{ v(t,x): v \in C(0,T;L_2(\Omega)), v_t \in L_2(0,T;W_2^{-1}(\Omega)) \right\},$$

$$|v|_{V(Q)} = \sup_{t_1 \leqslant t \leqslant t_2} |v(t,\cdot)|_{L_2(\Omega)} + |v_x|_{L_2(Q)} +$$

$$+ |v_t|_{L_2(0,T;W_2^{-1}(\Omega))}.$$

The author is partially supported by RFFI Grant № 93-012-606.

1. Unique solvability of the problem $(1^m)-(4^m)$.

Let us assume that there exist positive constants Λ_0, Λ_1,

132

K_1, K_2 g_0 such that uniformly with respect to $m=1,2,\ldots\infty$

$$\Lambda_0|\xi|^2 \leqslant \sum_{i,j=1}^n a_{ij}^m(t,x)\xi_i\xi_j \leqslant \Lambda_1|\xi|^2, \tag{5}$$

$$\forall\xi\equiv(\xi_1,\ldots,\xi_n)\in\mathbb{R}^n;$$

$$\left|a_i^m\right|_{L_\infty(Q)}, \left|b^m\right|_{L_\infty(Q)}, \left|g^m\right|_{L_\infty(0,T;L_2(\Omega))},$$

$$\left|h_j^m\right|_{L_2(Q)}, \left|u_0^m\right|_{L_2(\Omega)}, \left|w^m\right|_{W(Q)} \leqslant K_1, \quad i,j=1,2,\ldots,n; \tag{6}$$

$$|G^m(t)|\geqslant g_0 \quad \forall t\in[0,T]; \tag{7}$$

$$\left|\phi^m\right|_{W_2^1([0,T])} \leqslant K_2. \tag{8}$$

We suppose that $a_{ij}^m=a_{ji}^m$, and the compatibility condition

$$\phi^m(0)=\int_\Omega u_0^m(x)w^m(0,x)dx, \quad m=1,2,\ldots,\infty \tag{9}$$

is satisfied.

Definition 1. A pair of functions $\{u^m(t,x), f^m(t)\}$ is a generalized solution of the inverse problem $(1^m)-(4^m)$ if $u^m\epsilon V(Q)$, $f^m(t)\epsilon L_2([0,T])$, the following integral identity

$$\int_\Omega u^m(T,x)\Phi(T,x)dx - \int_\Omega u_0^m(x)\Phi(0,x)dx - \int_Q u^m\Phi_t dxdt +$$

$$+ \sum_{i,j=1}^{n} \int_Q a_{ij}^m(t,x) u_{x_i}^m \Phi_{x_j} \, dxdt + \sum_{i=1}^{n} \int_Q a_i^m(t,x) u_{x_i}^m \Phi \, dxdt +$$

$$+ \int_Q b^m(t,x) u^m \Phi \, dxdt = \int_Q f^m(t) g^m(t,x) \Phi \, dxdt -$$

$$- \sum_{j=1}^{n} \int_Q h_j^m(t,x) \Phi_{x_j} \, dxdt \qquad\qquad (10^m)$$

holds for any $\Phi(t,x) \in W_2^{1,1}(Q)$ $\left[\Phi\big|_{x\in\partial\Omega} = 0\right]$ and the overdetermination condition (4^m) is satisfied.

Let $\{u^m(t,x), f^m(t)\}$ – be a generalized solution of the problem $(1^m)-(4^m)$. Let us choose $\Phi(t,x)$ in (10^m) is equal to $w^m(t,x)\chi(t)$, $\chi(t) \in \overset{o}{C}{}^\infty([0,T])$. Taking into account the condition (4^m) it is easy to show that the pair $\{u^m,f^m\}$ satisfies the relation

$$f^m(t) = \frac{\phi^{m'}(t)}{G^m(t)} + \frac{1}{G^m(t)} \int_\Omega \left\{ \sum_{i,j=1}^{n} a_{ij}^m(t,x) u_{x_i}^m(t,x) w_{x_j}^m(t,x) + \right.$$

$$+ \sum_{i=1}^{n} a_i^m(t,x) u_{x_i}^m(t,x) w^m(t,x) + b^m(t,x) u^m(t,x) w^m(t,x) -$$

$$\left. - u^m(t,x) w_t^m(t,x) + \sum_{j=1}^{n} h_j^m(t,x) w_{x_j}^m(t,x) \right\} dx \qquad (11)$$

for almost all $t \in [0,T]$.

Theorem 1. *Pair of functions* $\{u^m(t,x),\ f^m(t)\}$ *is a generalized solution of the problem* $(1^m)-(4^m)$ *if and only if it satisfies the relations* $(10^m),(11)$.

The proof of this theorem is standard [11,12].

Now we turn our attention to the demonstration of the unique solvability of the problem $(1^m)-(4^m)$.

In what follows let us agree to denote by c (possibly with index) any positive constants depending only on n, T, mes Ω, Λ_0, Λ_1, K_1, K_2, g_0.

Theorem 2. *Suppose that the conditions* $(5)-(9)$ *are satisfied. Then the inverse problem* $(1^m)-(4^m)$ *has unique solution* $\{u^m,\ f^m\}$ *for any m, and the estimates*

$$\left| f^m \right|_{L_2([0,T])} \leqslant c_1, \tag{12}$$

$$\left| u^m \right|_{V(Q)} \leqslant c_2, \tag{13}$$

hold uniformly with respect to m.

Remark 1. Existence theorem closed to theorem 2 for more general equation but under more strong assumptions was established in [5]. Below we give independent proof of unique solvability of the problem $(1^m)-(4^m)$.

Proof of Theorem 2. In order to simplify the representation we shall omit index m in the notations of the solutions and input data of the problem $(1^m)-(4^m)$. We divide the segment $[0,T]$ into a finite number of segments $[0,t_1]$, $[t_1,t_2],\ldots,$ $[t_{N-1},t_N{=}T]$ of length Δt whose size will be chosen below. We shall prove the existence of the solution of the problem

(1)-(4) on each of these segments.

Let us first consider segment $[0,t_1]$ and represent desired solution $\{u_1(t,x),f_1(t)\}$ on this segment in the form of a sum

$$\{u_1,f_1\} = \{v_1,0\} + \{z_1,f_1\}.$$

The function $v_1(t,x)$ is defined uniquely (see [9]) as a solution of direct problem on Q_{t_1} for the equation

$$v_t - \sum_{i,j=1}^{n} \frac{\partial}{\partial x_j}\left[a_{ij}(t,x)\frac{\partial v}{\partial x_i}\right] + \sum_{i=1}^{n} a_i(t,x)v_{x_i} +$$

$$+ b(t,x)v = \sum_{j=1}^{n} \frac{\partial h_j(t,x)}{\partial x_j} , \quad (14)$$

with boundary conditions (2),(3). This function satisfies the estimate (see [9])

$$\left|v_1\right|_{V(Q_{t_1})} \leqslant c. \quad (15)$$

Now we shall seek the pair $\{z_1(t,x),f_1(t)\}$ as a solution of inverse problem

$$z_t - \sum_{i,j=1}^{n} \frac{\partial}{\partial x_j}\left[a_{ij}(t,x)\frac{\partial z}{\partial x_i}\right] + \sum_{i=1}^{n} a_i(t,x)z_{x_i} +$$

$$+ b(t,x)z = f(t)g(t,x), \quad (t,x)\epsilon Q_{t_1}, \quad (16)$$

$$z\Big|_{t=0} = z\Big|_{x\epsilon\partial\Omega} = 0, \quad (17)$$

136

$$\int_\Omega z(t,x)w(t,x)dx = \phi(t) -$$

$$- \int_\Omega v_1(t,x)w(t,x)dx \equiv \phi_1(t), \quad t\in[0,t_1]. \quad (18)$$

Let us note that $\phi_1(t)\in W_2^1([0,t_1])$, and $\phi_1(0)=0$.

According to the Theorem 1 let us consider the relation

$$f(t) = \frac{\phi_1'(t)}{G(t)} + \frac{1}{G(t)} \int_\Omega \left\{ - z(t,x)w_t(t,x) + \right.$$

$$+ \sum_{1,j=1}^{n} a_{1j}(t,x)z_{x_1}(t,x)w_{x_j}(t,x) +$$

$$+ \sum_{1=1}^{n} a_1(t,x)z_{x_1}(t,x)w(t,x) +$$

$$\left. + b(t,x)z(t,x)w(t,x) \right\}dx, \quad t\in[0,T]. \quad (19)$$

Now we introduce the linear operator

$$L: L_2([0,t_1]) \rightarrow L_2([0,t_1])$$

by the formula

$$(Lf)(t) = \frac{1}{G(t)} \iint\limits_{\Omega} \Bigg\{ - z(t,x)w_t(t,x) +$$

$$+ \sum_{i,j=1}^{n} a_{ij}(t,x)z_{x_i}(t,x)w_{x_j}(t,x) +$$

$$+ \sum_{i=1}^{n} a_i(t,x)z_{x_i}(t,x)w(t,x) + b(t,x)z(t,x)w(t,x) \Bigg\} dx, \quad (20)$$

where $z(t,x)$ - is a solution in Q_{t_1} of direct problem (16),(17) with fixed $f(t) \epsilon L_2([0,t_1])$ in the right side of the equation (16).

Note, that in view of (5)-(8) and theorem 4.2 of [9, p.189]), the operator L can be defined by (20) for any function $f(t) \epsilon L_1([0,t_1])$ and it easy to see that then $(Lf)(t) \epsilon L_2([0,t_1])$.

By means of operator L it is possible to write the relation (19) in the form of second kind operator equation

$$f = \frac{\phi_1'}{G} + Lf \quad (21)$$

Now we show that the operator L is a contracting operator in $L_1([0,t_1])$ for some $\Delta t = t_1$.

In fact, for any $f(t) \epsilon L_1([0,t_1])$ using conditions (5)-(7), Cauchy inequality and energy inequality we obtain following estimates

$$\|Lf\|_{L_1([0,t_1])} \leqslant c_1 \int_0^{t_1} \Bigg\{ \int_{\Omega} |z|^2 dx \Bigg\}^{1/2} \cdot \Bigg\{ \int_{\Omega} |w_t|^2 dx \Bigg\}^{1/2} dt +$$

$$+ c_2 \int_0^{t_1} \left\{ \int_\Omega |z_x|^2 dx \right\}^{1/2} \cdot \left\{ \int_\Omega |w_x|^2 dx \right\}^{1/2} dt +$$

$$+ c_3 \int_0^{t_1} \left\{ \int_\Omega |z_x|^2 dx \right\}^{1/2} \cdot \left\{ \int_\Omega |w|^2 dx \right\}^{1/2} dt +$$

$$+ c_4 \int_0^{t_1} \left\{ \int_\Omega |z|^2 dx \right\}^{1/2} \cdot \left\{ \int_\Omega |w|^2 dx \right\}^{1/2} dt \leqslant c_5 t^{1/2} |z_x|_{L_2(Q_{t_1})} +$$

$$+ c_6 t^{1/2} |z|_{L_2(Q_{t_1})} \leqslant c_7 t^{1/2} |fg|_{L_1(0,T;L_2(\Omega))} \leqslant$$

$$\leqslant c^* t^{1/2} |f|_{L_1([0,t_1])}, \tag{22}$$

where the constant c^* is independent of $\phi_1(t)$. If we choose Δt such, that

$$c^* \Delta t^{1/2} < \frac{1}{2}, \tag{23}$$

then by virtue of (22) the operator L we be a contracting operator in $L_1([0,t_1])$. Consequently, there exist unique solution $f_1(t) \in L_1([0,t_1])$ of the equation (21), and the following estimate

$$|f_1|_{L_1([0,t_1])} \leqslant c \tag{24}$$

holds.

Now we show that actually $f_1(t) \in L_2([0,t_1])$. In fact, carring out the estimates similar to (22) and using (24), we obtain

$$\left|Lf_1\right|_{L_2([0,t_1])} \leq c_1\left|z_x\right|_{L_2(Q_{t_1})} + c_2\left|z\right|_{L_2(Q_{t_1})} \leq$$

$$c_3\left|f_1g\right|_{L_1(0,T;L_2(\Omega))} \leq c_4\left|f_1\right|_{L_1([0,t_1])} \leq c_5. \qquad (25)$$

As $\phi_1'(t)/G(t)\in L_2([0,t_1])$, then from (21) and (25) it follows that $f_1(t)\in L_2([0,t_1])$ with

$$\left|f_1\right|_{L_2([0,t_1])} \leq c. \qquad (26)$$

Let $z_1(t,x)$ be a (unique) solution of direct problem (16),(17) with $f_1(t)$ obtained above. Note, that for $z_1(t,x)$ the following estimate holds:

$$\left|z_1\right|_{V(Q_{t_1})} \leq c. \qquad (27)$$

According to Theorem 1 pair $\{z_1,f_1\}$ is a unique solution of inverse problem (16)-(18). Hence, there exists unique solution $\{u_1,f_1\}$ of the inverse problem (1)-(4) in the cylinder Q_{t_1}, and by virtue of (15) and (27) the estimate

$$\left|u_1\right|_{V(Q_{t_1})} \leq c \qquad (28)$$

is true.

Now we turn our attention to finding the solution $\{u_2,f_2\}$ of the problem (1)-(4) for $t\in[t_1,t_2]$. As above, we shall seek this solution in the form $\{u_2,f_2\} = \{v_2,0\} +$

$+ \{z_2, f_2\}$. The function $v_2(t.x)$ is a unique solution in $Q_{t_2}^{t_1}$ of the direct problem for the equation (14) with initial condition

$$v\big|_{t=t_1} = u_1(t_1, x)$$

and boundary condition (3), and this solution satisfies the estimate

$$\left|v_2\right|_{V(Q_{t_2}^{t_1})} \leqslant c. \tag{29}$$

We find the pair $\{z_2, f_2\}$ as a solution of the inverse problem in $Q_{t_2}^{t_1}$ for the equation (16) with boundary conditions (17) and overdetermination condition

$$\int_\Omega z(t,x)w(t,x)dx = \phi(t) -$$

$$- \int_\Omega v_2(t,x)w(t,x)dx = \phi_2(t), \quad t\in[t_1,t_2]. \tag{30}$$

As the constant c^* and consequently the choice of Δt in (23) do not depend on the overdetermination condition, then after repeating the preceding arguments we obtain the existence of a unique solution $\{z_2, f_2\}$ of the inverse problem (16),(17),(30) and the following estimates

$$\left|f_2\right|_{L_2([t_1,t_2])} , \quad \left|z_2\right|_{V(Q_{t_2}^{t_1})} \leqslant c.$$

Hence, there exists a unique solution $\{u_2, z_2\}$ of the problem (1)-(4) in $Q_{t_2}^{t_1}$, and the estimates

$$\left|f_2\right|_{L_2([t_1, t_2])}, \ \left|u_2\right|_{V(Q_{t_2}^{t_1})} \leqslant c \tag{31}$$

are valid.

After similar arguments we can prove the existence of the unique solution $\{u_1, f_1\}$ of the inverse problem (1)-(4) in each partial cylinder $Q_{t_1}^{t_{1-1}}$, $1=1,2,\ldots,N$. Moreover, these solutions will satisfy the estimates similar to (31).

Thus, we obtain that there exists a unique solution $\{u^m, f^m\}$ of the problem $(1^m)-(4^m)$, $m=1,2,\ldots,\infty$ in the whole cylinder Q and the estimates (12),(13) are valid uniformly with respect to m. Theorem 2 is proved.

Remark 2. Let in addition to the hypotheses of the theorem 2 it is known that

$$\left|g^m\right|_{L_\infty(Q)}, \ \left|h_j^m\right|_{L_\infty(Q)} \leqslant K_3 \tag{32}$$

uniformly with respect to m. Then using theorem 10.1 of [9, chapter 3] it is easy to obtain that $u^m(t,x) \in C^{0,\alpha}(Q^\delta)$ (for any $\delta > 0$) and the following estimate

$$\left|u^m\right|_{C^{0,\alpha}(Q^\delta)} \leqslant c(\delta) \tag{33}$$

is valid uniformly with respect to m.

2. Passage to the Limit.

In this section we consider a question of passage to the limit in the problems $(1^m)-(4^m)$ as $m \to \infty$. In addition to the conditions (6)-(9) we shall assume that as $m \to \infty$

$$a_{ij}^m(t,x) \to a_{ij}^\infty(t,x), \quad i,j=1,2,\ldots,n, \tag{34}$$

$$a_i^m(t,x) \to a_i^\infty(t,x), \quad i=1,2,\ldots,n, \tag{35}$$

$$b^m(t,x) \to b^\infty(t,x), \tag{36}$$

$$h_j^m(t,x) \to h_j^\infty(t,x), \quad j=1,2,\ldots,n, \tag{37}$$

weakly in $L_2(Q)$;

$$u_0^m(x) \to u_0^\infty(x) \tag{38}$$

weakly in $L_2(\Omega)$;

$$g^m(t,x) \to g^\infty(t,x), \tag{39}$$

weakly in $L_2(\Omega)$ for any fixed $t \in [0,T]$;

$$w^m(t,x) \to w^\infty(t,x), \tag{40}$$

uniformly on Q;

$$\phi^m(t) \to \phi^\infty(t), \tag{41}$$

in the norm of $L_2([0,T])$.

In the theorem to be proved below we establish

conditions providing the problem $(1^\infty)-(4^\infty)$ (which is the limit problem for $(1^m)-(4^m)$ in the sense of convergence of the coefficients) to be the limit problem for $(1^m)-(4^m)$ in the sense of G-convergence.

Theorem 3. *Let the conditions* (6)-(9) *as well as the relations* (34)-(41) *are fulfilled. Assume that there exists a sequence* $\mu(m) \rightarrow 0$ *(as* $m \rightarrow \infty$*) such that for any* $\sigma(t,x)\in$ *$\in V(Q)$ the following inequalities hold:*

$$\left| \iint_Q \sum_{i=1}^{n} \left[a^m_{ij}(t,x) - a^\infty_{ij}(t,x) \right] \sigma_{x_i}(t,x) dxdt \right| \leqslant$$

$$\leqslant \mu(k) |\sigma|_{V(Q)}, \quad j=1,2,\ldots,n; \quad (42)$$

$$\left| \iint_Q \sum_{i=1}^{n} \left[a^m_i(t,x) - a^\infty_i(t,x) \right] \sigma_{x_i}(t,x) dxdt \right| \leqslant$$

$$\leqslant \mu(k) |\sigma|_{V(Q)}. \quad (43)$$

Then the following limit relations (as $m \rightarrow \infty$*) are satisfied:*

$$u^m(t,x) \rightarrow u^\infty(t,x) \qquad (44)$$

in the norm of $L_2(Q)$;

$$u^m_x(t,x) \rightarrow u^\infty_x(t,x) \qquad (45)$$

weakly in $L_2(Q)$;

$$u_t^m(t,x) \longrightarrow u_t^\infty(t,x) \tag{46}$$

weakly in $L_2(0,T;W_2^{-1}(\Omega))$;

$$f^m(t) \longrightarrow f^\infty(t) \tag{47}$$

weakly in $L_2([0,T])$.

Here the pairs $\{u^m,f^m\}$ - are solutions of the problems $(1^m)-(4^m)$, m=1,2,...., and pair $\{u^\infty,f^\infty\}$ is a solution of the problem $(1^\infty)-(4^\infty)$.

Proof. Estimates (12),(13) imply existing the subsequence $m_\nu \to \infty$, $\nu \to \infty$, and the pair $\{u^*(t,x),f^*(t)\}$ ($u^* \epsilon V(Q)$, $f^* \epsilon L_2([0,T])$) also satisfying the estimates (12),(13) such that for $\{u^{m_\nu}(t,x),f^{m_\nu}(t)\}$ and $\{u^*(t,x),f^*(t)\}$ the limit relations (44)-(47) are true while m, u^∞, f^∞ are replaced by m_ν, u^*, f^* correspondingly.

By virtue of definition of the solution $u^*(t,x)$ it satisfies boundary condition (3). Let us show that pair $\{u^*,f^*\}$ satisfies the integral identity (10^∞). To do this wi fix function $\Phi(t,x) \epsilon C^\infty(Q)$, $\Phi(T,x)=0$, $\Phi\big|_{x \epsilon \partial\Omega} =0$.

Letting

$$J \equiv \left| -\int_\Omega u_0^\infty(x)\Phi(0,x)dx - \int_Q u^*\Phi_t dxdt + \right.$$

$$+ \sum_{i,j=1}^n \int_Q a_{ij}^\infty(t,x)u_{x_i}^*\Phi_{x_j} dxdt + \sum_{i=1}^n \int_Q a_i^\infty(t,x)u_{x_i}^*\Phi \, dxdt +$$

$$+ \int_Q b^\infty(t,x)u^*\Phi \, dxdt - \int_Q f^*g^\infty(t,x)\Phi \, dxdt +$$

$$+ \sum_{j=1}^n \int_Q h_j^\infty(t,x)\Phi_{x_j} \, dxdt \Bigg|$$

and using the fact that $\{u^{m_\nu}, f^{m_\nu}\}$ is a solution of the problem $(1^{m_\nu})-(4^{m_\nu})$ we have

$$J \leqslant \Bigg| \int_\Omega \Big[u_0^{m_\nu}(x) - u_0^\infty(x) \Big] \Phi(0,x) dx \Bigg| + \Bigg| \int_Q \Big[u^{m_\nu} - u^* \Big] \Phi_t \, dxdt \Bigg| +$$

$$+ \Bigg| \int_Q \sum_{i,j=1}^n \Big[a_{ij}^{m_\nu} u_{x_i}^{m_\nu} - a_{ij}^\infty u_{x_i}^* \Big] \Phi_{x_j} \, dxdt \Bigg| +$$

$$+ \Bigg| \int_Q \sum_{i=1}^n \Big[a_i^{m_\nu} u_{x_i}^{m_\nu} - a_i^\infty u_{x_i}^* \Big] \Phi \, dxdt \Bigg| +$$

$$+ \Bigg| \int_Q \Big[b^{m_\nu} u^{m_\nu} - b^\infty u^* \Big] \Phi \, dxdt \Bigg| +$$

$$+ \Bigg| \int_Q \Big[f^{m_\nu}(t)g^{m_\nu}(t,x) - f^*(t)g^\infty(t,x) \Big] \Phi \, dxdt \Bigg| +$$

$$+ \Bigg| \sum_{j=1}^n \int_Q \Big[h_j^{m_\nu}(t,x) - h_j^\infty(t,x) \Big] \Phi_{x_j} \, dxdt \Bigg| =$$

$$\equiv J_0^\nu + J_t^\nu + J_{a_{1j}}^\nu + J_{a_1}^\nu + J_b^\nu + J_f^\nu + J_g^\nu. \qquad (48)$$

Assumptions (36)–(38) and condition (44) imply convergence to zero of the integrals J_0^ν, J_t^ν, J_b^ν, J_g^ν as $\nu \to \infty$.

For the integral $J_{a_{1j}}^\nu$ we write

$$J_{a_{1j}}^\nu \leqslant \sum_{j=1}^{n} \left| \iint_Q \sum_{i=1}^{n} \left[a_{ij}^{m_\nu} - a_{ij}^{\infty} \right] \left(u^{m_\nu} \Phi_{x_j} \right)_{x_i} dx dt \right| +$$

$$+ \sum_{i,j=1}^{n} \left| \iint_Q \left[a_{ij}^{m_\nu} - a_{ij}^{\infty} \right] \left[u^{m_\nu} - u^* \right] \Phi_{x_i x_j} dx dt \right| +$$

$$+ \sum_{i,j=1}^{n} \left| \iint_Q \left[a_{ij}^{m_\nu} - a_{ij}^{\infty} \right] u^* \Phi_{x_i x_j} dx dt \right| +$$

$$+ \sum_{i,j=1}^{n} \left| \iint_Q a_{ij}^{\infty} \left[u_{x_i}^{m_\nu} - u_{x_i}^* \right] \Phi_{x_j} dx dt \right|. \qquad (49)$$

By virtue of (44),(45) and (34) the last three terms in the right-hand side of inequality (49) vanish as $\nu \to \infty$. In accordance with (42) the first term does not exceed the quantity $\mu(m_\nu) \sum_{j=1}^{n} \left| u^{m_\nu} \Phi_{x_j} \right|_{V(Q)}$. It is easy to check that by virtue of estimate (13) the norms $\left| u^{m_\nu} \Phi_{x_j} \right|_{V(Q)}$ are bounded uniformly with respect to ν. Therefore, first term in the

147

right-hand side of (49) and hence the whole integral $J^{\nu}_{a_{1j}}$ tend to zero as $\nu \to \infty$.

Similarly, using condition (43) one proves that integral $J^{\nu}_{a_1}$ also tends to zero as $\nu \to \infty$.

Finally, integral J^{ν}_f satisfies inequality

$$J^{\nu}_f \leq \left| \int_0^T \left[f^{m_{\nu}}(t) - f^*(t) \right] \left\{ \int_{\Omega} g^{\infty}(t,x) \Phi(t,x) dx \right\} dt \right| +$$

$$+ \left| \int_0^T f^{m_{\nu}}(t) \, \Phi^{\nu}(t) dt \right| = S^{\nu}_1 + S^{\nu}_2,$$

where $\Phi^{\nu}(t) = \int_{\Omega} \left[g^{m_{\nu}}(t,x) - g^{\infty}(t,x) \right] \Phi(t,x) dx$.

The term $S^{\nu}_1 \to 0$ as $\nu \to \infty$ according to (47). On the other hand, by virtue of (39) $\Phi^{\nu}(t)$ tends to zero as $\nu \to \infty$ for any $t \in [0,T]$. Moreover, $\Phi^{\nu}(t)$ is bounded on $[0,T]$ uniformly with respect to ν. Consequently, according to the Lebesque theorem on passage to the limit we have that $\left| \Phi^{\nu} \right|_{L_2([0,T])} \to 0$, as $\nu \to \infty$. With regard to the estimate (12) it means that S^{ν}_2 and therefore J^{ν}_f tend to zero as $\nu \to \infty$.

The arguments above imply the equality $J = 0$, hence the pair $\{u^*, f^*\}$ satisfies the integral identity (10^{∞}).

Now we show that $u^*(t,x)$ satisfies the condition of

overdetermination (4^∞). In fact, for arbitrary function $\chi(t) \in \overset{O}{C}{}^\infty([0,T])$ we write the equality

$$\int_0^T \int_\Omega u^{m_\nu}(t,x) w^{m_\nu}(t,x) \chi(t) dx dt = \int_0^T \phi^{m_\nu}(t) \chi(t) dt,$$

which is true by virtue of (4^{m_ν}). Passage to the limit as $\nu \to \infty$ in this equality which is possible according to (40), (41), and (44), implies the identity

$$\int_0^T \int_\Omega u^*(t,x) w^*(t,x) \chi(t) dx dt = \int_0^T \phi^\infty(t) \chi(t) dt.$$

As $\chi(t)$ is arbitrary, we obtain that $u^*(t,x)$ satisfies the identity (4^∞).

Thus the pair $\{u^*, f^*\}$ is a generalized solution of the inverse problem (1^∞)-(4^∞). By virtue of the uniqueness of such solution (see Theorem 2) we obtain that $u^*(t,x)=u^\infty(t,x)$, $f^*(t)=f^\infty(t)$, and moreover that limit relations (44)-(47) are true. That completes the proof of the theorem.

Remark 3. If it is known that the assumptions (32) are satisfied then by virtue of Remark 2 $u^m(t,x) \in C^{0,\alpha}(Q^\delta)$ (for any $\delta > 0$). In this case it is easy to prove that in the Theorem 3 we obtain the following limit relation instead of (44): as $m \to \infty$

$$u^m(t,x) \to u^\infty(t,x) \quad \text{uniformly on } Q^\delta \text{ for any } \delta > 0.$$

BIBLIOGRAFIA

1.   N.S.BAKHVALOV - G.P.PANASENKO, *Homogenization of processes in periodic media*, Nauka, Moscow (1984)(Russian).

2.   A.BENSOUSSAN - J.L.LIONS - G.PAPANICOLAOU, *Asymptotic analysis for periodic structures*, North-Holland, Amsterdam (1978).

3. G.BUTTAZZO, G. DAL MASO. Γ-*convergence and optimal control problems*, J. Optim. Theory Appl. **38** (1982), pp. 385-407.

4. G.BUTTAZZO, L.FREDDI. *Sequences of optimal control problems with measures as controls*, Adv. Math. Sci. Appl. 2 (1993), pp. 215-230.

5. J.R.CANNON, Y.LIN. *Determination of a parameter* $p(t)$ *in some quasilinear parabolic differential equations*, Inverse Problems **4**, n. 1 (1988), pp. 35-45.

6.   V.L.KAMYNIN. *Passage to the limit in the inverse problem for nondivergence parabolic equations with the condition of final overdetermination*, Differentsial'nye Uravneniya, **28**, n.2, (1992), pp. 247-253; English transl. in Diff. Equations, **28**,(1992). pp. 213-218.

7.   V.L.KAMYNIN, I.A.VASIN. *Inverse problems for linearized Navier-Stokes equations with integral overdetermination. Unique solvability and passage to the limit*, Ann. Univ. Ferrara, Ser. 7. **38**, (1992), pp. 326-332.

8.   S.N.KRUZHKOV - V.L.KAMYNIN. *Convergence of the solutions of quasilinear parabolic equations with weakly converging coefficients*, Dokl. Akad. Nauk SSSR, **270**, n.3 (1983),   pp.533-536; English transl. in Soviet Math. Dokl. **27**, n.3   (1983).

9. O.A.LADYZHENSKAYA, V.A.SOLONNIKOV, N.N.URALTSEVA. *Linear and Quasilinear Equations of Parabolic Type*, Nauka, Moscow (1967) (Russian).

10. A.I.PRILEPKO – D.G.ORLOVSKY. *On determination of a parameter in evolution equation and on inverse problems of mathematical physics*.II, Differentsial'nye Uravneniya, 21, n. 4 (1985), pp. 694–700 (Russian).

11. A.I.PRILEPKO – A.B.KOSTIN – I.V.TIKHONOV. *Inverse problems for evolution equation*, Ill-posed Problems in Natural Sciences, Proc. of the Intern. Conf., Moscow, August 19–25, 1991, TVP-VSP Sci Publ, Moscow (1992), pp. 379–389.

12. I.A.VASIN, *Some inverse problems of viscous fluid dynamics with integral overdetermination*. Zhurnal Vychislit. Matematiki i Matematich. Fiziki 32, n.7 (1992), pp. 1071–1079 (Russian).

13. V.V.ZHIKOV – S.M.KOZLOV – OLEINIK, *G–convergence of parabolic operators*. Uspekhi Mat. Nauk 36, n.1 (1981), pp. 11–56; English transl. in Russian Math. Surveys, 36, (1981).

J POPIOLEK

On the Airy equation with a coefficient depending on t

Equations of the type $D_t u = D_x^3 u + G(u, D_x u, D_x^2 u)$ describe η – pseudospherical surfaces (see [11]). In paper [3] has been examined the equation $D_t u = a D_x^3 u$ which is called the Airy equation and is a linear version of the Korteweg – de Vries (KdV) equation. It arises in the description of the slow variation of a wave front in coordinates moving with the wave. It also describes the developments of long waves in various physical contexts, for example, plasma, water waves or nonlinear lattice.

In papers [5], [6] it is proved that equation $D_t u = D_x^3 u$ is one of the canonical forms of third–order partial differential equations and it is called the equation with characteristics multiple (see [4], p. 132). Boundary–value problems for that equation have been studied by a number of authors ([1], [2], [3], [4], [9] and [10]).

The paper concerns the boundary–value problem for the Airy equation with a coefficient depending on t. Using the methods of the theory of integral equations we prove the existence of a solution of the problem considered.

To the best of our knowledge, boundary–value problems for Airy equation with a variable coefficient have not been examined so far.

1 Fundamental solutions

We shall present the construction of fundamental solutions for the equation

$$\mathcal{L}\,[u(x,t)] \equiv a(t)\,D_x^3\,u(x,t) - D_t\,u(x,t) = 0\,, \quad (x,t) \in \mathcal{D}\,, \tag{1}$$

where $\mathcal{D} = \{(x,t) \in \mathbb{R}^2 : 0 < x < 1, 0 < t \le T\}$, $T = \text{const} > 0$, $D_x = \frac{\partial}{\partial x}$, $D_t = \frac{\partial}{\partial t}$.

We assume that the coefficient a is defined for all $t \in [0, T]$ and satisfies the following assumptions

$\mathcal{A}.\,1$ *there exist numbers A_0, A_1 such that*

$$0 < A_0 \le a(t) \le A_1, \quad t \in [0, T],$$

$\mathcal{A}.\,2$ *the function a satisfies the Hölder condition of the form*

$$|a(t) - a(s)| \le A|t - s|^{h_a}, \quad s, t \in [0, T],$$

where $A = const > 0$, $0 < h_a \le 1$.

Let

$$\vartheta(t) = \int_0^t a(r)\, dr.$$

Having performed the transfomation

$$\xi = x \quad \tau = \vartheta(t), \tag{2}$$

we reduce equation (1) to the following one

$$D_\xi^3 u(\xi, \tau) - D_\tau u(\xi, \tau) = 0.$$

The fundamental solutions of this equation are of the form (see [4], p. 133)

$$\mathcal{U}(\xi, \tau; \eta, \nu) = (\tau - \nu)^{-\frac{1}{3}}\, \mathbf{Ai}\left[\, (\xi - \eta)\,(\tau - \nu)^{-\frac{1}{3}}\,\right], \tag{3}$$

$$\mathcal{V}(\xi, \tau; \eta, \nu) = (\tau - \nu)^{-\frac{1}{3}}\, \mathbf{Bi}\left[\, (\xi - \eta)\,(\tau - \nu)^{-\frac{1}{3}}\,\right], \tag{4}$$

where

$$\mathbf{Ai}(z) = \frac{\pi\sqrt{z}}{3\sqrt{3}}\left[\mathbf{J}_{\frac{1}{3}}\left(\frac{2}{3\sqrt{3}}\,z^{\frac{3}{2}}\right) + \mathbf{J}_{-\frac{1}{3}}\left(\frac{2}{3\sqrt{3}}\,z^{\frac{3}{2}}\right)\right], \tag{5}$$

$$\mathbf{Bi}(z) = \frac{\pi\sqrt{z}}{3\sqrt{3}}\left[\mathbf{J}_{\frac{1}{3}}\left(\frac{2}{3\sqrt{3}}\,z^{\frac{3}{2}}\right) - \mathbf{J}_{-\frac{1}{3}}\left(\frac{2}{3\sqrt{3}}\,z^{\frac{3}{2}}\right)\right] \tag{6}$$

and $\mathbf{J}_\mu$ is the Bessel's function. The function $\mathbf{Ai}$ is called the Airy's function and the function $\mathbf{Bi}$ is called the associated Airy's function.

Thus, the fundamental solutions of equation (2) are the functions

$$\mathcal{U}(x, t; y, s) = a(s)\,(\vartheta(t) - \vartheta(s))^{-\frac{1}{3}}\, \mathbf{Ai}\left[\,(x - y)\,(\vartheta(t) - \vartheta(s))^{-\frac{1}{3}}\,\right], \tag{7}$$

$$\mathcal{V}(x, t; y, s) = a(s)\,(\vartheta(t) - \vartheta(s))^{-\frac{1}{3}}\, \mathbf{Bi}\left[\,(x - y)\,(\vartheta(t) - \vartheta(s))^{-\frac{1}{3}}\,\right]. \tag{8}$$

For the functions $\mathcal{U}$ and its derivatives the following inequalities hold (see [2], [4])

$$|D_x^m\, \mathcal{U}(x, t; y, s)| \le C\,|x - y|^{\frac{2m-1}{4}}\,(t - s)^{-\frac{2m+1}{4}} \tag{9}$$

when $(x - y)\,(y - t)^{-\frac{1}{3}} \to +\infty$, $m = 0, 1, \ldots$, $C = const > 0$,

$$|D_x^m\, \mathcal{U}(x, t; y, s)| \le C\,(t - s)^{-\frac{m+1}{3}}\,\exp\left[\,-c\,|x - y|^{\frac{3}{2}}\,(t - s)^{-\frac{1}{2}}\,\right] \tag{10}$$

when $(x - y)\,(t - s)^{-\frac{1}{3}} \to -\infty$, $m = 0, 1, \ldots$, $C, c = const > 0$.

For the functions $\mathcal{V}$ and its derivatives the following inequality holds (see [2], [4])

$$|D_x^m\, \mathcal{V}(x, t; y, s)| \le C\,|x - y|^{\frac{2m-1}{4}}\,(t - s)^{-\frac{2m+1}{4}} \tag{11}$$

when $(x - y)\,(y - t)^{-\frac{1}{3}} \to +\infty$, $m = 0, 1, \ldots$, $C = const > 0$.

2 Properties of certain integrals

In this section we give some lemmas that will be used in the further considerations.

Let us consider the integral

$$\mathfrak{I}(x,t) \; = \; \frac{1}{\pi} \int_0^1 \mathfrak{U}(x,t;y,0)\,\varphi(y)\,dy\,, \tag{12}$$

where the function φ is defined in the interval $[\,0\,;1\,]$, $\varphi(0) = \varphi(1) = 0$.

We have the following

Lemma 1 *If the function φ is continuous in the interval $[\,0\,;1\,]$ and its the first derivative is continuous in the interval $(\,0\,;1\,)$ and*

$$|\varphi'(x)| \; \leq \; M_\varphi\,, \quad x \in (\,0\,;1\,)\,,$$

where M_φ is a positive constant, then

$$\lim_{t\to 0} \mathfrak{I}(x,t) \; = \; a(0)\cdot\varphi(x)\,, \quad x \in (\,0\,;1\,)\,. \tag{13}$$

Now we consider the integral

$$\mathfrak{J}(x,t;s) \; = \; \frac{1}{\pi} \int_0^1 \mathfrak{U}(x,t;y,s)\,\phi(y,s)\,dy\,, \tag{14}$$

where the function ϕ is defined in $\overline{\mathcal{D}}$ and $\phi(0,t) = \phi(1,t) = 0$, $0 \leq t \leq T$.

The following lemma is valid

Lemma 2 *If the function ϕ is continuous in the domain $\overline{\mathcal{D}}$ and its the first derivative with respect to the variable x is continuous in the domain $\mathcal{D}$ and*

$$|D_x\,\phi(x,t)| \; \leq \; M_\phi\,, \quad (x,t) \in \mathcal{D}\,,$$

where M_ϕ is a positive constant, then

$$\lim_{s\to t} \mathfrak{J}(x,t;s) \; = \; a(t)\cdot\phi(x,t)\,, \quad (x,t) \in \mathcal{D}\,. \tag{15}$$

Let us introduce the integral

$$\mathcal{W}(x,t) \; = \; \frac{1}{\pi} \int_0^t \int_0^1 \mathfrak{U}(x,t;y,s)\,\psi(y,s)\,dy\,ds\,, \tag{16}$$

where the function ψ is defined in $\overline{\mathcal{D}}$ and $\psi(0,t) = \psi(1,t) = 0$, $0 \leq t \leq T$.

In further reasonoing we will also need the following lemma that is a consequence of Lemma 2.

Lemma 3 *If the function ψ is continuous in the domain $\overline{\mathcal{D}}$ and its the first derivative with respect to the variable x is continuous in the domain $\mathcal{D}$ and*

$$|D_x \psi(x,t)| \leq M_\psi, \quad (x,t) \in \mathcal{D},$$

where M_ψ is a positive constant, then

$$\mathcal{L}[\mathcal{W}(x,t)] = -a(t) \cdot \psi(x,t), \quad (x,t) \in \mathcal{D}. \tag{17}$$

Let $\gamma(t)$ be a continuous function for $t \in [0;T]$. Consider the integral

$$\mathcal{K}_\gamma(x,t) = \frac{1}{\pi} \int_0^t \mathcal{U}(x,t;0,s)\,\gamma(s)\,ds. \tag{18}$$

The integral $\mathcal{K}_\gamma$ has properties similar to those of the heat potential of first kind (see [8], p. 480). We shall call this integral the Airy's potential of first kind of the segment $x = 0$, $0 \leq t \leq T$, with the density $\gamma(s)$.

We have the following

Lemma 4 *If the function γ is continuous in the interval $[0;T]$ and*

$$|\gamma(t) - \gamma(s)| \leq M_\gamma |t - s|^{h_\gamma}, \tag{19}$$

where $M_\gamma = const > 0$, $\frac{1}{4} < h_\gamma \leq 1$, then

$$\lim_{x \to 0} D_x^2 \mathcal{K}_\gamma(x,t) = -\frac{2}{3}\gamma(t), \quad t \in (0;T]. \tag{20}$$

3 Boundary – value problem

We consider the equation

$$\mathcal{L}[u(x,t)] \equiv a(t) D_x^3 u(x,t) - D_t u(x,t) = f(x,t), \quad (x,t) \in \mathcal{D}. \tag{21}$$

We pose the following boundary–value problem: find a function $u \in \mathcal{C}_{x,t}^{3,1}(\mathcal{D}) \cap \mathcal{C}_{x,t}^{1,0}(\overline{\mathcal{D}})$ satisfying the conditions

$$u(x,0) = \psi(x), \quad 0 \leq x \leq 1, \tag{22}$$

$$u(0,t) = \varphi_0(t), \quad 0 \leq t \leq T, \tag{23}$$

$$u(1,t) = \varphi_1(t), \quad 0 \leq t \leq T, \tag{24}$$

$$D_t u(0,t) = \varphi_2(t), \quad 0 \leq t \leq T. \tag{25}$$

Let us introduce the function w such that

$$w(x,t) = u(x,t) - \psi(x), \quad (x,t) \in \overline{\mathcal{D}}.$$

The problem (21) – (25) is equivalent the following one

$$\mathcal{L}\left[w(x,t)\right] = f(x,t) - a(t)\,\psi'''(x) \equiv g(x,t), \quad (x,t) \in \mathcal{D}, \tag{26}$$

$$w(x,0) = 0, \quad 0 \le x \le 1, \tag{27}$$

$$w(0,t) = \varphi_0(t) - \psi(0) \equiv \phi_0(t), \quad 0 \le t \le T, \tag{28}$$

$$w(1,t) = \varphi_1(t) - \psi(1) \equiv \phi_1(t), \quad 0 \le t \le T, \tag{29}$$

$$D_t\,w(0,t) = \varphi_2(t) - \psi'(0) \equiv \phi_2(t), \quad 0 \le t \le T. \tag{30}$$

We shall seek a solution of problem (26) – (30) in the form

$$\pi\,w(x,t) = \int_0^t \mathcal{U}\left(x,t;0,s\right)\beta_0(s)\,ds + \int_0^t \mathcal{V}\left(x,t;0,s\right)\beta_1(s)\,ds + \tag{31}$$

$$+ \int_0^t \mathcal{U}\left(x,t;1,s\right)\beta_2(s)\,ds - \int_0^t \int_0^1 \mathcal{U}\left(x,t;y,s\right)g(y,s)\,dy\,ds,$$

where functions $\mathcal{U}$ and $\mathcal{V}$ are defined by formulas (7) and (8) respectively; β_0, β_1 and β_2 are unknown functions.

The function w, given by formula (31) satisfies equation (26) and initial condition (27). Imposing boundary conditions (28) – (30), we get

$$\pi\,\phi_0(t) = \mathbf{Ai}(0) \int_0^t \left[\vartheta(t) - \vartheta(s)\right]^{-\frac{1}{3}} a(s)\,\beta_0(s)\,ds + \tag{32}$$

$$+ \mathbf{Bi}(0) \int_0^t \left[\vartheta(t) - \vartheta(s)\right]^{-\frac{1}{3}} a(s)\,\beta_1(s)\,ds +$$

$$+ \int_0^t \mathcal{U}\left(0,t;1,s\right)\beta_2(s)\,ds - \mathcal{W}(0,t;g),$$

$$\pi\,\phi_1(t) = \int_0^t \mathcal{U}\left(1,t;0,s\right)\beta_0(s)\,ds + \int_0^t \mathcal{V}\left(1,t;0,s\right)\beta_1(s)\,ds + \tag{33}$$

$$+ \mathbf{Ai}(0) \int_0^t \left[\vartheta(t) - \vartheta(s)\right]^{-\frac{1}{3}} a(s)\,\beta_2(s)\,ds - \mathcal{W}(1,t;g),$$

$$\pi\,\phi_2(t) = \mathbf{Ai}'(0) \int_0^t \left[\vartheta(t) - \vartheta(s)\right]^{-\frac{2}{3}} a(s)\,\beta_0(s)\,ds + \tag{34}$$

$$+ \mathbf{Bi}'(0) \int_0^t \left[\vartheta(t) - \vartheta(s)\right]^{-\frac{2}{3}} a(s)\,\beta_1(s)\,ds +$$

$$+ \int_0^t D_x\,\mathcal{U}\left(0,t;1,s\right)\beta_2(s)\,ds - D_x\,\mathcal{W}(0,t;g),$$

where

$$\mathcal{W}(0,t;g) = \int_0^t \int_0^1 \mathcal{U}\left(0,t;y,s\right)g(y,s)\,dy\,ds,$$

$$\mathcal{W}(1,t;g) = \int_0^t \int_0^1 \mathcal{U}\left(1,t;y,s\right)g(y,s)\,dy\,ds.$$

Equations (32) – (34) are Volterra integral equations of the first kind. We shall reduce these equations to Volterra integral equations of the second kind. In this aim we introduce the operator

$$\mathcal{R}_\sigma [\varphi(t)] = D_t \int_0^t [\vartheta(t) - \vartheta(z)]^{-\sigma} \varphi(z)\, dz, \qquad (35)$$

where $0 < \sigma < 1$ and φ is continuous function.

Let us consider equation (32). It can be written in the form

$$\pi \phi_0(t) = \mathbf{Ai}(0) \int_0^t [\vartheta(t) - \vartheta(s)]^{-\frac{1}{3}} a(t) \beta_0(s)\, ds + \qquad (36)$$

$$+ \mathbf{Ai}(0) \int_0^t [\vartheta(t) - \vartheta(s)]^{-\frac{1}{3}} [a(s) - a(t)] \beta_0(s)\, ds +$$

$$+ \mathbf{Bi}(0) \int_0^t [\vartheta(t) - \vartheta(s)]^{-\frac{1}{3}} a(t) \beta_1(s)\, ds +$$

$$+ \mathbf{Bi}(0) \int_0^t [\vartheta(t) - \vartheta(s)]^{-\frac{1}{3}} [a(s) - a(t)] \beta_1(s)\, ds +$$

$$+ \int_0^t \mathcal{U}(0, t; 1, s) \beta_2(s)\, ds - \mathcal{W}(0, t; g),$$

Lemma 5 *If functions a and β_0 are continuous, then the following relation*

$$\mathcal{R}_{2/3}\left[\int_0^t [\vartheta(t) - \vartheta(s)]^{-\frac{1}{3}} a(t) \beta_0(s)\, ds \right] = \frac{2\pi}{\sqrt{3}} \beta_0(t)$$

holds good.

Lemma 6 *If the function β_0 is continuous and the function a satisfies assumptions $\mathcal{A}.1$ and $\mathcal{A}.2$, then the relation holds*

$$\mathcal{R}_{2/3}\left[\int_0^t [\vartheta(t) - \vartheta(s)]^{-\frac{1}{3}} [a(s) - a(t)] \beta_0(s)\, ds \right] = \int_0^t \mathcal{K}_{00}(t, s) \beta_0(s)\, ds,$$

where

$$\mathcal{K}_{00} = D_t \int_s^t [\vartheta(t) - \vartheta(z)]^{-\frac{2}{3}} [\vartheta(z) - \vartheta(s)]^{-\frac{1}{3}} [a(s) - a(z)]\, dz.$$

Lemma 7 *If functions a and β_1 are continuous, then the following relation*

$$\mathcal{R}_{2/3}\left[\int_0^t [\vartheta(t) - \vartheta(s)]^{-\frac{1}{3}} a(t) \beta_1(s)\, ds \right] = \frac{2\pi}{\sqrt{3}} \beta_1(t)$$

holds good.

Lemma 8 *If the function β_1 is continuous and the function a satisfies assumptions $\mathcal{A}.1$ and $\mathcal{A}.2$, then the relation holds*

$$\mathcal{R}_{2/3}\left[\int_0^t \left[\vartheta(t) - \vartheta(s)\right]^{-\frac{1}{3}}\left[a(s) - a(t)\right]\beta_1(s)\,ds\right] = \int_0^t \mathcal{K}_{01}(t,s)\,\beta_1(s)\,ds\,,$$

where

$$\mathcal{K}_{01} = D_t \int_s^t \left[\vartheta(t) - \vartheta(z)\right]^{-\frac{2}{3}}\left[\vartheta(z) - \vartheta(s)\right]^{-\frac{1}{3}}\left[a(s) - a(z)\right]dz\,.$$

Lemma 9 *If the function β_2 is continuous and the function a satisfies assumptions $\mathcal{A}.1$ and $\mathcal{A}.2$, then the relation holds*

$$\mathcal{R}_{2/3}\left[\int_0^t \mathcal{U}(0,t;1,s)\,\beta_2(s)\,ds\right] = \int_0^t \mathcal{K}_{02}(t,s)\,\beta_2(s)\,ds\,,$$

where

$$\mathcal{K}_{02} = D_t \int_s^t \left[\vartheta(t) - \vartheta(z)\right]^{-\frac{2}{3}}\mathcal{U}(0,z;1,s)\,dz\,.$$

Applying the operator $\mathcal{R}_{2/3}$, defined by formula (35) to the both sides of equation (36), in virtue of Lemmas 5 – 9, we get

$$\frac{2\,\pi}{\sqrt{3}}\,\mathbf{Ai}(0)\,\beta_0(t) + \mathbf{Ai}(0)\int_0^t \mathcal{K}_{00}(t,s)\,\beta_0(s)\,ds + \qquad (37)$$

$$+\frac{2\,\pi}{\sqrt{3}}\,\mathbf{Bi}(0)\,\beta_1(t) + \mathbf{Bi}(0)\int_0^t \mathcal{K}_{01}(t,s)\,\beta_1(s)\,ds + $$

$$+\int_0^t \mathcal{K}_{02}(t,s)\,\beta_2(s)\,ds = \pi\,\mathcal{R}_{2/3}\left[\phi_0(t)\right] + \mathcal{R}_{2/3}\left[\mathcal{W}(0,t;g)\right].$$

Now, we consider equation (33). It can be written in the form

$$\pi\,\phi_1(t) = \int_0^t \mathcal{U}(1,t;0,s)\,\beta_0(s)\,ds + \int_0^t \mathcal{V}(1,t;0,s)\,\beta_1(s)\,ds + \qquad (38)$$

$$+ \mathbf{Ai}(0)\int_0^t \left[\vartheta(t) - \vartheta(s)\right]^{-\frac{1}{3}}\left[a(s) - a(t)\right]\beta_2(s)\,ds + $$

$$+ \mathbf{Ai}(0)\int_0^t \left[\vartheta(t) - \vartheta(s)\right]^{-\frac{1}{3}}a(s)\,\beta_2(s)\,ds - \mathcal{W}(1,t;g)\,.$$

Lemma 10 *If the function β_0 is continuous and the function a satisfies assumptions $\mathcal{A}.1$ and $\mathcal{A}.2$, then the relation holds*

$$\mathcal{R}_{2/3}\left[\int_0^t \mathcal{U}(1,t;0,s)\,\beta_0(s)\,ds\right] = \int_0^t \mathcal{K}_{10}(t,s)\,\beta_0(s)\,ds\,,$$

where

$$\mathcal{K}_{10} = D_t \int_s^t \left[\vartheta(t) - \vartheta(z)\right]^{-\frac{2}{3}}\mathcal{U}(1,z;0,s)\,dz\,.$$

Lemma 11 *If the function β_1 is continuous and the function a satisfies assumptions $\mathcal{A}.1$ and $\mathcal{A}.2$, then the relation holds*

$$\mathcal{R}_{2/3}\left[\int_0^t \mathbf{V}(1,t;0,s)\,\beta_1(s)\,ds\,\right] = \int_0^t \mathcal{K}_{11}(t,s)\,\beta_1(s)\,ds\,,$$

where

$$\mathcal{K}_{11} = D_t \int_s^t \left[\vartheta(t) - \vartheta(z)\right]^{-\frac{2}{3}} \mathbf{V}(1,z;0,s)\,dz\,.$$

Lemma 12 *If functions a and β_2 are continuous, then the following relation*

$$\mathcal{R}_{2/3}\left[\int_0^t \left[\vartheta(t) - \vartheta(s)\right]^{-\frac{1}{3}} a(t)\,\beta_2(s)\,ds\,\right] = \frac{2\pi}{\sqrt{3}}\,\beta_2(t)$$

holds good.

Lemma 13 *If the function β_2 is continuous and the function a satisfies assumptions $\mathcal{A}.1$ and $\mathcal{A}.2$, then the relation holds*

$$\mathcal{R}_{2/3}\left[\int_0^t \left[\vartheta(t) - \vartheta(s)\right]^{-\frac{1}{3}}\left[a(s) - a(t)\right]\beta_2(s)\,ds\,\right] = \int_0^t \mathcal{K}_{12}(t,s)\,\beta_2(s)\,ds\,,$$

where

$$\mathcal{K}_{12} = D_t \int_s^t \left[\vartheta(t) - \vartheta(z)\right]^{-\frac{2}{3}}\left[\vartheta(z) - \vartheta(s)\right]^{-\frac{1}{3}}\left[a(s) - a(z)\right]dz\,.$$

Applying the operator $\mathcal{R}_{2/3}$, defined by formula (35) to the both sides of equation (37), in virtue of Lemmas $10-13$, we get

$$\int_0^t \mathcal{K}_{10}(t,s)\,\beta_0(s)\,ds + \int_0^t \mathcal{K}_{11}(t,s)\,\beta_1(s)\,ds + \tag{39}$$

$$+\frac{2\pi}{\sqrt{3}}\mathbf{Ai}(0)\,\beta_2(t) + \mathbf{Ai}(0)\int_0^t \mathcal{K}_{12}(t,s)\,\beta_2(s)\,ds =$$

$$= \pi\mathcal{R}_{2/3}\left[\phi_1(t)\right] + \mathcal{R}_{2/3}\left[\mathbf{W}(1,t;g)\right].$$

Now we consider equation (34). It can be written in the form

$$\pi\phi_2(t) = \mathbf{Ai}'(0)\int_0^t \left[\vartheta(t) - \vartheta(s)\right]^{-\frac{2}{3}} a(t)\,\beta_0(s)\,ds + \tag{40}$$

$$+\mathbf{Ai}'(0)\int_0^t \left[\vartheta(t) - \vartheta(s)\right]^{-\frac{2}{3}}\left[a(s) - a(t)\right]\beta_0(s)\,ds +$$

$$+\mathbf{Bi}'(0)\int_0^t \left[\vartheta(t) - \vartheta(s)\right]^{-\frac{2}{3}} a(t)\,\beta_1(s)\,ds +$$

$$+\mathbf{Bi}'(0)\int_0^t \left[\vartheta(t) - \vartheta(s)\right]^{-\frac{2}{3}}\left[a(s) - a(t)\right]\beta_1(s)\,ds +$$

$$+\int_0^t D_x\mathbf{U}(0,t;1,s)\,\beta_2(s)\,ds - D_x\mathbf{W}(0,t;g)\,.$$

159

Lemma 14 *If functions a and β_0 are continuous, then the following relation*

$$\mathcal{R}_{1/3}\left[\int_0^t [\vartheta(t) - \vartheta(s)]^{-\frac{2}{3}}\, a(t)\,\beta_0(s)\,ds \right] = \frac{2\,\pi}{\sqrt{3}}\,\beta_0(t)$$

holds good.

Lemma 15 *If the function β_0 is continuous and the function a satisfies assumptions $\mathcal{A}.1$ and $\mathcal{A}.2$, then the relation holds*

$$\mathcal{R}_{1/3}\left[\int_0^t [\vartheta(t) - \vartheta(s)]^{-\frac{2}{3}}\, [a(s) - a(t)]\,\beta_0(s)\,ds \right] = \int_0^t \mathcal{K}_{20}(t,s)\,\beta_0(s)\,ds\,,$$

where

$$\mathcal{K}_{20} = D_t \int_s^t [\vartheta(t) - \vartheta(z)]^{-\frac{1}{3}} [\vartheta(z) - \vartheta(s)]^{-\frac{2}{3}} [a(s) - a(z)]\,dz\,.$$

Lemma 16 *If functions a and β_1 are continuous, then the following relation*

$$\mathcal{R}_{1/3}\left[\int_0^t [\vartheta(t) - \vartheta(s)]^{-\frac{2}{3}}\, a(t)\,\beta_1(s)\,ds \right] = \frac{2\,\pi}{\sqrt{3}}\,\beta_1(t)$$

holds good.

Lemma 17 *If the function β_1 is continuous and the function a satisfies assumptions $\mathcal{A}.1$ and $\mathcal{A}.2$, then the relation holds*

$$\mathcal{R}_{1/3}\left[\int_0^t [\vartheta(t) - \vartheta(s)]^{-\frac{2}{3}}\, [a(s) - a(t)]\,\beta_1(s)\,ds \right] = \int_0^t \mathcal{K}_{21}(t,s)\,\beta_1(s)\,ds\,,$$

where

$$\mathcal{K}_{21} = D_t \int_s^t [\vartheta(t) - \vartheta(z)]^{-\frac{1}{3}} [\vartheta(z) - \vartheta(s)]^{-\frac{2}{3}} [a(s) - a(z)]\,dz\,.$$

Lemma 18 *If the function β_2 is continuous and the function a satisfies assumptions $\mathcal{A}.1$ and $\mathcal{A}.2$, then the relation holds*

$$\mathcal{R}_{1/3}\left[\int_0^t D_x\,\mathcal{U}(0,t;1,s)\,\beta_2(s)\,ds \right] = \int_0^t \mathcal{K}_{22}(t,s)\,\beta_2(s)\,ds\,,$$

where

$$\mathcal{K}_{22} = D_t \int_s^t [\vartheta(t) - \vartheta(z)]^{-\frac{1}{3}}\, D_x\,\mathcal{U}(0,z;1,s)\,dz\,.$$

Applying the operator $\mathcal{R}_{1/3}$, defined by formula (35) to the both sides of equation (40), in virtue of Lemmas $14 - 18$, we get

$$\frac{2\,\pi}{\sqrt{3}}\,\mathbf{Ai}^{'}(0)\,\beta_0(t) + \mathbf{Ai}^{'}(0) \int_0^t \mathcal{K}_{20}(t,s)\,\beta_0(s)\,ds + \tag{41}$$

$$+\frac{2\pi}{\sqrt{3}}\mathbf{Bi}'(0)\,\beta_1(t) + \mathbf{Bi}'(0)\int_0^t \mathcal{K}_{21}(t,s)\,\beta_1(s)\,ds +$$

$$+\int_0^t \mathcal{K}_{22}(t,s)\,\beta_2(s)\,ds = \pi\mathcal{R}_{2/3}[\phi_2(t)] + \mathcal{R}_{2/3}[D_x\mathcal{W}(0,t;g)]\,.$$

Equations (37), (39) and (41) can be writen in the form

$$\mathbf{Ai}(0)\,\beta_0(t) + \mathbf{Bi}(0)\,\beta_1(t) = \mathcal{F}_0(t) - \sum_{j=0}^{2} a_j \int_0^t \mathcal{K}_{0j}(t,s)\,\beta_j(s)\,ds\,, \qquad (42)$$

$$\mathbf{Ai}(0)\,\beta_2(t) = \mathcal{F}_1(t) - \sum_{j=0}^{2} b_j \int_0^t \mathcal{K}_{1j}(t,s)\,\beta_j(s)\,ds\,, \qquad (43)$$

$$\mathbf{Ai}'(0)\,\beta_0(t) + \mathbf{Bi}'(0)\,\beta_1(t) = \mathcal{F}_2(t) - \sum_{j=0}^{2} c_j \int_0^t \mathcal{K}_{2j}(t,s)\,\beta_j(s)\,ds\,, \qquad (44)$$

where

$$\mathcal{F}_0(t) = \frac{\sqrt{3}}{2\pi}\Big[\pi\mathcal{R}_{2/3}[\phi_0(t)] + \mathcal{R}_{2/3}[\mathcal{W}(0,t;g)]\Big]\,,$$

$$a_0 = \frac{\sqrt{3}}{2\pi}\mathbf{Ai}(0)\,,\quad a_1 = \frac{\sqrt{3}}{2\pi}\mathbf{Bi}(0)\,,\quad a_2 = \frac{\sqrt{3}}{2\pi}\,,$$

$$\mathcal{F}_1(t) = \frac{\sqrt{3}}{2\pi}\Big[\pi\mathcal{R}_{2/3}[\phi_1(t)] + \mathcal{R}_{2/3}[\mathcal{W}(1,t;g)]\Big]\,,$$

$$b_0 = \frac{\sqrt{3}}{2\pi}\,,\quad b_1 = \frac{\sqrt{3}}{2\pi}\,,\quad b_2 = \frac{\sqrt{3}}{2\pi}\mathbf{Ai}(0)\,,$$

$$\mathcal{F}_2(t) = \frac{\sqrt{3}}{2\pi}\Big[\pi\mathcal{R}_{2/3}[\phi_2(t)] + \mathcal{R}_{2/3}[D_x\mathcal{W}(0,t;g)]\Big]\,,$$

$$c_0 = \frac{\sqrt{3}}{2\pi}\mathbf{Ai}'(0)\,,\quad c_1 = \frac{\sqrt{3}}{2\pi}\mathbf{Bi}'(0)\,,\quad c_2 = \frac{\sqrt{3}}{2\pi}\,.$$

The determinant of this system is of the form

$$\Delta = \begin{vmatrix} \mathbf{Ai}(0) & \mathbf{Bi}(0) & 0 \\ 0 & 0 & \mathbf{Ai}(0) \\ \mathbf{Ai}'(0) & \mathbf{Bi}'(0) & 0 \end{vmatrix} = \mathbf{Ai}(0)\,\mathbf{Bi}(0)\,\mathbf{Ai}'(0) -$$

$$- [\mathbf{Ai}(0)]^2\,\mathbf{Bi}'(0) \neq 0\,.$$

By using the Cramer's formulae we obtain

$$\beta_0(t) = \lambda_0\,\mathcal{F}_2(t) - \lambda_0'\,\mathcal{F}_0(t) - \lambda_0 \sum_{j=0}^{2} c_j \int_0^t \mathcal{K}_{2j}(t,s)\,\beta_j(s)\,ds + \qquad (45)$$

$$+ \lambda_0' \sum_{j=0}^{2} a_j \int_0^t \mathcal{K}_{0j}(t,s)\,\beta_j(s)\,ds\,,$$

$$\beta_1(t) = \lambda_1 \, \mathcal{F}_2(t) - \lambda_1' \, \mathcal{F}_0(t) - \lambda_0 \sum_{j=0}^{2} c_j \int_0^t \mathcal{K}_{2j}(t,s)\,\beta_j(s)\,ds + \qquad (46)$$

$$+ \lambda_0' \sum_{j=0}^{2} a_j \int_0^t \mathcal{K}_{0j}(t,s)\,\beta_j(s)\,ds \, ,$$

$$\beta_2(t) = \lambda_2 \, \mathcal{F}_1(t) - \lambda_2 \sum_{j=0}^{2} b_j \int_0^t \mathcal{K}_{1j}(t,s)\,\beta_j(s)\,ds \, , \qquad (47)$$

where

$$\lambda_0 = \frac{\mathbf{Bi}(0)}{\mathbf{Ai}'(0)\,\mathbf{Bi}(0) - \mathbf{Ai}(0)\,\mathbf{Bi}'(0)} \, ,$$

$$\lambda_0' = \frac{\mathbf{Bi}'(0)}{\mathbf{Ai}'(0)\,\mathbf{Bi}(0) - \mathbf{Ai}(0)\,\mathbf{Bi}'(0)} \, ,$$

$$\lambda_1 = \frac{\mathbf{Ai}(0)}{\mathbf{Bi}'(0)\,\mathbf{Ai}(0) - \mathbf{Bi}(0)\,\mathbf{Ai}'(0)} \, ,$$

$$\lambda_1' = \frac{\mathbf{Ai}'(0)}{\mathbf{Bi}'(0)\,\mathbf{Ai}(0) - \mathbf{Bi}(0)\,\mathbf{Ai}'(0)} \, ,$$

$$\lambda_2 = \frac{1}{\mathbf{Ai}(0)} \, .$$

It is easy to see that system (45) – (47) can be written in the form

$$\beta_i(t) = \overline{\mathcal{F}}_i(t) - \sum_{j=0}^{2} \int_0^t \overline{\mathcal{K}}_{ij}(t,s)\,\beta_j(s)\,ds \, , \quad i = 0, 1, 2. \qquad (48)$$

It is also easily seen that (48) is the system of second-kind Volterra equations. Hence, we can assert that there exists a solution of the said system of the form

$$\beta_i(t) = \overline{\mathcal{F}}_i(t) - \sum_{j=0}^{2} \int_0^t \mathcal{N}_{ij}(t,s)\,\overline{\mathcal{F}}_i(s)\,ds \, , \quad i = 0, 1, 2, \qquad (49)$$

where $\mathcal{N}_{ij}$ being resolvent kernels of $\overline{\mathcal{K}}_{ij}$, $i, j = 0, 1, 2$.

As a result of the foregoing considerations we can formulate the following theorem:

Theorem 1 *If assumptions of Lemmas 1 – 18 are satisfied then there exists a function $u \in \mathcal{C}_{x,t}^{3,1}(\mathcal{D}) \cap \mathcal{C}_{x,t}^{1,0}(\mathcal{D})$ which is a solution of the problem (21) – (25).*

References

[1] S. ABDINAZAROV, *General boundary-value problems for a third-order equation with multiple charakteristic*, Diferencial'nyje Uravnenija 17(1991), No. 1, 1–8, transl. Differential Equations 17(1991), No.1, 1–8.

[2] L. CATTABRIGA, *Un problema al contorno per una equazione di ordine dispari*, Annali della scuola normale superiori di Pisa fis e mat. 13(1959), No. 1, 163–203.

[3] W. CRAIG and J. GOODMAN, *Linear dispersive equations of Airy type*, J. Differential Equations 87(1990), No. 1, 38–61.

[4] T. D. DZHURAEV, *The boundary-value problems for equations of mixed and mixed-composed types*, FAN, Tashkent, 1979 (in Russian).

[5] T. D. DZHURAEV and J. POPIOŁEK, *Canonical forms of third-order partial differential equations*, Uspekhi Mat. Nauk 44(1989), No. 4(268), 237–238, transl. Russian Math. Surveys 44(1989), No. 4, 203–204.

[6] T. D. DZHURAEV and J. POPIOŁEK, *Classification and reduction to canonical forms of third-order partial differential equations*, Diferencial'nyje Uravnenija 27(1991), No. 10, 1734–1745, transl. Differential Equations 27(1991), No. 10, 1225–1235.

[7] A. FRIEDMAN, *Partial differential equation of parabolic type*, Prentice-Hall, New York, 1964.

[8] M. KRZYŻAŃSKI, *Partial differential equations of second order*, vol I, PWN, Warsaw, 1971.

[9] J. POPIOŁEK, *On the contact problem for the partial differential equations of third order*, Fasc. Math. 23(1991), 91–96.

[10] J. POPIOŁEK, *On the contact problem for the Airy's equation*, Selected Invited Lectures and Short Communications Deliverd at the First International Colloquium on Numerical Analysis and the Third International Colloquium on Differential Equations, Plovdiv, Bulgaria, August 1992, 178–187.

[11] M. L. REBELO and K. TENENBLAT, *A classification of pseudospherical surface equations of type* $u_t = u_{sss} + G(u, u_s, u_{ss})$, J. Math. Phys. 33(1992), No. 2, 537–549.

INSTITUTE OF MATHEMATICS, WARSAW UNIVERSITY, AKADEMICKA 2, 15–267 BIAŁYSTOK, POLAND

W REICHEL

Radial symmetry by moving planes for semilinear elliptic BVPs on annuli and other non-convex domains

Abstract: We use a variant of the moving hyperplane method of Alexandroff to obtain radial symmetry for solutions of $\Delta u + f(x, u, \nabla u) = 0$ on non-convex, annulus-type domains. For an annulus A we prove a symmetry theorem for boundary conditions $u = 0$ on the outer sphere, $u = a > 0$ on the inner sphere and $0 \le u \le a$ on A. We then drop the restriction $0 \le u \le a$ and consider two arbitrary real numbers as boundary values on the limiting spheres. Using the first result we can in this case prove a characterization of the radially symmetric solutions by their sets of extremal points. We also consider an overdetermined BVP (i.e., constant Neumann and Dirichlet boundary values) on a general ring-domain $\Omega = \Omega_0 \backslash \overline{\Omega}_1$ and prove the radial symmetry of the solution and the domain. In a forthcoming paper a corresponding theorem for an exterior domain will be given.

1 The annulus

1.1 Introduction

In this section we study the symmetry of solutions of the following semilinear elliptic boundary value problem on an annulus $A = \{x \in \mathbb{R}^n : R_1 < |x| < R_2\}$

$$\Delta u + f(|x|, u, |\nabla u|) = 0 \qquad \text{in } \overline{A} \tag{1}$$
$$u = 0, \qquad |x| = R_2 \tag{2}$$
$$u = a \ge 0, \qquad |x| = R_1 \tag{3}$$

where $f : [R_1, R_2] \times \mathbb{R} \times [0, \infty) \to \mathbb{R}$ and $u \in C^2(\overline{A})$. There are some known results:

(I) Suppose the nonlinearity $f = f(r, p, q)$ satisfies the following Lipschitz condition for all $q_1, q_2 \ge 0$ and $p_1, p_2 \in \mathbb{R}$ with $p_1 \ge p_2$

$$f(r, p_1, q_1) - f(r, p_2, q_2) \le L_1(p_1 - p_2) + L_2|q_1 - q_2|$$

where L_1 is strictly less than the first eigenvalue of $-\Delta$ with zero boundary-conditions on A. By the maximum principle we have uniqueness of the solution of (1)-(3) and hence radial symmetry.

(II) More generally Aviles[3] and Lazer&McKenna[8] showed radial symmetry of all solutions for the case $a = 0$ and $f = f(r, p)$ continuously differentiable w.r.t. p

under the conditions $f_p < \lambda_2$ or $\lambda_n < f_p < \lambda_{n+1}$ for some $n \geq 3$, where λ_n $(n \geq 1)$ are the eigenvalues of $-\Delta$ on A with zero boundary-conditions.

(III) Brezis&Nirenberg[4] and many others (see Esteban[6]) showed for $a = 0$ that there are indeed non-symmetric positive solutions of (1)-(3). This illustrates the difference to the corresponding boundary value problem on balls, where due to a result of Gidas, Ni, Nirenberg [7] all positive solutions are indeed radially symmetric. (Their result was extended to non-negative solutions by Castro&Shivaji [5] and to fully non-linear equations by Li [9].)

1.2   The main result

We now state an additional condition on u and reasonable hypotheses on f, which guarantee the radial symmetry of the solutions of (1)-(3).

$$0 \leq u(x) \leq a \text{ for all } x \in A \tag{4}$$

$f(r, p, q)$ is decreasing in r (H1)

$f(r, p, q) = f_1(r, p, q) + f_2(r, p)$ with   (H2)

f_1 Lipschitz continuous in p, q, f_2 increasing in p

Theorem 1 *Let f satisfy* (H1), (H2). *Then every solution u of* (1)-(4) *is radially symmetric and decreasing in r.*

Remark: In the case $f(r, p, q) \leq 0$, the solutions of (1)-(3) are subharmonic and therefore attain their maximum and minimum on the boundary of A. Hence (4) holds automatically for this class of nonlinearities.

In order to be able to treat Dirichlet-boundary conditions with values different from $0, a$ we introduce

$$
\begin{array}{rcll}
u(x) & = & b, & |x| = R_2 \qquad (2') \\
u(x) & = & c, & |x| = R_1 \qquad (3') \\
b \leq u(x) & \leq & c, & \text{in } A \qquad (4'\alpha) \\
c \leq u(x) & \leq & b, & \text{in } A \qquad (4'\beta)
\end{array}
$$

$$f(r, p, q) \text{ is increasing in } r. \tag{$\widetilde{H1}$}$$

It is easy to derive the following corollaries from the above theorem

Corollary 1 *Let f satisfy* (H1), (H2). *Then every solution u of* (1), (2'), (3'), (4'α) *is radially symmetric and decreasing in r.*

165

Corollary 2 *Let f satisfy $(\widetilde{H1})$, $(H2)$. Then every solution u of (1), (2'), (3'), (4'β) is radially symmetric and increasing in r.*

We use the following notation: $x = (x_1, \ldots, x_n) = (x_1, x')$ denotes a point in $\mathbb{R}^n$ with $x' = (x_2, \ldots, x_n) \in \mathbb{R}^{n-1}$. $B_r(x)$ is the open ball with radius r centered at x. $C^2(\Omega)$ stands for the space of real valued functions on the open set Ω with continuous derivatives up to order 2. $C^2(\overline{\Omega})$ stands for all functions in $C^2(\Omega)$ which extend together with all their derivatives of order ≤ 2 continuously to the boundary of Ω. For partial derivatives of u in the k-th coordinate direction we use $\partial_{x_k} u = \partial_k u = u_k$ and for derivatives of u in direction $m \in \mathbb{R}^n$ we use $\frac{\partial u}{\partial m} = \partial_m u$. Sometimes it is convenient to write $f = f(r, p, q_1, \ldots, q_n)$ instead of $f = f(r, p, q)$ where $q = |(q_1, \ldots, q_n)|$. Note that w.r.t. Lipschitz continuity both notations are consistent, i.e. $f(r, p, q_1, \ldots, q_n)$ is Lipschitz continuous w.r.t. $q_1, \ldots, q_n$ iff $f(r, p, q)$ is Lipschitz w.r.t. q and in fact the corresponding constants can be chosen equally. Finally, for a real valued function c defined on a subset of $\mathbb{R}^n$ we use $c^-(x) = \min\{c(x), 0\}$.

1.3 The proof of Theorem 1

Observe that problem (1)-(4) is rotationally invariant, that is to say if u is a solution then for every rotation matrix R (i.e.; $R^T = R^{-1}$) $u(Rx)$ is also a solution of (1)-(4). Therefore it suffices for the proof of radial symmetry to establish the symmetry w.r.t. one coordinate axis, e.g. the x_1-axis. In fact, it is already enough to establish

$$u(x_1, x') \leq u(-x_1, x') \text{ for all } x = (x_1, x') \in A, x_1 > 0. \tag{5}$$

By using a rotation which replaces the x_1-axis of our coordinate frame by the $-x_1$-axis, we can then deduce equality in (5), i.e. symmetry in the x_1-direction.

In order to proof (5) we use the geometrical device of moving planes as in Gidas, Ni, Nirenberg [7]. Since our domain A is non-convex we have to extend the definition of 'right-hand-caps' to our needs ($\lambda \in (0, R_2)$).

$$
\begin{aligned}
T_\lambda &= \{x | x_1 = \lambda\}, \text{ the hyperplane}\\
x^\lambda &= \widetilde{(2\lambda - x_1, x')}, \text{ the reflection of } x \text{ at } T_\lambda\\
\overline{\Sigma(\lambda)} &= \{(x_1, x') \in A | x_1 > \lambda\}, \text{ the right hand cap}\\
\overline{B_{R_1}(0)} &= \text{ closed inner ball}\\
\Sigma(\lambda) &= \overline{\Sigma(\lambda)} \backslash \overline{B_{R_1}(0)}^\lambda, \text{ the reduced right hand cap}
\end{aligned}
$$

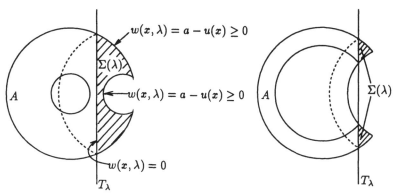

Figure 1: Thick and thin annuli

The main feature of the reduced right hand cap is that for all $\lambda \in (0, R_2)$ there is now a well defined comparison function

$$w(x, \lambda) = u(x^\lambda) - u(x) = v(x, \lambda) - u(x)$$

on $\Sigma(\lambda)$. Alessandrini [1] and Willms, Gladwell, Siegel [16] make similar use of reduced right hand caps. The plan of our proof is to establish the following properties of $w(x, \lambda)$ for all $\lambda \in (0, R_2)$:

$$w(x, \lambda) \geq 0 \quad \text{in } \Sigma(\lambda) \tag{i}$$
$$\partial_1 w(x, \lambda) > 0 \quad \text{on } A \cap T_\lambda \tag{ii}$$

(i) and (ii) will be proved by an initial step, i.e. for $\lambda \in (R_2 - \epsilon, R_2)$ and a continuation step, i.e. for all λ in a maximal interval (μ, R_2) with $\mu = 0$. Letting $\lambda \to 0$ in (i) will give (5) and hence symmetry in the x_1-direction while property (ii) will give the monotonicity since $\partial_1 w(x, \lambda) = -2\partial_1 u(x)$ on T_λ. Notice that the condition $0 \leq u(x) \leq a$ guarantees $w \geq 0$ on $\partial \Sigma(\lambda)$.

We start with geometric properties of the reduced right hand caps. For $n = 2$ the reduced right hand caps need not be connected (see Figure 1). But by the additional rotational symmetry for $n = 3$ they are indeed connected. For $n \geq 4$ we do not know about the connectedness of our reduced right hand caps. The following lemma states another important geometric property.

Lemma 1 *For every component Z of $\Sigma(\lambda)$ the following holds: $\partial Z \cap \{x : |x| = R_2, x_1 > \lambda\} \neq \emptyset$ and in particular there is a sequence $\{x^{(k)}\}_{k=1}^\infty$ in this set s.t. $x_1^{(k)} \searrow \lambda$ as $k \to \infty$.*

In the case $n = 2$ the lemma is obvious. For $n \geq 3$ the lemma is proved by considering the projections $Z_0, \Sigma_0(\lambda)$ of $Z, \Sigma(\lambda)$ rsp. into the $x_1, x_2, \ldots, x_{n-1}$ plane. Z_0 is a union of one or more components of the $n - 1$-dimensional right hand cap $\Sigma_0(\lambda)$. By an induction step, Z_0 and hence Z has the property, that on its boundary one can reach

167

T_λ from the right by points of radius R_2, which is the property stated in the lemma.

An examination of the boundary behaviour of the solutions of (1)-(4) shows, that u is strictly increasing (decreasing) along any inward direction starting from a boundary point of A on the outer (inner) sphere. This observation is stated in the next proposition. The proof will be given in Section 2 in a more general setting.

Proposition 1 *Let u be a solution of* (1)-(4). *For $z \in \partial A$ let $m \in \mathbb{R}^n$ be an inward unit direction at z, i.e. $\nu(z) \cdot m > 0$ where $\nu(z)$ is the interior normal to ∂A at z. Then there exists a radius $\rho = \rho(z)$ and a ball $B_\rho(z)$ s.t. $\partial_m u > 0$ in $B_\rho(z) \cap A$ if $|z| = R_2$ and $\partial_m u < 0$ in $B_\rho(z) \cap A$ if $|z| = R_1$.*

Remark: We always assume the ball $B_\rho(z)$ so small, that m is an inward direction for all boundary points in $\partial A \cap B_\rho(z)$.

Corollary 3 *There exists $\epsilon > 0$ s.t. $u > 0$ on $\{x \in A : \text{dist}(x, \{|x| = R_2\}) < \epsilon\}$ and $u < a$ on $\{x \in A : \text{dist}(x, \{|x| = R_1\}) < \epsilon\}$.*

Proof: We consider the situation on the outer boundary. Let us fix x with $|x| = R_2$ and take the ball $B_\rho(x)$ according to Proposition 1 with $m = \nu(x)$. Next we take $0 < \rho_1 < \rho$ so small, that for every point $y \in B_{\rho_1}(x) \cap A$ the straight line $y - tm, t > 0$ stays inside $B_\rho(x) \cap A$ until it hits a point $y_0 \in \partial A \cap B_\rho(x)$. Integrating $\partial_m u(x) > 0$ along this straight line connecting y_0 to y we get $u(y) > 0$ for every $y \in B_{\rho_1}(x) \cap A$. Considering all balls $B_{\rho_1}(x)$ for all x with $|x| = R_2$ and using a compactness argument we obtain the claim. $\qquad\square$

Remark: A corollary of the above form was found by Castro&Shivaji [5] in the ball case.

We can now prove the **initial step:** Fix $x = (R_2, 0, \ldots, 0)$ and take $B_\rho(x)$ with $m = (-1, 0, \ldots, 0)$ as in Proposition 1. Then observe that $\overline{\Sigma(\lambda) \cup \Sigma(\lambda)^\lambda} \subset B_\rho(x)$ for all λ close to R_2. Hence in $\Sigma(\lambda)$ we have $\partial_1 u < 0$ for all $\lambda \in (R_2 - \epsilon, R_2)$ and some $\epsilon > 0$. If we take $\lambda \in (R_2 - \frac{\epsilon}{2}, R_2)$ and $x \in \Sigma(\lambda)$ we see that the straight line $[x, x^\lambda]$ lies in $\Sigma(R_2 - \epsilon)$ and hence by integration of $\partial_1 u$ as before we get $u(x^\lambda) > u(x)$, i.e. (i) holds for $\lambda \in (R_2 - \frac{\epsilon}{2}, R_2)$. For the same range of λ (ii) holds, since $\partial_1 w(x, \lambda) = -2\partial_1 u(x) > 0$ on $T_\lambda \cap A$, which finishes the proof of the initial step. $\qquad\square$

Continuation step: (I) By the initial step we know the existence of $\lambda \in (0, R_2)$ s.t. $w(x, \lambda) \geq 0$ in $\Sigma(\lambda)$. For such λ we show, that $w(x, \lambda) = u(x^\lambda) - u(x)$ is a supersolution of a linear, uniformly elliptic operator with bounded coefficients. The function $v(x, \lambda) = u(x^\lambda)$ satisfies in $\Sigma(\lambda)$

$$\Delta v + f(|x^\lambda|, v, |\nabla v|) = 0.$$

168

Hence we find

$$
\begin{aligned}
0 &= \Delta w + f(|x^\lambda|, v, |\nabla v|) - f(|x|, u, |\nabla u|) \\
&\geq \Delta w + f(|x|, v, |\nabla v|) - f(|x|, u, |\nabla u|) \qquad \text{by (H1)} \qquad (6) \\
&\geq \Delta w + f_1(|x|, v, |\nabla v|) - f_1(|x|, u, |\nabla u|)
\end{aligned}
$$

where the last inequality holds by (H2) and the assumption $v(x, \lambda) \geq u(x)$. Next we define

$$
\begin{aligned}
c(x) &= (f_1(x, v, \nabla v) - f_1(x, u, \nabla v))/(v - u) \\
b_i(x) &= \frac{f_1(x, u, u_1, \ldots, u_{i-1}, v_i, \ldots, v_n) - f_1(x, u, u_1, \ldots, u_i, v_{i+1}, \ldots, v_n)}{v_i - u_i}
\end{aligned}
$$

where the quotients are supposed to be zero, if the denominator is zero. With these definitions, the above differential inequality (6) reads

$$
0 \geq \Delta w + b_i \partial_i w + c w \geq \Delta w + b_i \partial_i w + c^- w \qquad (7)
$$

where b_i, c^- are bounded functions by the Lipschitz continuity of f_1.

(II) For λ as in (I) with $\lambda > 0$ we want to strengthen the inequality $w(x, \lambda) \geq 0$ to $w(x, \lambda) > 0$ in $\Sigma(\lambda)$ and $\partial_1 w(x, \lambda) > 0$ in $A \cap T_\lambda$. This can be deduced from (7) by means of the strong maximum principle and it boundary version if we can exclude $w \equiv 0$ on a component Z of $\Sigma(\lambda)$. Suppose for contradiction $w \equiv 0$ in Z. By Lemma 1 there is a sequence $x^{(k)}$ in ∂Z with $|x^{(k)}| = R_2$ and $x_1^{(k)} \searrow \lambda$. By the boundary condition $u(x^{(k)}) = 0$ and by the hypothesis $w(x^{(k)}, \lambda) = 0$ we get $u(x^{(k),\lambda}) = 0$. The reflected point $x^{(k),\lambda}$ is inside the annulus because $\lambda > 0$. Since $|x^{(k),\lambda} - x^{(k)}| \to 0$ implies $|x^{(k),\lambda}| \to R_2$ we obtain $u(x^{(k),\lambda}) > 0$ by Corollary 3 - contradiction!

(III) Due to the initial step we can now define

$$
\mu = \inf\{\alpha > 0 : \text{(i) holds } \forall \lambda \in (\alpha, R_2)\}.
$$

By the observation in (II) we have that (i) and (ii) hold for all $\lambda \in (\mu, R_2)$. We want to show $\mu = 0$ and suppose the contrary $\mu > 0$. Hence there exist sequences λ_k and $x^{(k)} \in \Sigma(\lambda_k)$ s.t. $\lambda_k \nearrow \mu$ and $w(x^{(k)}, \lambda_k) < 0$. Next we choose $x^{(k)}$ s.t. $w(x, \lambda_k)$ attains its negative minimum over $\overline{\Sigma(\lambda_k)}$ in $x^{(k)}$. $x^{(k)} \notin \partial\Sigma(\lambda_k)$ since on $\partial\Sigma(\lambda_k) \cap T_{\lambda_k}$ we have $w = 0$ and on the remaining parts of $\partial\Sigma(\lambda_k)$ the following holds:

$$
w(x, \lambda_k) = \left\{ \begin{array}{ll} u(x^{\lambda_k}) - 0 & \text{if } |x| = R_2 \\ a - u(x) & \text{if } |x^{\lambda_k}| = R_1 \end{array} \right\} \geq 0.
$$

Hence $x^{(k)} \in \Sigma(\lambda_k)$ and $\nabla_x w(x^{(k)}, \lambda_k) = 0$. Choosing a convergent subsequence we obtain a point $\overline{x} \in \overline{\Sigma(\mu)}$ with $\nabla_x w(\overline{x}, \mu) = 0$ and $w(\overline{x}, \mu) \leq 0$. Since (i) holds at

$\lambda = \mu$ we have $w(\overline{x}, \mu) = 0$. Because $w(x, \mu)$ is supersolution of an elliptic operator we get by the strengthened inequalities in (II) that $\overline{x} \in \partial \Sigma(\mu)$. Also because of the boundary version of the strong maximum principle $\overline{x}$ can not lie on the smooth part of $\partial \Sigma(\mu)$. Hence $\overline{x} \in (\partial A \cap T_\mu) \cup (\partial B_{R_1}(0)^\mu \cap \partial A)$. Since on the latter of these two sets $w(x, \mu) = a - 0 > 0$, it follows that $\overline{x} \in \partial A \cap T_\mu$. Let us denote the reflection of $x^{(k)}$ at T_{λ_k} with $y^{(k)}$ rather than $x^{(k), \lambda_k}$. Since $\overline{x}$ lies on the hyperplane, we find $x^{(k)}, y^{(k)} \to \overline{x}$. By Proposition 1 there exists a ball $B_\rho(\overline{x})$ s.t. $\partial_1 u < 0$ in $A \cap B_\rho(\overline{x})$. For k big enough we have $x^{(k)}, y^{(k)} \in B_\rho(\overline{x})$ and hence

$$u(x^{(k)}) - u(y^{(k)}) = \int_{2\lambda^{(k)} - x_1^{(k)}}^{x_1^{(k)}} \partial_1 u(t, x^{(k)'}) dt < 0$$

in contradiction to $w(x^{(k)}, \lambda_k) < 0$. This shows $\mu = 0$ and finishes the proof. $\qquad \square$

1.4   A characterization theorem for radial symmetry

The following application of Theorem 1 gives a characterization of the radially symmetric solutions of (1) with arbitrary Dirichlet boundary conditions and no restriction of type (4). To be precise we consider the boundary value problem

$$\begin{aligned} \Delta u + f(u, |\nabla u|) &= 0 & &\text{in } \overline{A}, & (8) \\ u &= b, & &|x| = R_2, & (9) \\ u &= c, & &|x| = R_1. & (10) \end{aligned}$$

Theorem 2 *Let u be a solution of (8)-(10) and suppose $f(p, q)$ is Lipschitz continuous in p, q. Then u is radially symmetric iff the set of its local extremal points in A is radially symmetric.*

Remark: It follows from the proof that the above theorem is also true for the corresponding Dirichlet problem on balls. In this case our result is complementary to a result of Min [10], who charaterized the radially symmetric solutions on balls by the radial symmetry of their nodal set. By the symmetry-breaking result of Brezis&Nirenberg [4], Min's result cannot be true for annuli. Our proof follows some ideas of Min.

Proof: Let us denote the set of points of local extrema of u by E. The necessity of the criterium is trivial. Let us prove the sufficiency. If u attains neither supremum nor infimum in A then E is empty and the claim follows from one of the Corollaries 1 or 2. To avoid these trivial situations let us in the following suppose that u is non-constant and $E \neq \emptyset$. If we introduce $S(t) = \{x : |x| = t\}$ then E can be written as $E = \bigcup_{t \in T} S(t)$ where $T \subset (R_1, R_2)$. We shall show that T has no accumulation points. Then $E = \bigcup_{i=1}^{N} S(t_i)$; that is to say E is the union of N annuli $A_1, \ldots, A_N$. On every A_i the values of u are strictly between the values of u on the limiting spheres of A_i. Since the non-linearity f is independent on r one of the Corollaries 1, 2 applies to A_i and gives the radial symmetry of u on A_i. As this is true for all $i = 1, \ldots, N$ the claim follows. It remains to show that T has no accumulation points. Suppose for contradiction that there exists $\{t_k\}_{k=1}^{\infty} \subset T$ with $t_k \to \bar{t}$. On $S(t_k)$ we have $u = \text{const.} = u_{t_k}$ and $\nabla u = 0$. Hence by continuity $u = u_{\bar{t}}$, $\nabla u = 0$ on $S(\bar{t})$. Let now $x \in S(\bar{t})$. For a coordinate direction x_i tangent to $S(\bar{t})$ at x we find $\partial_{x_i}^2 u(x) = 0$ since u is constant on $S(\bar{t})$. If the x_i–direction is non-tangential at x, then along a ray in the x_i-direction starting from x we choose a sequence of points $x^{(k)}$ on $S(t_k)$. Since $x^{(k)} \to x$ we use the definition of the second derivative and find

$$\partial_{x_i}^2 u(x) = \lim_{k \to 0} \frac{\partial_{x_i} u(x^{(k)}) - \partial_{x_i} u(x)}{(x^{(k)} - x)_i} = 0.$$

Hence $\Delta u(x) = 0$ on $S(\bar{t})$ which implies $f(u_{\bar{t}}, 0) = 0$. Simliar to the proof of Theorem 1 we define

$$
\begin{aligned}
c(x) &= (f(u, 0) - f(u_{\bar{t}}, 0))/(u - u_{\bar{t}}) \\
b_i(x) &= \frac{f(u, 0, \ldots, 0, u_i, \ldots, u_n) - f(u, 0, \ldots, 0, u_{i+1}, \ldots, u_n)}{u_i}
\end{aligned}
$$

with b_i, c defined zero if the corresponding denominator is zero. By hypothesis on f the functions b_i, c are bounded and we obtain the following differential equality on A

$$
\begin{aligned}
\Delta(u(x) - u_{\bar{t}}) &+ b_i(x)\partial_i(u(x) - u_{\bar{t}}) + c(x)(u(x) - u_{\bar{t}}) \\
&= \Delta u(x) + f(u(x), \nabla u(x)) - f(u_{\bar{t}}, 0) \\
&= 0.
\end{aligned}
$$

Since $u(x) - u_{\bar{t}} = 0$ and $\nabla(u(x) - u_{\bar{t}}) = 0$ on $S(\bar{t})$ we get $u \equiv u_{\bar{t}}$ on A by the unique continuation property for linear, uniformly elliptic equations, see e.g. Miranda [11], Theorem 19,II. $u \equiv u_{\bar{t}}$ contradicts the assumption of u being non-constant. $\square$

Remark: For the corresponding Dirichlet problem on balls the proof is essentially the same as above apart from the possible accumulation of the spheres $S(t)$ at $\bar{t} = 0$. Corollaries 1, 2 are then to be applied to a sequence of anulli which shrink to the origin.

2   Overdetermined BVPs in ring shaped domains

2.1   The problem and some old and new results

In the preceding section we concluded that a solution of the Dirichlet-problem (8)-(10) is radially symmetric if and only if the set E of local extrema is radially symmetric. This characterization is somewhat redundant as the following simple case $n = 2$ suggests: Suppose E is a simple closed C^2-curve winding round 0 once and hence dividing the annulus into two ring-shaped domains A_1, A_2. On both A_1, A_2 the original boundary value problem now becomes overdetermined since $u = $ const. and $\nabla u = 0$ on E. Also on A_1, A_2 the values of u are strictly between the values on the boundaries. In a famous paper Serrin [15] considered a similar setting where he showed that an overdetermined boundary value problem determines the geometry of the underlying set. Using Serrin's ideas we shall show that in the above case E is radial and hence Theorem 1 applies. And in fact our observation is true in a much more general setting.

Definition 1 *Let Ω_0, Ω_1 be two simply connected C^2-domains s.t. $\overline{\Omega}_1 \subset \Omega_0$. Then $\Omega = \Omega_0 \backslash \overline{\Omega}_1$ is called a ring-shaped domain.*

We study the solutions $u \in C^2(\overline{\Omega})$ of the following boundary-value problem

$$
\begin{aligned}
\Delta u + f(u) &= 0 & &\text{in } \overline{\Omega} & &(11) \\
u(x) &= 0 & &\text{on } \partial\Omega_0 & &(12) \\
u(x) &= a > 0 & &\text{on } \partial\Omega_1 & &(13) \\
0 < u(x) &< a, & &\text{in } \Omega & &(14)
\end{aligned}
$$

with three types of Neumann boundary conditions

$$
\begin{aligned}
\partial u/\partial \nu|_{\partial\Omega_1} &= \text{const.} = c_1 & &\text{if } \Omega_0 \text{ is a ball} & &(15a) \\
\partial u/\partial \nu|_{\partial\Omega_0} &= \text{const.} = c_0 & &\text{if } \Omega_1 \text{ is a ball} & &(15b) \\
\partial u/\partial \nu|_{\partial\Omega_i} &= c_i, (i = 0, 1) & &\text{if } \Omega \text{ is a general ring domain} & &(15c)
\end{aligned}
$$

Here and in the following $\nu(x)$ denotes always the unit inner normal with respect to Ω. Note that consistency requires $c_1 \leq 0 \leq c_0$. Note also, that the additional Neumann-boundary values are imposed on those parts of the boundary of Ω where the radial symmetry is not a priori known.

Theorem 3 *If $f = f_1 + f_2 : [0, a] \to I\!R$ with f_1 Lipschitz continuous and f_2 increasing then in all three case of the boundary value problem (11)-(15) Ω is an annulus and all solutions are radially symmetric and decreasing in r.*

Remark: (a) In a forthcoming paper we shall prove Theorem 3 in the case of an exterior domain $\Omega = I\!R^n \backslash \overline{\Omega}_1$ with constant Dirichlet and Neumann boundary values on $\partial\Omega_1$ and $u \to 0$ at ∞.
(b) It is straight forward to generalize the above theorem to boundary values a_1, a_2 with $a_1 < u(x) < a_2$ (see also Corollaries 1 and 2).
(c) In the same way as in the annulus-case we can also consider nonlinearities $f(u, |\nabla u|)$ with $f(p, q) = f_1(p, q) + f_2(p)$ and f_1 Lipschitz in p, q and f_2 increasing in p.
(d) Finally, we can admit domains with a finite number of simply connected cavities: $\Omega = \Omega_0 \backslash \bigcup \overline{\Omega}_i$ where $\overline{\Omega}_i$ are mutually disjoint subsets of Ω_0. In this case we have to assume the following boundary values: $u = 0$ on $\partial\Omega_0$, $u = a$ on $\partial\Omega_i$ for $i = 1, \ldots, K$, $0 < u < a$ and $\partial u/\partial \nu = c_j$ on $\partial\Omega_j$ for $j = 0, \ldots, K$.

(e) We conjecture that the above theorem is still true with the relaxed condition $0 \le u(x) \le a$ instead of (14). But unlike in the case of the annulus, where the geometry of the right hand caps is well understood, we could not find a proof. In the annulus case Lemma 1 showed that one can reach T_λ from the right by points on the boundary of any component Z of $\Sigma(\lambda)$. This need not be true for the general ring domain.

(f) It is well known (see Gidas, Ni, Nirenberg [7]) that the relaxed condition $0 \le u \le a$ together with $f(0) \ge 0$ and $f(a) \le 0$ imply the strict inequalities $0 < u < a$ in Ω. A quick look at (16) of Proposition 2 will convince the reader.

Some results related to the problem are know to the author. It is important to say that in theses works condition (14) is not supposed a priori but deduced for the particular problems.

(I) In generalization of earlier works of Philippin [13] and Philippin&Payne [14], Alessandrini [1] proved Theorem 3 with $f \equiv 0$ and Δ replaced by a quasilinear, possibly degenerate elliptic operator of divergence type. The author considers more than one cavity as long as the Dirichlet boundary data is the same on all cavities. He also uses moving hyperplanes and reduced right hand caps. (14) is obtained by direct application of a comparison theorem for elliptic equations.

(II) Complementing another paper of Payne and Philippin [12], recently Wilms, Gladwell and Siegel [16] proved Theorem 3 in $I\!R^2$ for the torsion problem ($f \equiv 1$) also by using moving planes and reduced right hand caps. They show that a violation of (14) results in the existence of a saddle point which they exclude by exploiting $f \equiv 1$ in the differential equation and by supposing the inner boundary to be convex.

2.2 Definition of right hand caps

In order to prove radial symmetry, i.e. symmetry with respect to all directions, it is (by the rotation invariance of the problem) sufficient to do so for one coordinate direction, e.g. the x_1−direction. We shall again employ the method of moving planes as in Serrin [15]. Like in the case of an annulus, the key step is to define proper right hand caps. As before $T_\lambda = \{x | x_1 = \lambda\}$ denotes the hyperplane and $x^\lambda = (2\lambda - x_1, x')$ is the reflection of $x = (x_1, x')$ at T_λ. Furthermore let

$$
\begin{aligned}
\Sigma_0(\lambda) &= \{x \in \Omega_0 | x_1 > \lambda\} \text{ the outer cap,} \\
\Sigma_1(\lambda) &= \{x \in \Omega_1 | x_1 > \lambda\} \text{ the inner cap,} \\
\Gamma_i(\lambda) &= \{x \in \partial\Omega_i | x_1 > \lambda\}, (i = 0, 1) \text{ the right hand boundaries,} \\
M_i &= \sup\{x_1 | x_1 \in \Omega_i\}, (i = 0, 1).
\end{aligned}
$$

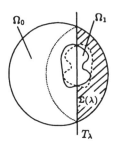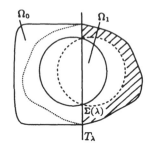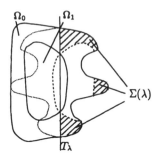

Figure 2: Three types of domains

For the geometry of right hand caps it is well known (see Amick&Fraenkel [2], Lemma A.1) that for values of λ a little less than M_i the reflection of $\Sigma_i(\lambda)$ is lying in Ω_i and the x_1-direction is external at every point of Γ_i. Also it is well known that this stays true for decreasing values of λ until $\Sigma_i(\lambda)^\lambda$ becomes internally tangent to $\partial\Omega_i$ or the normal on $\Gamma_i(\lambda)$ becomes perpendicular to the x_1-direction at a point $x \in \overline{\Gamma_i(\lambda)} \cap T_\lambda$. We denote this critical value of λ by m_i and set $m = \max\{m_1, m_2\}$. The admissible range for λ is then (m, M_0). The definition of the reduced right hand caps is now analogous to the annulus-case. For $\lambda \in (m, M_0)$ let

$$\Sigma(\lambda) = \Sigma_0(\lambda) \backslash \overline{\Omega}_1^\lambda.$$

On $\Sigma(\lambda)$ the comparison function $w(x, \lambda) = v(x, \lambda) - u(x) = u(x^\lambda) - u(x)$ is well defined.

2.3 Proof of Theorem 3

We start our analysis with the investigation of the behaviour of u near the boundary of Ω. In the next lemma, proposition and corollary we only make use of the Dirichlet boundary data at $\partial\Omega$. In particular Proposition 1 will be included.

Lemma 2 *Suppose D is a C^2-domain and $u \in C^2(\overline{D})$. Let $z \in \partial D$ and $m \in \mathbb{R}^n$ a unit, non-tangent vector (i.e. $\nu(z) \cdot m \neq 0$), s.t. for a ball $B(z)$ we have $u = $ const., $\partial_m u \leq 0$ or ≥ 0 on $\partial D \cap B(z)$ and $\partial_m u(z) = 0$. Then $\nabla u(z) = 0$ and $\partial_m^2 u(z) = \Delta u(z)(m \cdot \nu(z))^2$.*

Proof: See Appendix.

Proposition 2 *Let u be a solution of (11)-(14). For $z \in \partial\Omega$ let $m \in \mathbb{R}^n$ be an inward unit direction, i.e. $\nu(z) \cdot m > 0$. Then there exists a ball $B_\rho(z)$ s.t. $\partial_m u > 0$ in $B_\rho(z) \cap \Omega$ if $z \in \partial\Omega_0$ and $\partial_m u < 0$ in $B_\rho(z) \cap \Omega$ if $z \in \partial\Omega_1$.*

Remark: Again we assume ρ so small, that m is an inward direction for all boundary points in $\partial\Omega \cap B_\rho(z)$.

174

Proof: We only give the proof for $z \in \partial\Omega_1$. Clearly $\partial_m u(z) \leq 0$ since $u < a$ in Ω by (14). In the case $\partial_m u(z) < 0$ the statement follows by continuity; so let us suppose $\partial_m u(z) = 0$.

<u>Case 1:</u> $f(a) \leq 0$ Then the following differential inequality holds:

$$\begin{aligned}
0 &= \Delta u + f(u) \\
&\leq \Delta u + f(u) - f(a) \\
&\leq \Delta(u-a) + f_1(u) - f_1(a),
\end{aligned} \tag{16}$$

where the last inequality holds since $u \leq a$ and f_2 increasing. By the Lipschitz continuity of f_1 we see that $\bar{u} = u - a$ satisfies

$$\Delta\bar{u} + c^-(x)\bar{u} \geq \Delta\bar{u} + c(x)\bar{u} \geq 0$$

with bounded functions $c(x) = (f_1(u(x)) - f_1(a))/(u(x) - a)$, i.e. $\bar{u}$ is a subsolution. Since $u(z) = a$ we get $\partial_m u(z) < 0$ by the Hopf-version of the strong maximum principle. Hence we have a contradiction to our assumption. This shows that case 1 can not occur and we turn to

<u>Case 2:</u> $f(a) > 0$ Applying the preceding lemma we find

$$\partial_m^2 u(z) = -f(a)(m \cdot \nu(z))^2 < 0$$

so that $\partial_m^2 u < 0$ in $B_\rho(z) \cap \Omega$. Using a smaller ball $B_{\rho_1}(z)$ as in Corollary 3 we can integrate $\partial_m^2 u(x)$ along lines $y - tm, t > 0$ which stay in $B_\rho(z)$ and connect $y \in B_{\rho_1}(z) \cap \Omega$ to a boundary point y_0 where $\partial_m u(y_0) \geq 0$. By this we obtain $\partial_m u(y) > 0$ for all $y \in B_{\rho_1}(z) \cap \Omega$. This finishes the proof of the proposition. For points $z \in \partial\Omega_0$ all inequality signs are reversed and a is substituted by 0. $\qquad\square$

Remark: To include Proposition 1 in Proposition 2 ($\Omega = A$) we have to write (16) in the case $f(R_1, a, 0) \leq 0$ as

$$\begin{aligned}
0 &\leq \Delta u + f(|x|, u, |\nabla u|) - f(R_1, a, 0) \\
&\leq \Delta u + f(R_1, u, |\nabla u|) - f(R_1, a, 0) \\
&\leq \Delta u + f_1(R_1, u, \nabla u) - f_1(R_1, a, 0)
\end{aligned}$$

and linearize this (as seen before in the continuation step for Theorem 1) to $\Delta\bar{u} + b_i\partial_i\bar{u} + c^-\bar{u} \geq \Delta\bar{u} + b_i\partial_i\bar{u} + c\bar{u} \geq 0$.

With these prelimnary considerations in hand we can now prove Theorem 3. Our plan is to show the following properties for $w(x, \lambda)$ for all $\lambda \in (m, M_0)$:

$$\begin{aligned}
w(x, \lambda) &\geq 0 \quad &&\text{in } \Sigma(\lambda) &&\text{(i)} \\
\partial_1 w(x, \lambda) &> 0 \quad &&\text{on } \Omega \cap T_\lambda &&\text{(ii)}
\end{aligned}$$

175

This will be done as before by an initial step for $\lambda \in (M_0 - \epsilon, M_0)$ and by a continuation step for all $\lambda \in (m, M_0)$. We then have $w(x, m) \geq 0$ in $\Sigma(m)$. From this we shall conclude $w(x, m) \equiv 0$ on a component Z of $\Sigma(m)$ and finally we shall show $\overline{Z \cup Z^m} = \overline{\Omega}$, which is the desired symmetry of Ω.

Let us start with the initial step: We consider the compact set K of all points $x \in \partial \Omega_0$ with $x_1 = M_0$. For any of those points there is - according to Proposition 2 - a radius $\rho = \rho(x)$ s.t. $\partial_1 u < 0$ on $B_\rho(x) \cap \Omega$. Using the compactness of K we can find a radius σ independent on x s.t. $\partial_1 u < 0$ in $B_\sigma(x) \cap \Omega$ for all $x \in K$. With K_σ we denote $\bigcup_{x \in K} B_\sigma(x) \cap \Omega$. Now we proceed as in the initial step of the annulus-case. For λ close to M_0 the reflected cap $\Sigma(\lambda)^\lambda$ lies in K_σ. Integration of $\partial_1 u$ along straight lines connecting x and x^λ gives $u(x^\lambda) > u(x)$ for all $x \in \Sigma(\lambda)$. Thus (i) holds and for the same range of λ (ii) holds since $\partial_1 w(x, \lambda) = -2\partial_1 u(x) > 0$ in $T_\lambda \cap \Omega$. $\square$

Continuation step: Some parts of the proof, which are very much like in the annulus case, we only sketch.

(I) By the initial step we know that there exist $\lambda \in (m, M_0)$ s.t. $w(x, \lambda) \geq 0$ in $\Sigma(\lambda)$. For such λ there are bounded functions b_i, c, c^- such that

$$0 \geq \Delta w + b_i \partial_i w + cw \geq \Delta w + b_i \partial_i w + c^- w. \tag{17}$$

(II) Next we show, that $w \equiv 0$ in Z implies $\overline{Z \cup Z^\lambda} = \overline{\Omega}$ which is the desired symmetry of Ω. So suppose $w \equiv 0$ in Z. The boundary of Z decomposes into three disjoint parts

$$
\begin{aligned}
P &= \{x \in \partial Z | x_1 = \lambda, x \in \Omega\}, \\
Q &= \{x \in \partial Z | x_1 \geq \lambda, x \in \partial \Omega_0\}, \\
R &= \{x \in \partial Z | x_1 \geq \lambda, x \in \partial \Omega_1^\lambda\}.
\end{aligned}
$$

By the assumption $w \equiv 0$ we have $u = 0$ on Q^λ and $u = a$ on R and by (14) this implies $Q^\lambda \subset \partial \Omega_0$ and $R \subset \partial \Omega_1$. In the Appendix will shall show that this implies

$$\Omega = Z \cup Z^\lambda \cup (\partial Z \cap \Omega) \cup (\partial Z^\lambda \cap \Omega).$$

(III) If, for λ as in (I) with $\lambda > m$, we apply the strong maximum principle to a component Z of $\Sigma(\lambda)$ we deduce from (17) that either $w \equiv 0$ or $w > 0$ in Z. So we could strengthen $w \geq 0$ to $w > 0$ in $\Sigma(\lambda)$ if we can exclude $w \equiv 0$ on a component Z. Suppose for contradiction $w \equiv 0$ on Z. In (II) it was shown that this implies $\overline{Z \cup Z^\lambda} = \overline{\Omega}$ which is impossible for $\lambda > m$.

(IV) By the initial step the following quantity μ is well defined

$$\mu = \inf\{\alpha > m : \text{(i) holds } \forall \lambda \in (\alpha, M)\}$$

and both (i) and (ii) hold for all λ in (μ, M) by (I), (III) and the maximum principle. Our intention is to show $\mu = m$. Suppose $\mu > m$. By this assumption, the x_1–direction is non-tangent on $\partial\Sigma(\mu)$. Like in the annulus-case there are sequences $\lambda_k \nearrow \mu$ and $x^{(k)} \in \Sigma(\lambda_k)$ s.t. $w(x, \lambda_k)$ attains is negative minimum over $\Sigma(\lambda_k)$ in $x^{(k)}$ and $x^{(k)} \to \bar{x}$ with $w(\bar{x}, \mu) = 0$, $\nabla_x w(\bar{x}, \mu) = 0$. By (I), (III) and the Hopf Lemma $\bar{x}$ lies on the non-smooth part of $\partial\Sigma(\mu)$, that is $\bar{x} \in \partial\Omega \cap T_\mu$ or $\bar{x} \in \partial\Omega_0 \cap \partial\Omega_1^\lambda$. The latter case is excluded since there $w(\bar{x}, \mu) = a - 0 > 0$. Hence $\bar{x} \in \partial\Omega \cap T_\mu$ and $x^{(k)}, x^{(k), \lambda_k} \to \bar{x}$. By Proposition 2, in the vicinity of $\bar{x}$ we have $\partial_1 u < 0$. Integration of $\partial_1 u$ along straight lines from $x^{(k)}$ to $x^{(k), \lambda_k}$ yields $w(x^{(k)}, \lambda_k) > 0$ for k big enough, a contradiction. This shows $\mu = m$ and finishes the continuation step.

So far we have the following conclusion: $w(x, m) \geq 0$ in $\Sigma(m)$ and in particular $w > 0$ or $w \equiv 0$ for any component Z of $\Sigma(m)$ by the strong maximum principle. If $w \equiv 0$ on a component on Z then Ω is symmetric. To finish the proof of Theorem 3 it remains to show that there is such a component Z, i.e. that there is one point x in $\Sigma(m)$ with $w(x, m) = 0$. In this final step the different Neumann condition will come into the game.

(V) Suppose for contradiction $w > 0$ in $\Sigma(m)$. Remember that the critical position of T_m originates from either one of the $\Sigma_i(m)^m$ becoming internally tangent to $\partial\Omega_i$ or from the tangent on one of the right hand boundaries Γ_i becoming orthogonal to the x_1-direction.

Let us think about the critical position of T_m in the case where one of the Ω_i is a ball. If T_m becomes critical because one of the above conditions is fulfilled on the ball, then T_m goes through the center of that ball and both conditions are fulfilled. If we choose the $-x_1$-direction instead of the x_1-direction as the approach-direction of the reflecting hyperplane then the hyperplane cannot become critical because of the ball-part of the boundary. Hence by this possible change in the approach direction of T_λ we can suppose that the hyperplane sits in the critical position because one of the two above conditons is fulfilled on that part of the boundary where constant Neumann-conditions are prescribed.

(a) **Internal tangency:** Let p^m be a point where $\partial\Sigma_i(m)^m$ meets $\partial\Omega_i$, $p \notin T_m$. Clearly $w(p, m) = 0$ so that by the differential inequality in (I) and by the Hopf Lemma we find $\partial_{\nu(p)} w(p, m) \neq 0$ where $\nu(p)$ is the common normal to $\partial\Sigma_i(m)$ and $\partial\Sigma(m)$ at p. But by the internal tangency one can easily compute for the constant Neumann boundary condition that $\partial_{\nu(p)} w(p, m) = 0$; contradiction.

(b) **There exists a point $p \in T_m \cap \partial\Omega_i$ with $\nu_1(p) = o$:** Observe that p is a right-angled corner of $\Sigma(m)$ so that a direct application of the Hopf Lemma as in (a) is not at hand. Instead we show that w has a zero of second order at p which contradicts the corner-version of the Hopf Lemma due to Serrin [15] (see Appendix). To calculate the derivatives of w we use a rectangular coordinate frame with origin at p, ξ_n–axis along the inward normal $\nu(p)$ and ξ_1–axis collinear to the x_1–axis. In this coordinate frame $\partial\Omega$ is locally expressed by

$$\xi_n = g(\xi_1, \ldots, \xi_{n-1}) = g(\xi'), \qquad g \in C^2$$

and the inward normal $\nu(\xi')$ at $(\xi', g(\xi'))$ is given by

$$\nu(\xi') = \frac{(\nabla g(\xi'), 1)}{(|\nabla g(\xi')|^2 + 1)^{\frac{1}{2}}}.$$

The fact that the inward normal at p coincides with the ξ_n−axis means that $\nabla g(0) = 0$. The point $x = p$ corresponds to $\xi = 0$. Using the ξ-coordinates we introduce new functions $\tilde{u}, \tilde{v}, \tilde{w}$ by

$$\tilde{u}(\xi) = u(x), \qquad \tilde{v}(\xi) = v(x, m) = u(x^m), \qquad \tilde{w}(\xi) = w(x, m)$$

and find the relations

$$
\begin{aligned}
\tilde{v}(\xi) &= \tilde{u}(-\xi_1, \xi_2, \ldots, \xi_n) \\
\tilde{w}(\xi) &= \tilde{u}(-\xi_1, \xi_2, \ldots, \xi_n) - \tilde{u}(\xi_1, \ldots, \xi_n).
\end{aligned}
$$

The Dirichlet and Neumann boundary conditions now read

$$\tilde{u}(\xi) = \text{const.}, \qquad \frac{\partial \tilde{u}}{\partial \nu}(\xi) = \nabla_\xi \tilde{u}(\xi) \cdot \nu(\xi) = \text{const.}$$

and with help of the parametrization of $\partial\Omega_i$ we find

$$\tilde{u}(\xi', g(\xi')) = \text{const.},$$

$$\sum_{l=1}^{n-1} \tilde{u}_l(\xi', g(\xi'))g_l(\xi') + \tilde{u}_n(\xi', g(\xi')) = \text{const.} \cdot (|\nabla g(\xi')|^2 + 1)^{\frac{1}{2}}.$$

Differentiation w.r.t. ξ_j for $j = 1, \ldots, n-1$ gives (we use summation convention for $l = 1, \ldots, n-1$)

$$\tilde{u}_j + \tilde{u}_n g_j = 0$$

$$\tilde{u}_{lj}g_l + \tilde{u}_{ln}g_jg_l + \tilde{u}_l g_{lj} + \tilde{u}_{nj} + \tilde{u}_{nn}g_j = \text{const.} \cdot \frac{g_l g_{lj}}{(|\nabla g|^2 + 1)^{\frac{1}{2}}}$$

and evaluation at $\xi' = 0$ results in (remember $\nabla g(0) = 0$)

$$
\begin{aligned}
\tilde{u}_j(0) &= 0 && \text{for } j = 1, \ldots, n-1 && (18) \\
\tilde{u}_{nj}(0) &= 0 && \text{for } j = 1, \ldots, n-1. && (19)
\end{aligned}
$$

Let us now collect our results for the derivatives of $\tilde{w}$ at $\xi = 0$. Notice that all first and second derivatives of $\tilde{w}$ in directions other than the ξ_1−direction vanish at $\xi = 0$ since it is a point on the hyperplane where ξ and its reflected point coincide. Also $\tilde{w}_{11}(0) = 0$.

178

- $\partial_\alpha \tilde{w}(0) = 0$ for $\alpha = 2, \ldots, n$ and for $\alpha = 1$ by (18).

- $\partial_{11} \tilde{w}(0) = \partial_{\alpha\beta} \tilde{w}(0)$ for $\alpha, \beta = 2, \ldots, n$.

- $\partial_{1n} \tilde{w}(0) = 0$ by (19).

To obtain our contradiction to Serrin's version of the boundary point lemma we need to show that also the remaining derivatives $\partial_{1\alpha} \tilde{w}(0)$ vanish for $\alpha = 2, \ldots, n$. We do this by using the following Taylor-expansion of $\tilde{w}$:

$$\tilde{w}(\xi) = \tilde{w}(0) + \sum_{j=1}^{n} \partial_j \tilde{w}(0)\xi_j + \sum_{j,k=1}^{n} \partial_{jk}\tilde{w}(0)\frac{\xi_j\xi_k}{2} + o(|\xi|^2) \tag{20}$$

$$= \sum_{\alpha=1}^{n-1} \partial_{1\alpha}\tilde{w}(0)\xi_1\xi_\alpha + o(|\xi|^2).$$

Fix $\alpha \in \{2, \ldots, n\}$, let $\sigma, \rho > 0$ and define $\xi = \sigma(\rho, 0, \ldots, 0, \overset{\alpha}{\pm\rho}, 0, \ldots, 0, 1)$, where the plus sign is taken if $\partial_{1\alpha}\tilde{w}(0) \leq 0$ and the minus sign if $\partial_{1\alpha}\tilde{w}(0) > 0$. The point $\xi(\rho, \sigma)$ is moving along a straight line through the origin, which has the angle

$$\cos \theta(\rho) = 1/\sqrt{2\rho^2 + 1}$$

with the inner normal $\nu(p)$. Since $\Sigma(m)$ has a rectangular corner at $x = p$ resp. $\xi = 0$ and since the ξ_1-component of $\xi(\rho, \sigma)$ is positive we finde that for σ, ρ small the point $\xi(\sigma, \rho)$ is in $\Sigma(\lambda)$. After this choice of σ, ρ it follows from (20) that

$$\tilde{w}(\xi) = -\sigma^2\rho^2|\tilde{w}_{1\alpha}(0)| + o(\sigma^2) \qquad \text{for } \sigma \to 0.$$

Since also $\tilde{w} > 0$ by our assumption we see that this is only possible if $\tilde{w}_{1\alpha}(0) = 0$. Therefore $\tilde{w}$ and w have a second order zero at p in contradiction to Serrin's corner lemma. This finishes the proof of Theorem 3. $\qquad \square$

Appendix

Proof of Lemma 2: Since $u = $ const. locally around z on ∂D it follows that every tangential derivative is 0 at z. Furthermore $\partial_m u(z) = 0$ for the non-tangential direction m. Therefore $\nabla u(z) = 0$. The Laplacien is rotation invariant, i.e. $\Delta u(z)$ can be calculated by summing the second derivatives w.r.t. n arbitrary, orthogonal unit vectors, e.g. $\nu(z), \xi_2, \ldots, \xi_n$ where $\xi_2, \ldots, \xi_n$ are tangent at z. Hence $\Delta u(z) = \partial_\nu^2 u(z)$. In the following expression we use summation convention for $i, j = 2, \ldots, n$

$$\frac{\partial}{\partial m} = (m \cdot \nu)\frac{\partial}{\partial \nu} + m_i \frac{\partial}{\partial \xi_i}$$

$$\frac{\partial^2}{\partial m^2} = (m \cdot \nu)^2 \frac{\partial^2}{\partial \nu^2} + 2(m \cdot \nu)m_i \frac{\partial}{\partial m}\frac{\partial}{\partial \xi_i} + m_i m_j \frac{\partial}{\partial \xi_i}\frac{\partial}{\partial \xi_j}$$

We apply this operator to u and evaluate at z. Observe that $u = $ const. on ∂D implies $\partial_{\xi_i}\partial_{\xi_j}u(z) = 0$. Also since $\partial_m u$ has a local extremum at z on ∂D we find $\partial_{\xi_i}\partial_m u(z) = 0$. Hence the result:

$$\frac{\partial^2}{\partial m^2}u(z) = (m \cdot \nu(z))^2 \frac{\partial^2}{\partial \nu^2}u(z) = (m \cdot \nu(z))^2 \Delta u(z).$$

$\square$

Serrin's corner lemma *Let D be a domain lying to the right of the hyperplane T_λ and let $Q \in \partial D \cap T_\lambda$ be a point where ∂D intersects T_λ orthogonally. Suppose w is a C^2-function in $\overline{D}$ satisfying the elliptic differential inequality*

$$\Delta w + b_i \partial_i w + c^- w \le 0, \quad \text{in } D$$

while also $w \ge 0$ in D and $w(Q) = 0$. Let $m \in \mathbb{R}^n$ be a direction which enters D at Q non-tangentially. Then either

$$\frac{\partial w}{\partial m}u(Q) > 0 \qquad \text{or} \qquad \frac{\partial^2}{\partial m^2}w(Q) > 0$$

unless $w \equiv 0$.

Continuation of part (II) of the proof of Theorem 3: So far we know $\partial Z \backslash T_\lambda \subset \partial \Omega$ and $\partial Z^\lambda \backslash T_\lambda \subset \partial \Omega$. We define

$$X = Z \cup Z^\lambda \cup (\partial Z \cap \Omega) \cup (\partial Z^\lambda \cap \Omega)$$

and show that X is open. Then it is easy to calculate

$$\partial X \subset (\partial Z \cup \partial Z^\lambda) \backslash (\partial Z \cap \Omega) \backslash (\partial Z^\lambda \cap \Omega) \subset \partial \Omega$$

which implies $\Omega = X$ since X is open and Ω connected. Let us show the openness of Ω. Clearly $Z \cup Z^\lambda$ is a subset of the interior of X.
(a) Take $x \in \partial Z \cap \Omega$. Since $\partial Z \backslash T_\lambda \subset \partial \Omega$ we find $x \in T_\lambda \cap \Omega$. By the openness of Ω we can use a ball $B_\rho(x) \subset \Omega$ to define

$$
\begin{aligned}
B_> &= B_\rho(x) \cap \{x_1 > \lambda\}, \\
B_< &= B_\rho(x) \cap \{x_1 < \lambda\}, \\
E &= B_\rho(x) \cap \{x_1 = \lambda\}.
\end{aligned}
$$

Cleary $Z \cap B_> \ne \emptyset$. If also $Z^C \cap B_> \ne \emptyset$ then $\partial Z \cap B_> \ne \emptyset$ which is impossible since $\partial Z \cap B_> \subset (\partial \Omega \cup T_\lambda) \cap B_> = \emptyset$. Hence $B_> \subset Z$, $B_< \subset Z^\lambda$ and $E \subset \Omega \cap (\overline{Z} \cap T_\lambda) \subset \Omega \cap \partial Z$ altogether result in $B_\rho(x) \subset X$.
(b) Take $x \in \partial Z^\lambda \cap \Omega$. As before we see that $\partial Z^\lambda \backslash T_\lambda \subset \partial \Omega$ implies $x \in T_\lambda$. Hence there is a sequence $\{x_\nu\}_1^\infty \subset Z^\lambda \cap \Omega$ with $x_\nu \to x$. Since $x_\nu^\lambda \in Z$ also converge to $x = x^\lambda$ we find $x \in \overline{Z} \cap T_\lambda \cap \Omega \subset \partial Z \cap \Omega$, i.e. x lies in the interior of X by (a). $\square$

References

[1] G. Alessandrini. A symmetry theorem for condensers. *Math. Meth. Appl. Sci.*, 15:315–320, 1992.

[2] C.J. Amick and L.E. Fraenkel. Uniqueness of Hill's spherical vortex. *Arch. Rat. Mech. Anal.*, 92:91–119, 1986.

[3] P. Aviles. Symmetry theorems related to Pompeiu's problem. *Am. J. Math.*, 108:1023–1036, 1986.

[4] H. Brezis and L. Nirenberg. Positive solutions of nonlinear elliptic equations involving critical Sobolev exponents. *Comm. Pure Appl. Math.*, 36:437–477, 1983.

[5] A. Castro and R. Shivaji. Nonnegative solutions to a semilinear Dirichlet problem in a ball are positive and radially symmetric. *Comm. PDE*, 14(8&9):1091–1100, 1989.

[6] M.J. Esteban. About the symmetry of positive solutions of nonlinear elliptic problems in symmetric domains. *Pitman Res. Notes*, 249:14–25, 1991.

[7] B. Gidas, Wei-Ming Ni, and L. Nirenberg. Symmetry and related problems via the maximum principle. *Comm. Math. Phys.*, 68:209–243, 1979.

[8] A. Lazer and P.J. McKenna. A symmetry theorem and applications to nonlinear PDEs. *J. Diff. Eqns.*, 72:95–106, 1988.

[9] Congming Li. Monotonicity and symmetry of solutions of fully nonlinear elliptic equations on bounded domains. *Comm. PDE*, 16(2&3):491–526, 1991.

[10] Ji Min. A remark on the symmetry of solutions to nonlinear elliptic equations. *Pac. J. Math.*, 153(1):157–162, 1992.

[11] C. Miranda. *Partial Differential Equations of Elliptic Type*. Springer Verlag, 2nd edition, 1970.

[12] L.E. Payne and G.A. Philippin. On two free boundary problems in potential theory. *J. Math. Anal. Appl.*, 161:332–342, 1991.

[13] G.A. Philippin. On a free boundary problem in electrostatics. *Math. Meth. Appl. Sci.*, 12:387–392, 1990.

[14] G.A. Philippin and L.E. Payne. On the conformal capacity problem. In G. Talenti, editor, *Geometry of Solutions to Partial Differential Equations*, London, 1989. Academic Press.

[15] J. Serrin. A symmetry theorem in potential theory. *Arch. Rat. Mech. Anal.*, 43:304–318, 1971.

[16] N.B. Willms, G. Gladwell, and D. Siegel. Symmetry theorems for some overde-termined boundary value problems on ring domains. *ZAMP*, 45:556–579, 1994.

Wolfgang Reichel
Mathematisches Institut I
Universität Karlsruhe
D-76128 Karlsruhe
Germany

N SATO, J SHIROHZU AND N KENMOCHI

Large-time behavior of the solution to a phase change problem with constraint

1. Introduction

We consider the following nonlinear system:

$$\frac{\partial \rho(u)}{\partial t} + \frac{\partial w}{\partial t} - \triangle u = f(t, x) \qquad \text{in } Q := (0, +\infty,) \times \Omega, \tag{1.1}$$

$$\nu \frac{\partial w}{\partial t} + \beta(w) + g(w) \ni u \quad \text{in } Q \tag{1.2}$$

with lateral boundary condition:

$$\frac{\partial u}{\partial n} + \alpha_N(x)u = h_N(t, x) \quad \text{on } \Sigma := (0, +\infty) \times \Gamma, \tag{1.3}$$

and initial conditions:

$$u(0, \cdot) = u_0, \quad w(0, \cdot) = w_0 \quad \text{in } \Omega. \tag{1.4}$$

Here Ω is a bounded domain in R^N ($N \geq 1$) with smooth boundary $\Gamma := \partial\Omega$; ρ is a monotone increasing and bi-Lipschitz continuous function on R; ν is a positive constant; β is a maximal monotone graph in $R \times R$; g is a smooth function defined on R; α_N is a non-negative, bounded and measurable function on Γ such that $\alpha_N > 0$ on a subset of Γ with positive measure; f, h_N, u_0 and w_0 are given data.

For simplicity problem (1.1)-(1.4) is denoted by (CP). This is a simplified model for a class of solid-liquid phase change problems, and in this context u represents a function related to temperature and w a non-conserved order parameter (the state variable characterizing phase). For instance, we have the following examples:

(1) Stefan problem with phase relaxation, in which β is the subdifferential of the indicator function of the interval $[0, 1]$ and $g \equiv 0$. This case was discussed as a melting problem with supercooling and superheating effect in [12,5].

(2) Phase-field model with constraint, in which β is the same as in (1), $\rho(u) = u$, $g(w) = w^3 - cw$ with a positive constant c, and a diffusion term $-\kappa \triangle w$ is added to the left side of (1.2). This is a phase-field model with constraint $0 \le w \le 1$ and was discussed in [6,9,11]. We may consider system (1.1)-(1.4) as an approximation of this problem with small $\kappa > 0$.

Furthermore we refer to [2,1] for papers dealing with similar problems.

In this paper, we discuss the large-time behavior of the solution $\{u, w\}$. In fact, under the condition that $f(t, x) \to f^\infty(x)$ and $h_N(t, x) \to h_N^\infty(x)$ in an appropriate sense as $t \to +\infty$, it will be shown that as $t \to +\infty$, $u(t, \cdot)$ and $w(t, \cdot)$ converge to a solution $\{u^\infty, w^\infty\}$ of the corresponding steady-state problem

$$
\begin{cases}
-\triangle u^\infty = f^\infty(x) & \text{in } \Omega, \\[2mm]
\dfrac{\partial u^\infty}{\partial n} + \alpha_N(x) u^\infty = h_N^\infty(x) & \text{on } \Gamma, \\[2mm]
\beta(w^\infty) + g(w^\infty) \ni u^\infty & \text{in } \Omega.
\end{cases}
$$

2. Existence and uniqueness result for (CP)

Problem (CP) is discussed under the following assumptions (A1)-(A6):

(A1) $\rho : R \longrightarrow R$ is an increasing and bi-Lipschitz continuous function.

(A2) β is a maximal monotone graph in $R \times R$ such that for some numbers σ_*, σ^* with $-\infty < \sigma_* < \sigma^* < +\infty$

$$\overline{D(\beta)} = [\sigma_*, \sigma^*];$$

note in this case that $R(\beta) = R$, so that there is a non-negative proper l.s.c. convex function $\hat{\beta}$ on R whose subdifferential $\partial\hat{\beta}$ coincides with β in R, and in the context of solid-liquid system we can consider that $w = \sigma_*$ (resp. σ^*) indicates the pure solid (resp. liquid) phase and any intermediate value w indicates a state of mixture.

(A3) $g : R \to R$ is a Lipschitz continuous function with compact support in R; in this case note that there is a non-negative primitive $\hat{g}$ of g.

(A4) $f \in L^2_{loc}(R_+; L^2(\Omega))$.

(A5) $h_N \in W^{1,2}_{loc}(R_+; L^2(\Gamma))$ with $\sup_{t \ge 0} |h_N|_{W^{1,2}(t,t+1;L^2(\Gamma))} < +\infty$.

(A6) $u_0 \in L^2(\Omega)$ and $w_0 \in L^2(\Omega)$ with $\hat{\beta}(w_0) \in L^1(\Omega)$.

We introduce some function spaces and a convex function in order to discuss (CP) in the framework of abstract evolution equations of the form

$$U'(t) + \partial \varphi^t(U(t)) + G(U(t)) \ni \tilde{f}(t).$$

Let $V := H^1(\Omega)$ with norm

$$|z|_V := \{|\nabla z|^2_{L^2(\Omega)} + \int_\Gamma \alpha_N |z|^2 d\Gamma\}^{\frac{1}{2}},$$

and denote by V^* the dual space of V and by $\langle \cdot, \cdot \rangle$ the duality pairing between V^* and V. Then, identifying $L^2(\Omega)$ with its dual space by means of the usual inner product

$$(v, z) := \int_\Omega vz dx,$$

we see that

$$V \subset L^2(\Omega) \subset V^*$$

with compact injections.

Let F be the duality mapping from V onto V^* which is given by the formula

$$\langle Fv, z \rangle = \int_\Omega \nabla v \cdot \nabla z dx + \int_\Gamma \alpha_N vz d\Gamma \quad \text{for any } v, z \in V.$$

It is easy to see that V^* becomes a Hilbert space with inner product $(\cdot, \cdot)_*$ given by

$$(v, z)_* := \langle v, F^{-1}z \rangle \ (= \langle z, F^{-1}v \rangle) \quad \text{for any } v, z \in V^*.$$

Now, consider the product space

$$X := V^* \times L^2(\Omega),$$

which becomes a Hilbert space with inner product $(\cdot, \cdot)_X$ given by

$$([e_1, w_1], [e_2, w_2])_X := (e_1, e_2)_* + \nu(w_1, w_2) \quad \text{for any } [e_i, w_i] \in X \ (i = 1, 2).$$

Next, given the boundary data h_N, choose $h : R_+ \longrightarrow H^1(\Omega)$ such that for each $t \geq 0$

$$\int_\Omega \nabla h(t) \cdot \nabla z dx + \int_\Gamma \alpha_N h(t) z d\Gamma = \int_\Gamma h_N(t) z d\Gamma \quad \text{for all } z \in V;$$

note from (A5) that $\sup_{t \geq 0} |h|_{W^{1,2}(t, t+1; H^1(\Omega))} < +\infty$.

Also, using h and $\hat{\beta}$, for each $t \geq 0$, define a proper l.s.c. convex function φ^t on X by

$$\varphi^t(U) = \begin{cases} \int_\Omega \rho^*(e - w) dx + \int_\Omega \hat{\beta}(w) dx - (h(t), e) \\ \qquad \text{if } U = [e, w] \in L^2(\Omega) \times L^2(\Omega) \text{ with } \hat{\beta}(w) \in L^1(\Omega), \\ +\infty \quad \text{otherwise}, \end{cases}$$

where ρ^* is a non-negative primitive of ρ^{-1}. We denote by $\partial\varphi^t$ the subdifferential of φ^t in X and its characterization is given by the following theorem.

Theorem 2.1. *(cf. [5,9]) Let $t \geq 0$, $[e^*, w^*] \in X$ and $[e, w] \in D(\partial\varphi^t)$. Then $[e^*, w^*] \in \partial\varphi^t([e, w])$ if and only if conditions (a) and (b) below are satisfied:*

(a) $e^* = F(\rho^{-1}(e - w) - h(t))$, *that is,* $\rho^{-1}(e - w) - h(t) \in V$ *and*

$$\langle e^*, z \rangle = \int_\Omega \nabla(\rho^{-1}(e - w) - h(t)) \cdot \nabla z dx + \int_\Gamma \alpha_N(\rho^{-1}(e - w) - h(t))z d\Gamma$$

for all $z \in V$;

(b) *there exists a function* $\xi \in L^2(\Omega)$ *such that* $\xi \in \beta(w)$ *a.e. on* Ω *and*

$$\nu w^* = \xi - \rho^{-1}(e - w) \quad in \ L^2(\Omega).$$

Moreover, for $U_i^* = [e_i^*, w_i^*] \in \partial\varphi^t(U_i)$ *with* $U_i = [e_i, w_i] \in D(\partial\varphi^t)$ *$(i = 1, 2)$,*

$$(U_1^* - U_2^*, U_1 - U_2)_X = |(e_1 - w_1) - (e_2 - w_2)|_{L^2(\Omega)} + (\xi_1 - \xi_2, w_1 - w_2),$$

where $\xi_i \in L^2(\Omega)$ *is as any function* ξ *in (b) for each* $i = 1, 2$.

A weak formulation for (CP) is given as follows.

Definition 2.1. A couple $\{u, w\}$ of functions $u : R_+ \longrightarrow V^*$ and $w : R_+ \longrightarrow L^2(\Omega)$ is called a (weak) solution of (CP) on R_+, if the following conditions (w1)-(w3) are fulfilled for any finite $T > 0$:

(w1) $\rho(u) \in C([0, T]; V^*) \cap W_{loc}^{1,2}((0, T]; V^*) \cap L^2(0, T; L^2(\Omega))$, $u \in L_{loc}^2((0, T]; H^1(\Omega))$, $w \in C([0, T]; L^2(\Omega)) \cap W_{loc}^{1,2}((0, T]; L^2(\Omega))$, and $\hat{\beta}(w) \in L^1(0, T; L^1(\Omega))$.

(w2) $\rho(u)(0) = \rho(u_0)$ and

$$\langle u'(t) + w'(t), z \rangle + \int_\Omega \nabla(u(t) - h(t)) \cdot \nabla z dx + \int_\Gamma \alpha_N(u(t) - h(t))z d\Gamma = (f(t), z)$$

for all $z \in V$ and a.e. $t \in [0, T]$, where the prime $'$ denotes the derivative in time.

(w3) there exists $\xi \in L_{loc}^2((0, T]; L^2(\Omega))$ such that $\xi \in \beta(w)$ a.e. on $Q_T := (0, T) \times \Omega$ and

$$\nu(w'(t), z) + (\xi(t) + g(w(t)), z) = (u(t), z)$$

for all $z \in L^2(\Omega)$ and a.e. $t \in [0, T]$.

186

According to Theorem 2.1, (CP) can be reformulated as an evolution equation in X in the following form:

$$\begin{cases} U'(t) + \partial\varphi^t(U(t)) + G(U(t)) \ni \tilde{f}(t), & \text{in } X,\ t \geq 0, \\ U(0) = [\rho(u_0) + w_0, w_0], \end{cases}$$

where $U(t) = [\rho(u(t)) + w(t), w(t)]$, $G(U(t)) = [0, \frac{1}{\nu}g(w(t))]$ and $\tilde{f}(t) = [f(t), 0]$.

As to the solvability of (CP) we have:

Theorem 2.2. *(cf. [6,9]) Assume that (A1)-(A6) hold. Then, for any $T > 0$, (CP) admits one and only one solution $\{u, w\}$ on $[0, T]$ such that*

$$\begin{cases} t^{\frac{1}{2}}\rho(u)' \in L^2(0, T; V^*), & t^{\frac{1}{2}}u \in L^2(0, T; H^1(\Omega)), \\ t\rho(u)' \in L^2(0, T; L^2(\Omega)), & tu \in L^\infty(0, T; H^1(\Omega)), \end{cases}$$

$$t^{\frac{1}{2}}w' \in L^2(0, T; L^2(\Omega)), \quad t\hat{\beta}(w) \in L^\infty(0, T; L^1(\Omega)),$$

$$t^{\frac{1}{2}}\xi \in L^2(0, T; L^2(\Omega))$$

where ξ is the function in condition (w3).

3. Large-time behavior of the solution

Further suppose that there are $h_N^\infty \in L^2(\Gamma)$ and $f^\infty \in L^2(\Omega)$ such that

$$h_N - h_N^\infty \in L^2(R_+; L^2(\Gamma)), \quad f - f^\infty \in L^2(R_+; L^2(\Omega)), \tag{3.1}$$

and consider the steady-state problem (3.2)-(3.3):

$$- \triangle u^\infty = f^\infty(x) \text{ in } \Omega, \quad \frac{\partial u^\infty}{\partial n} + \alpha_N(x)u^\infty = h_N^\infty(x) \quad \text{on } \Gamma, \tag{3.2}$$

$$\beta(w^\infty) + g(w^\infty) \ni u^\infty \quad \text{in } \Omega. \tag{3.3}$$

We should note that problem (3.2) does not include w^∞, and it has a unique solution $u^\infty \in H^1(\Omega)$ in the variational sense, i.e.,

$$\int_\Omega \nabla(u^\infty - h^\infty) \cdot \nabla z\, dx + \int_\Gamma \alpha_N(u^\infty - h^\infty)z\, d\Gamma = (f^\infty, z) \quad \text{for all } z \in V, \tag{3.4}$$

where $h^\infty \in H^1(\Omega)$ such that

$$\int_\Omega \nabla h^\infty \cdot \nabla z\, dx + \int_\Gamma \alpha_N h^\infty z\, d\Gamma = \int_\Gamma h_N^\infty z\, d\Gamma \quad \text{for all } z \in V.$$

We see from (3.1) that $h - h^\infty \in L^2(R_+; H^1(\Omega))$.

In the sequel we mean by (P^∞) the algebraic relation (3.3) with the solution $u^\infty \in H^1(\Omega)$ of (3.4), and $w^\infty = w^\infty(x)$ is called a solution of (P^∞).

As the following example shows, the steady-state problem (P^∞) has in general infinitely many solutions.

Example 3.1. Consider the case when

$$f^\infty(x) \equiv 0, \quad h_N^\infty(x) \equiv l_0, \quad \alpha_N(x) \equiv 1, \quad \beta = \partial I_{[-1,1]} \text{ and } g(w) = w^3 - w$$

where l_0 is a constant. Then, clearly $u^\infty \equiv l_0$ and we have the following three possibilities:

(i)   when $l_0 > \frac{2}{3\sqrt{3}}$ (resp. $l_0 < -\frac{2}{3\sqrt{3}}$), the algebraic relation

$$\beta(r) + g(r) \ni u^\infty (= l_0) \tag{3.5}$$

has exactly one solution $r = 1$ (resp. -1).

(ii)   when $l_0 = \frac{2}{3\sqrt{3}}$ (resp. $-\frac{2}{3\sqrt{3}}$), (3.5) has exactly two solutions $r = -\frac{1}{\sqrt{3}}$ (resp. $\frac{1}{\sqrt{3}}$), 1 (resp. -1).

(iii)   when $|l_0| < \frac{2}{3\sqrt{3}}$, (3.5) has exactly three solutions $r = \xi_-, \xi_0, \xi_+$ with $-1 \leq \xi_- < \xi_0 < \xi_+ \leq 1$.

Physically (i) means that if the temperature is kept high (resp. low) enough, then the limit state (as $t \longrightarrow +\infty$) will be of pure liquid (resp. solid). On the other hand, (ii) and (iii) mean that if the temperature is kept near the phase transition temperature, then the limit state possibly includes a mushy region. In particular, in the case of (iii), all step functions w^∞ with range in $\{\xi_-, \xi_0, \xi_+\}$ are solutions of (P^∞) and hence (P^∞) has in general infinitely many solutions.

Our main result is stated in the following theorem.

Theorem 3.1. *Suppose that conditions (A1)-(A6) and (3.1) hold, and let $\{u, w\}$ be the solution to (CP) on R_+. Further, suppose that for each $p \in R$ the (algebraic) inclusion*

$$\beta(r) + g(r) \ni p$$

has a finite number of solutions r in $\overline{D(\beta)}$. Then,

$$u(t) \longrightarrow u^\infty \text{ weakly in } H^1(\Omega) \text{ as } t \longrightarrow +\infty, \tag{3.6}$$

where u^∞ is the unique solution of (3.4), and there exists a function $w^\infty \in L^\infty(\Omega)$ such that

$$\beta(w^\infty(x)) + g(w^\infty(x)) \ni u^\infty(x) \quad \text{for a.e. } x \in \Omega$$

and

$$w(t,x) \longrightarrow w^\infty(x) \text{ for a.e. } x \in \Omega \text{ as } t \longrightarrow +\infty.$$

We prove the theorem by the following four lemmas.

Lemma 3.1. *Under the same assumptions of Theorem 3.1, for the solution $\{u, w\}$ to (CP) on R_+, we have*

$$u - u_\infty \in L^2(R_+; H^1(\Omega)), \ w' \in L^2(R_+; L^2(\Omega)) \text{ and } \hat\beta(w) \in L^\infty(R_+; L^1(\Omega)), \quad (3.7)$$

$$u \in L^\infty([1, +\infty); H^1(\Omega)). \tag{3.8}$$

Proof. Multiplying the difference of (1.1) and (3.2) by $u(t) - u^\infty$ and (1.2) by $w'(t)$, we get

$$\frac{d}{dt}\{\int_\Omega \rho^*(\rho(u(t)))dx - (\rho(u(t)) + w(t), u^\infty)\} + (w'(t), u(t)) +$$

$$+ |\nabla(u(t) - u^\infty)|^2_{L^2(\Omega)} + \int_\Gamma \alpha_N |u(t) - u^\infty|^2 d\Gamma$$

$$= (f(t) - f^\infty, u(t) - u^\infty) + \int_\Gamma (h(t) - h^\infty)(u(t) - u^\infty)d\Gamma$$

and

$$\nu |w'(t)|^2_{L^2(\Omega)} + \frac{d}{dt}\{\int_\Omega \hat\beta(w(t))dx + \int_\Omega \hat g(w(t))dx\} = (u(t), w'(t))$$

for a.e. $t \geq 0$. Adding these two equalities we have

$$\frac{d}{dt}\{\int_\Omega \rho^*(\rho(u(t)))dx - (\rho(u(t)) + w(t), u^\infty) + \int_\Omega \hat\beta(w(t))dx + \int_\Omega \hat g(w(t))dx\} +$$

$$+ \nu |w'(t)|^2_{L^2(\Omega)} + |\nabla(u(t) - u^\infty)|^2_{L^2(\Omega)} + \int_\Gamma \alpha_N |u(t) - u^\infty|^2 d\Gamma$$

$$= (f(t) - f^\infty, u(t) - u^\infty) + \int_\Gamma (h(t) - h^\infty)(u(t) - u^\infty)d\Gamma$$

for a.e. $t \geq 0$, so that there are positive constants C_1 and C_2 such that

$$\frac{d}{dt}\{\int_\Omega \rho^*(\rho(u(t)))dx - (\rho(u(t)) + w(t), u^\infty) + \int_\Omega \hat{\beta}(w(t))dx + \int_\Omega \hat{g}(w(t))dx\}+$$

$$+\nu|w'(t)|^2_{L^2(\Omega)} + C_1|u(t) - u^\infty|^2_{H^1(\Omega)}$$

$$\leq C_2\{|f(t) - f^\infty|^2_{L^2(\Omega)} + |h(t) - h^\infty|^2_{L^2(\Gamma)}\}$$

for a.e. $t \geq 0$.

Therefore, for all $T > 0$, we have

$$\int_\Omega \rho^*(\rho(u(T)))dx - (\rho(u(T)) + w(T), u^\infty) + \int_\Omega \hat{\beta}(w(T))dx + \int_\Omega \hat{g}(w(T))dx+$$

$$+\nu\int_0^T |w'(t)|^2_{L^2(\Omega)}dt + C_1\int_0^T |u(t) - u^\infty|^2_{H^1(\Omega)}dt$$

$$\leq C_2\{\int_0^T |f(t) - f^\infty|^2_{L^2(\Omega)}dt + \int_0^T |h(t) - h^\infty|^2_{L^2(\Gamma)}dt\}+$$

$$+ \int_\Omega \rho^*(\rho(u_0))dx - (\rho(u_0) + w_0, u^\infty) + \int_\Omega \hat{\beta}(w_0)dx + \int_\Omega \hat{g}(w_0))dx.$$

Hence (3.7) is obtained. Also, (3.8) is a direct consequence of (3.7) and a standard regularity result for parabolic equations. □

Lemma 3.2. *Under the same assumptions of Theorem 3.1, put*

$$U^t(x) := \int_t^{t+1} |w'(\tau, x)|^2 d\tau \quad \text{for } x \in \Omega.$$

Then, $U^t(x) \longrightarrow 0$ as $t \longrightarrow +\infty$ for a.e. $x \in \Omega$.

Proof. By Lemma 3.1, we have

$$\lim_{T \nearrow +\infty} \int_T^{+\infty} dt \int_\Omega |w'(t, x)|^2 dx = 0,$$

so that

$$\lim_{T \nearrow +\infty} \int_\Omega dx \int_T^{+\infty} |w'(t, x)|^2 dt = 0.$$

Hence,

$$\int_T^{+\infty} |w'(t, x)|^2 dt \longrightarrow 0 \text{ as } T \longrightarrow +\infty \text{ for a.e. } x \in \Omega.$$

This implies the lemma. □

Lemma 3.3. *Under the same assumptions of Theorem 3.1, (3.6) holds.*

190

Proof. Let $\{u, w\}$ be a solution to (CP) and u^∞ be the solution to (3.4). Let $\{t_n\}$ be any sequence with $t_n \longrightarrow +\infty$ as $n \longrightarrow +\infty$, and put

$$u_n(t) := u(t_n + t), \quad w_n(t) := w(t_n + t), \quad f_n(t) := f(t_n + t), \quad h_n(t) := h(t_n + t)$$

$$\text{for } 0 \leq t \leq 1.$$

Since by Lemma 3.1, $u - u^\infty$ and w' are in $L^2(R_+; H^1(\Omega))$ and $L^2(R_+; L^2(\Omega))$, respectively, we see that

$$u_n \longrightarrow u^\infty \text{ in } L^2(0, 1; H^1(\Omega)), \tag{3.9}$$

and

$$w'_n \longrightarrow 0 \text{ in } L^2(0, 1; L^2(\Omega)), \tag{3.10}$$

as $n \longrightarrow +\infty$. Moreover since by Lemma 3.1, u is bounded in $H^1(\Omega)$ on $[1, +\infty)$, we may assume that for a function $\tilde{u}^\infty$ in $H^1(\Omega)$

$$u_n(0) = u(t_n) \longrightarrow \tilde{u}^\infty \text{ weakly in } H^1(\Omega) \tag{3.11}$$

as $n \longrightarrow +\infty$. Now, consider the Cauchy problem for each n

$$\begin{cases} \rho(u_n)'(t) + \partial \Phi_n^t(u_n(t)) = f_n(t) - w'_n(t) & \text{in } L^2(\Omega),\ 0 \leq t \leq 1, \\ u_n(0) = u(t_n) \end{cases}$$

where Φ_n^t is a proper l.s.c. and convex function on $L^2(\Omega)$ such that for each n and $t \in [0, 1]$

$$\Phi_n^t(z) := \begin{cases} \dfrac{1}{2}|z - h_n(t)|_V^2 & \text{if } z \in V, \\ +\infty & \text{otherwise}, \end{cases}$$

and $\partial \Phi_n^t$ is the subdifferential of Φ_n^t in $L^2(\Omega)$. From (3.9), (3.10) and (3.11), we see that

$$\Phi_n^t \longrightarrow \Phi_\infty \quad \text{on } L^2(\Omega) \text{ in the sense of Mosco for every } t \in [0, 1],$$

$$f_n - w'_n \longrightarrow f^\infty \quad \text{in } L^2(0, 1; L^2(\Omega))$$

and

$$u_n(0) \longrightarrow \tilde{u}^\infty \quad \text{in } L^2(\Omega)$$

as $n \longrightarrow +\infty$, where Φ_∞ is a proper l.s.c. and convex function on $L^2(\Omega)$ such that

$$\Phi_\infty(z) := \begin{cases} \dfrac{1}{2}|z - h^\infty|_V^2 & \text{if } z \in V, \\ +\infty & \text{otherwise}. \end{cases}$$

191

Therefore, by a general theory in [8],

$$u_n \longrightarrow \tilde{u} \quad \text{in } C([0,1]; L^2(\Omega)) \text{ as } t \longrightarrow +\infty, \tag{3.12}$$

where $\tilde{u}$ is the solution of

$$\begin{cases} \rho(\tilde{u})'(t) + \partial\Phi_\infty(\tilde{u}(t)) = f^\infty & \text{in } L^2(\Omega), 0 \le t \le 1, \\ \tilde{u}(0) = \tilde{u}^\infty. \end{cases}$$

From (3.9) and (3.12) it follows that $\tilde{u} = u^\infty$ on $[0,1]$. Consequently (3.6) holds. $\square$

Lemma 3.4. *Under the same assumptions of Theorem 3.1, put*

$$V(x) := \{r \in \overline{D(\beta)}; w(t_n, x) \to r \text{ for some } t_n \text{ with } t_n \to +\infty\} \quad \text{for } x \in \Omega.$$

Then,

(1) $V(x) \neq \emptyset$ for a.e. $x \in \Omega$;

(2) $\beta(r) + g(r) \ni u^\infty(x)$ for all $r \in V(x)$ and a.e. $x \in \Omega$;

(3) $V(x)$ is a singleton for a.e. $x \in \Omega$.

Proof. (1) is clear by the boundedness of $w(t,x)$ on R.

Let $x \in \Omega$ with $\displaystyle\lim_{t \to +\infty} \int_t^{t+1} |w'(\tau, x)|^2 d\tau = 0$ (cf. Lemma 3.2) and $r \in V(x)$. Then, there exists a sequence $\{t_n\}$ such that

$$t_n \longrightarrow +\infty \text{ and } w(t_n, x) \longrightarrow r \quad \text{as } n \longrightarrow +\infty.$$

Fixing x, put

$$w_n(t) := w(t_n + t, x), \quad u_n(t) := u(t_n + t, x) \quad \text{for } 0 \le t \le 1.$$

By Lemma 3.3 and (A2), we may assume that

$$u_n - g(w_n) \longrightarrow u^\infty(x) - g(r) \quad \text{in } L^2(0,1) \text{ as } t \longrightarrow +\infty$$

Now consider a sequence of ODEs:

$$\begin{cases} w_n'(t) + \beta(w_n(t)) + g(w_n(t)) \ni u_n(t) & \text{for } 0 \le t \le 1, \\ w_n(0) = w(t_n, x). \end{cases}$$

By a general theory in [8] again, w_n converges in $C([0,1])$ to the solution $\tilde{w}$ of

$$\begin{cases} \tilde{w}'(t) + \beta(\tilde{w}(t)) + g(\tilde{w}(t)) \ni u^\infty(x) & \text{for } 0 \le t \le 1, \\ \tilde{w}(0) = r. \end{cases} \tag{3.13}$$

But, since $\tilde{w}' \equiv 0$ i.e. $\tilde{w} \equiv r$ by assumption, we see from (3.13) that

$$\beta(r) + g(r) \ni u^\infty(x).$$

Thus, (2) is proved. At last, we show (3). Suppose that $V(x)$ has more than two elements for some $x \in \Omega$, say $r_1, r_2 \in V(x)$, $r_1 < r_2$. By definition, there exists two sequence $\{s_n\}$ and $\{t_n\}$ such that

$$s_n \longrightarrow +\infty, \ w(s_n, x) \longrightarrow r_1,$$

$$t_n \longrightarrow +\infty, \ w(t_n, x) \longrightarrow r_2$$

as $n \longrightarrow +\infty$, and

$$s_n < t_n < s_{n+1} < t_{n+1} \ \text{for } n = 1, 2, 3, \cdots.$$

From the continuity of w with respect to t, for any $r \in (r_1, r_2)$, there exists a sequence $\{\tau_n\}$ with $\tau_n \longrightarrow +\infty (n \longrightarrow +\infty)$ such that

$$s_n < \tau_n < t_n \ \text{for } n = 1, 2, 3, \cdots \text{ and } w(\tau_n, x) = r \ \text{for large } n.$$

This implies that $r \in V(x)$ and hence $[r_1, r_2] \subset V(x)$. This contradicts the assumption that $\beta(r) + g(r) \ni u^\infty(x)$ has a finite number of solutions r in $\overline{D(\beta)}$. Thus, $V(x)$ must be a singleton for a.e. $x \in \Omega$. $\qquad\square$

In particular, (2) and (3) of Lemma 3.4 imply that $w(t, x)$ converges to a solution $w^\infty(x)$ for a.e. $x \in \Omega$ as $t \longrightarrow +\infty$ and the limit w^∞ is a solution of (P^∞). Thus we complete the proof of Theorem 3.1.

References

1. D. Blanchard, A. Damlamian and H. Ghidouche, A nonlinear system for phase change with dissipation, Diff. Int. Eq. **2**(3)(1989), 344-362.

2. D. Blanchard and H. Ghidouche, A nonlinear system for irreversible phase change, Euro. J. Appl. Math. **1**(1990), 91-100.

3. G. Caginalp, An analysis of a phase field model of a free boundary, Arch. Rat. Mech. Anal., **92**(1986), 205-245.

4. A. Damlamian and N. Kenmochi, Asymptotic behavior of solutions to a multi-phase Stefan problem, Japan J. Appl. Math., **3**(1986), 15-35.

5. A. Damlamian, N. Kenmochi and N. Sato, Subdifferential operator approach to a class of nonlinear systems for Stefan problems with phase relaxation, to appear in Nonlinear Anal. TMA.

6. A. Damlamian, N. Kenmochi and N. Sato, Phase field equations with constraints, *"Nonlinear Mathematical Problems in Industry"*, pp. 391-404, Gakuto. Inter. Ser. Math. Sci. Appl. Vol.2, Gakkōtosho, Tokyo, 1993.

7. G. J. Fix, Phase field models for free boundary problems, *Free Boundary Problems: Theory and Applications*, pp.580-589, Pitman Reserch Notes in Math. Ser. Vol. 79, 1983.

8. N. Kenmochi, Solvavility of nonlinear evolution equations with time-dependent constraints and applications, Bull. Fac. Education, Chiba Univ., **30**(1981), 1-87.

9. N. Kenmochi, Systems of nonlinear PDEs arising from dynamical phase transitions, to appear in Lecture Notes Math., Springer.

10. N. Kenmochi and M. Niezgódka, Systems of nonlinear parabolic equations for phase change problems, Adv. Math. Sci. Appl., **3**(1994), 89-117.

11. Ph. Laurençot, A double obstacle problem, to appear in J. Math. Anal. Appl.

12. A. Visintin, Stefan problems with phase relaxation, IMA J. Math., **34**(1985), 225-245.

Naoki SATO, Jun SHIROHZU
Department of Mathematics
Graduate School of Science and Technology
Chiba University
263 Chiba, Japan

Nobuyuki KENMOCHI
Department of Mathematics, Faculty of Education
Chiba University
263 Chiba, Japan

A SIMON
Entire positive solutions for semi-linear elliptic functions

1. Introduction

We consider the following problem

$$\begin{cases} Lu = f(x,u) \\ u \xrightarrow{|x|\to\infty} 0 \qquad (x \in \mathbb{R}^n, \ n > 2), \\ u > 0 \end{cases} \tag{1}$$

where $L = -\Delta + c^2$, $c > 0$, $f(x,0) = 0$, f is superlinear and subcritical. In [4] it was proved for peculiar f, the existence of a non-trivial C^2-solution of (1). The method is based upon an a-priori estimate and a fixed-point theorem in cones of Banach spaces. In this paper the nonlinearity is of a more general form, and we allow to the coefficients to go to zero at infinity in an arbitrary way and not necessarily exponentially. We use the notion of weight suitable to the integral operator associated to the Green function of L introduced in [5].

We recall the principal definitions and properties which are needed here.

We denote by $G(x,y)$ the Green function of L. We have

$$G(x,y) \overset{|x-y|\to\infty}{\sim} d_n \frac{e^{-c|x-y|}}{|x-y|^{(n-1)/2}}.$$

The first derivatives $\dfrac{\partial G}{\partial x_i}(x,y)$ have the same asymptotic behavior as $G(x,y)$. Let us denote by $\underline{G}$ the integral operator with kernel $G(x,y)$, i.e.,

$$\underline{G}\varphi(x) = \int G(x,y)\varphi(y)\,dy.$$

Let $\omega : \mathbb{R}^+ \to [1,+\infty[$ be a C^∞ increasing function such that $\lim\limits_{t\to+\infty} \omega(t) = +\infty$. We put

$$\omega_\delta(x) = [\omega(|x|)]^\delta, \qquad \delta > 0.$$

Definition 1. ω is a weight suitable to $\underline{G}$ if there exist three positive constants a, γ_1, γ_2 such that for every $\delta \in]0, a[$

(i) $0 < \gamma_1 \leq \displaystyle\int G(x,y) \frac{\omega_\delta(x)}{\omega_\delta(y)} \, dy \leq \gamma_2,$

(ii) $\displaystyle\int \left| \frac{\partial G}{\partial x_i}(x,y) \right| \frac{\omega_\delta(x)}{\omega_\delta(y)} \, dy \leq \gamma_2, \ i \in \{1, \ldots, n\}.$

Examples. 1) $\omega(t) = e^t$. ω is suitable to $\underline{G}$ with $a = c$ (cf. [4]).

2) $\omega(t) = 1 + t$. ω is suitable to $\underline{G}$ with a arbitrary.

Let ω be a weight suitable to $\underline{G}$ and $\delta < a$. We consider the Banach space

$$C^k_{\omega_\delta} = \left\{ u \in C^k(\mathbb{R}^n), \ \sup_{x \in \mathbb{R}^n} |D^\alpha u(x)| \omega_\delta(x) < \infty, \ |\alpha| \leq k \right\}.$$

The associated norm is

$$\|u\|_{k,\delta} = \sum_{|\alpha| \leq k} \sup_{x \in \mathbb{R}^n} |D^\alpha u(x)| \omega_\delta(x).$$

We have the fundamental properties:

Proposition 1. 1) *The injection* $C^k_{\omega_{\delta_1}} \to C^l_{\omega_{\delta_2}}$ *is continuous for* $k \geq l$ *and* $\delta_1 \geq \delta_2$.

2) *The injection* $C^{k+1}_{\omega_{\delta_1}} \to c^k_{\omega_{\delta_2}}$ *is compact for* $\delta_1 > \delta_2$.

Proposition 2. *The map* $\underline{G} : C^0_{\omega_\delta} \to C^1_{\omega_\delta}$ *is continuous.*

2. The method

First we give the assumptions on the nonlinearity.

(H_1) $f : \left\{ \begin{array}{l} (x,t) \mapsto f(x,t) \\ \mathbb{R}^n \times \mathbb{R}^+ \to \mathbb{R}^+ \end{array} \right.$ is locally höldercontinuous.

(H_2) $f(x,0) = 0$.

(H_3) $t \mapsto f_t(x,t)$ is continuous and increasing.

(H_4) There exist a weigth ω suitable to $\underline{G}$ and $\delta < a/2$, $\varphi_1, \varphi_2 \in C^0_{\omega_\delta}$, $\alpha_\tau > 0$ such that

$$0 \leq \varphi_1(x)t^{k-1} \leq f_t(x,t) \leq \varphi_2(x)t^{k-1}, \quad 1 < k < \frac{n+2}{n-2},$$

$0 < \alpha_\tau \leq \omega_\delta(x)f(x,\tau)$, τ constant, $\neq 0$.

(H_5) $f(x,t) = f(|x_1| \searrow, \ldots, |x_n| \searrow, t)$.

(H_6) $\displaystyle\lim_{t \to \infty} \frac{f(x,t)}{t^k} = h(x)$, h continuous.

(H) denotes the whole set of hypotheses (H_1)–(H_6).

Example. $f(x,t) = a(x)t^k$, $1 < k < (n+2)/(n-2)$, with $a(x)$ locally hölderconti-nuous, $a(x) > 0$, $a \in C^0_{\omega_\delta}$ and $0 < \alpha \leq a(x)\omega_\delta(x)$, a satisfies (H_5).

Let $\delta' < \delta$ fixed (where δ is given by (H_4)), $E = C^0_{\omega_{\delta'}}$ and K the cone of nonnegative functions in E. With (H), the problem (1) is equivalent to

$$u = \Phi(u), \qquad u \in K, \tag{2}$$

with

$$\Phi u(x) = \int G(x,y)f(y,u(y))\, dy.$$

We want to prove that Φ has a non-trivial fixed point in K.

In [5] we prove the existence of an a-priori estimate for the C^2-solutions of (1).

Proposition 3. *With* (H), *there exists a constant* $M > 0$ *such that for every* C^2-*solution of* (1), *we have*

$$u \leq M.$$

We need also the following problem

$$\begin{cases} Lu = f(x, u + \tau) \\ u \xrightarrow{|x| \to \infty} 0 \\ u > 0 \end{cases} \qquad (\tau \text{ positive constant}). \tag{1'}$$

One proves

Proposition 3'. *With* (H) *there exists a constant* $M_1 > 0$ *such that for every* C^2-*solution of* (1'), *we have*

$$u \leq M_1.$$

With the propositions 3 and 3' we get estimates in the space E for the solutions of (2) and (2'), with (2') defined by

$$u = \Phi(u + \tau), \qquad u \in K. \tag{2'}$$

Proposition 4. *With* (H) *there exists a constant* $N > 0$ *such that for every solution of* (2)

$$\|u\|_E \leq N.$$

Proof. Let u be a solution of (2). Then

$$
\begin{aligned}
(u\omega_{\delta'})(x) &= \int G(x,y)\omega_{\delta'}(x)f(y,u(y))\,dy \\
&\leq \int \omega_{\delta'}(x)G(x,y)\varphi_2(y)[u(y)]^k\,dy \\
&\leq \int \frac{\omega_{\delta'}(x)}{\omega_\delta(y)}G(x,y)[\varphi_2(y)\omega_\delta(y)][u(y)]^k\,dy \\
&\leq \|\varphi_2\|_{0,\delta}M^k\gamma_2
\end{aligned}
$$

(by the properties of ω, the definition 1, (H) and the proposition 3). Then we have

$$\|u\|_E \leq \|\varphi_2\|_{0,\delta}M^k\gamma_2 = N.$$

Proposition 4'. *With* (H), *there exists a constant* $N_1 > 0$ *such that for every solution of* (2') *we have*

$$u \leq N_1.$$

Proof. The proof is similar to the preceding one:

$$
\begin{aligned}
(u\omega_{\delta'})(x) &= \int \omega_{\delta'}(x)G(x,y)f(y,u(y)+\tau)\,dy \\
&\leq \int \omega_{\delta'}(x)G(x,y)\{\varphi_2(y)\tau^k + u(y)\varphi_2(y)[u(y)+\tau]^{k-1}\}\,dy \\
&\leq \|\varphi_2\|_{0,\delta}[\tau^k + M(M+\tau)^{k-1}]\gamma_2 = N_1.
\end{aligned}
$$

We recall the fixed-point theorem that we use. Let E be any Banach space, and K a cone in E. Let us write

$$\overline{B}_\rho = \{u \in K, \|u\| \leq \rho\}, \qquad \forall\, \rho > 0,$$

and finally let r and R be two positive numbers with $r < R$.

Theorem 1 [4][5]. *Let $\Phi : \overline{B}_R \to K$ a compact map such that*

(i) $u \neq s\Phi(u)$, $s \in [0,1]$, $u \in K$, $\|u\| = r$,

(ii) *there exist $\tau_0 > 0$ and $\Psi : [0, \tau_0] \times K \to K$ compact sucht that $\Psi(0, u) = \Phi(u)$,* $u \in K$, $\|u\| \leq R$,

and

(a) $u \neq \Psi(\tau, u)$ *for* $\tau = \tau_0$, $u \in K$, $\|u\| \leq R$,

(b) $u \neq \Psi(\tau, u)$ *for* $0 \leq \tau \leq \tau_0$, $u \in K$, $\|u\| = R$.

Then Φ has, at least, two fixed points u_1, u_2 with $\|u_1\| < r < \|u_2\| < R$.

3. Existence result

Theorem 2. *If f satifies* (H), *the problem* (1) *has a C^2 nontrivial solution. This solution is exponentially decreasing.*

Proof. We apply the theorem 1 to the nonlinear integral equation (2). In fact, we have to prove that we can choose r, R, τ_0, Ψ such that all the conditions of this theorem are fulfilled.

First $\Phi : \overline{B}_\rho \to K$ is compact for every $\rho > 0$. We have $\Phi(K) \subset K$ (by the proof of the proposition 4), and Φ can be written as

$$\Phi : u \mapsto f(x, u(x)) \overset{G}{\mapsto} \underline{G}[f(x, u(x)] \overset{i}{\mapsto} \underline{G}[f(x, u(x)]$$

$$K \longrightarrow C^0_{\omega_\delta} \longrightarrow C^1_{\omega_\delta} \longrightarrow C^0_{\omega_{\delta'}} = E.$$

All the maps are continuous, the last one being compact (by the propositions 1 and 2); thus every bounded set in K is transformed in a relatively compact set in E.

We prove the following

(i)* There exists $r > 0$ such that

$$u \neq s\Phi(u), \quad s \in [0,1], \quad u \in K, \quad \|u\|_E = r.$$

Proof of (i)*. We suppose the contrary. Then

$$(u\omega_{\delta'})(x) = s\int \omega_{\delta'}(x)G(x,y)f(y,u(y))\,dy$$

$$\leq \int \frac{\omega_{\delta'}(x)}{\omega_{\delta}(y)}G(x,y)[\varphi_2(y)\omega_{\delta}(y)][u(y)]^k\,dy$$

$$\leq \gamma_2\|\varphi_2\|_{0,\delta}r^k.$$

We choose r so small that

$$r^k \leq \frac{r}{2\|\varphi_2\|_{0,\delta}\gamma_2}.$$

So we get $r \leq r/2$. This gives the contradiction.

Now r is fixed such that (i)* is satisfied.

Next we prove

(ii)* There exist $R > r$, $\tau_0 > 0$ such that

    (a) $u \neq \Phi(u + \tau)$, $u \in K$, $\tau \geq \tau_0$,

    (b) $u \neq \Phi(u + \tau)$, $u \in K$, $0 \leq \tau \leq \tau_0$, $\|u\| = R$.

Obviously (ii)* implies (ii) of theorem 1 with

$$\psi(\tau, u) = \Phi(u + \tau).$$

Proof of (a). Let us notice that the solutions of (2′) are in fact in $C^0_{\omega_\delta}$. We have

$$(u\omega_\delta)(x) = \int \omega_\delta(x)G(x,y)f(y,u(y)+\tau)\,dy$$

$$\leq \gamma_2\|\varphi_2\|_{0,\delta}[\tau^k + M(M+\tau)^{k-1}]$$

by the proof of the proposition 4.

Now we prove that there exists no positive solution of $u = \Phi(u + \tau)$ in $C^0_{\omega_\delta}$, for τ "large". We assume the contrary. Then

$$u(x) = \int G(x,y)f(y,u(y)+\tau)\,dy$$

$$\geq G(x,y)[f(y,\tau) + u(y)\varphi_1(y)\tau^{k-1}]\,dy$$

or

$$u \geq \underline{G}f(x,\tau) + \tau^{k-1}\underline{G}(\varphi_1 u),$$

where $\geq$ is the order induced in $C^0_{\omega_\delta}$ by $\widetilde{K}$, cone of nonnegative functions. We put

$$\underline{G}_{\varphi_1}(u) = \underline{G}(\varphi_1 u).$$

Then $\underline{G}_{\varphi_1}$ is a linear positive compact operator from $C^0_{\omega_\delta}$ into itself. One can see this fact by writing the decomposition

$$u \mapsto \varphi_1 u \overset{\underline{G}}{\mapsto} \underline{G}(\varphi_1 u) \overset{i}{\mapsto} \underline{G}(\varphi_1 u) = \underline{G}_{\varphi_1}(u)$$

$$C^0_{\omega_\delta} \longrightarrow C^0_{\omega_{2\delta}} \longrightarrow C^1_{\omega_{2\delta}} \longrightarrow C^0_{\omega_\delta}.$$

Every map is continuous (as $2\delta < a$); the last is compact.

Let $(\widetilde{K})^*$ be the dual cone of K in $(C^0_{\omega_\delta})^*$, topological dual of $C^0_{\omega_\delta}$. By definition

$$(\widetilde{K})^* = \{w \in (C^0_{\omega_\delta})^*;\ (w, z) \geq 0,\ \forall z \in \widetilde{K}\},$$

where $(\ ,\ )$ denotes the duality between $C^0_{\omega_\delta}$ and its dual. We can prove that there exists e_1 in $(\widetilde{K})^*$ and $\lambda_1 > 0$ such that

$$\underline{G}^*_{\varphi_1} e_1 = \frac{1}{\lambda_1} e_1,$$

where $\underline{G}^*_{\varphi_1}$ is the conjugate operator of $\underline{G}_{\varphi_1}$. This results from a theorem of Krein-Rutman ([2], [5]). Now we choose $\tau_0 > (1/\lambda_1)^{1/(k-1)}$. Then, for $\tau \geq \tau_0$, we would have

$$(u, e_1) \geq (\underline{G}f(x, \tau), e_1) + \tau^{k-1}(\underline{G}_{\varphi_1} u_1, e_1).$$

Then

$$(u, e_1)\left[1 - \frac{\tau^{k-1}}{\lambda_1}\right] \geq (\underline{G}f(x, \tau), e_1). \tag{3}$$

But $f(x, \tau)$ is in the interior of $\widetilde{K}$ by (H$_4$) and $\underline{G}f(x, \tau)$ has the same property by the minoration in the definition 1. Then we have

$$(\underline{G}f(x, \tau), e_1) > 0.$$

But the left-hand side of (3) is nonpositive because $(u, e_1) \geq 0$ and $1 - \tau^{k-1}/\lambda_1 < 0$. Thus we get a contradiction.

Corollary. *There exists no solution of $u = \Phi(u + \tau)$ in K for $\tau \geq \tau_0$.*

We have proved (a).

Proof of (b). Now τ is fixed in $[0, \tau_0]$. By the proposition 4', there exists an a-priori estimate for the solutions of (2'), namely $\|u\|_{0,\delta'} \leq N_1$. We prove that this estimate is uniform with respect to τ. Then let R be a number strictly larger than N_1. Then (b) is satisfied.

Finally, Φ has two fixed points. As $u \equiv 0$ is a trivial solution in B_r, we get that the fixed point u_2 gives a non-trivial solution of (2), which gives a non-trivial C^2-solution of (1) in $C^0_{\omega_\delta}$. To get the exponential decay of the solution, one can give a proof similar to this of the proposition 4.1 in [1].

References

1. B. Gidas, W.M. Ni, L. Nirenberg: *Symmetry of positive solutions of nonlinear elliptic equations in* $\mathbb{R}^n$. Advances in Math. Studies A **7** (1981), 369–402.

2. M.G. Krein, M.A. Rutman: *Opérateurs linéaires laissant un cône invariant dans un espace de Banach*. Uspehi Mat. Nauk **3**, n.1 (23) (1948), 3–95. Translations of AMS **10** (1962), 199–325.

3. A. Simon: *An elliptic problem in* $\mathbb{R}^n$ *with a small perturbation*, in Progress in Partial Differential Equations, The Metz Surveys 2, Pitman Research Notes in Mathematics **296**, Longman, Harlow (1993), 194–200.

4. A. Simon (Chaljub-Simon), P. Volkmann: *Existence of ground states with exponential decay for semilinear elliptic equations in* $\mathbb{R}^n$. J. Diff. Equ. **76** (1988), 374–390.

5. A. Simon, P. Volkmann: *Existence de deux solutions positives pour un problème elliptique à paramètre dans* $\mathbb{R}^n$. A paraître.

Département de Mathématiques, Université d'Orléans,

B.P. 6759, 45067 Orléans Cedex 2, France

I V SKRYPNIK

On quasilinear parabolic higher order equations with Hölder continuous solutions

It is introduced a class of quasilinear parabolic higher order equations whose generalized solutions are Hölder continuous functions. A model equation of this class is
the equation

$$\frac{\partial u}{\partial t} + \sum_{|\alpha|=m} (-1)^{|\alpha|} D^\alpha [|D^m u|^{p-2} D^\alpha u] - \sum_{i=1}^{n} \frac{\partial}{\partial x_i} [|\frac{\partial u}{\partial x}|^{q-2} \frac{\partial u}{\partial x_i}] = f(x,t) \quad (1)$$

if $p \geq 2$, $q > mp$. For quasilinear parabolic equation with Caratheodory's coefficients they are obtained interior regularity of solutions, regularity of solutions near
the boundary for Dirichlet or Neumann boundary conditions. It is established that
the solutions belong to a class of functions $B_{q,s}$ which generalizes the corresponding
classes' of O.A. Ladyzhenskaya and N.N.Uraltzeva ($q = 2$, $s = 1$) [3] and of E. di
Benedetto ($s = 1$) [2]. The imbeddings of classes $B_{q,s}$ in spaces of Hölder continuous functions are established. Corresponding results about interior regularity were
published in papers [4,5], regularity near the boundary the author proved together
with prof. F.Nicolosi.

1. Statement of results about regularity of solutions

In the sequel Ω is a bounded open set in n- dimensional Euclidean space R^n, $n \geq 1$.
We will consider the regularity of bounded solutions of equation

$$\frac{\partial u}{\partial t} + \sum_{|\alpha|\leq m} (-1)^{|\alpha|} D^\alpha A_\alpha(x,t,u,\ldots,D^m u) = 0, \quad (x,t) \in \Omega \times [0,T). \quad (2)$$

Here $x = (x_1,\ldots,x_n)$, $\alpha = (\alpha_1,\ldots,\alpha_n)$ is a vector with nonnegative integer compo-
nents, $|\alpha| = \alpha_1 + \cdots + \alpha_n$ and

$$D^\alpha = (\frac{\partial}{\partial x_1})^{\alpha_1} \ldots (\frac{\partial}{\partial x_n})^{\alpha_n}, \quad D^k u = \{D^\alpha u : |\alpha| = k\}. \quad (3)$$

We suppose that $A_\alpha(x,t,\xi)$ are defined for $(x,t) = Q_T = \Omega \times (0,T)$, $\xi = \{\xi_\gamma :
|\gamma| \leq m\}$, $\xi_\gamma \in R^1$ and they are satisfied for all values of arguments next conditions:

1) $A_\alpha(x,t,\xi)$ are Caratheodory's functions, i.e. measurable functions with respect to x,t for all ξ and they are continuous functions with respect to ξ for almost all values of x,t;

2) the inequalities

$$\sum_{1\leq|\alpha|\leq m} A_\alpha(x,t,\xi)\xi_\alpha \geq C' \sum_{|\alpha|=m} |\xi_\alpha|^p + C' \sum_{|\alpha|=1} |\xi_\alpha|^q - C'' \sum_{|\alpha|=2}^{m-1} |\xi_\alpha|^{p_\alpha} - f(x,t) \quad (4)$$

$$|A_\alpha(x,t,\xi)| \leq C'' \sum_{1\leq|\beta|\leq m} |\xi_\beta|^{p_{\alpha\beta}} + f_\alpha(x,t), \ |\alpha| \leq m \quad (5)$$

are valid with some positive constants C', C'' for $p \geq 2$, $q > mp$. By this the numbers p_α and $p_{\alpha\beta}$ in (4),(5) are defined by equalities

$$\frac{1}{p_\alpha} = \frac{|\alpha|-1}{m-1} \cdot \frac{1}{p} + \frac{m-|\alpha|}{m-1} \cdot \frac{1}{q_1} \text{ for } 1 < |\alpha| \leq m,$$

$$p_\alpha = q \text{ for } |\alpha| = 1, \ p_{\alpha\beta} = p_\beta(1 - \frac{1}{p_\alpha}) \text{ if } |\alpha| \geq 1, \ p_{\alpha\beta} = p_\beta \text{ for } |\alpha| = 0 \quad (6)$$

and the number q_1 satisfies the inequality $mp < q_1 < q$.

The functions $f(x,t)$ and $f_\alpha(x,t)$ in (4),(5) satisfy the condition

$$F(x,t) = |f(x,t)| + \sum_{1\leq|\alpha|\leq m} |f_\alpha(x,t)|^{\frac{p_\alpha}{p_\alpha-1}} + |f_0(x,t)| \in L_{\rho_0,r_0}(Q_T) \quad (7)$$

where $\rho_0 \geq 1$, $r_0 \geq 1$ and

$$\frac{1}{r_0} + \frac{n}{q\rho_0} = 1 - \kappa_1; \quad (8)$$

here $\kappa_1 \in (0, \frac{q-1}{q})$ if $n = 1$, $\kappa_1 \in (0,1)$ if $n > 1$, $q \leq n$ and $\kappa_1 \in (\frac{q-n}{q},1)$ if $1 < n \leq q$.

Under these counditions we study the Hölder continuity of generalized solution of equation (2). By such solution we mean a function

$$u(x,t) \in V_{2,p,q}^{m,1}(Q_T) = C(0,T;L_2(\Omega)) \cap L_p(0,T;W_p^m(\Omega)) \cap L_q(0,T;W_q^1(\Omega))$$

satisfying the integral identity

$$\int_\Omega u(x,\tau)\varphi(x,\tau)dx\Big|_{t_1}^{t_2} +$$

$$+ \int_{t_1}^{t_2} \int_\Omega \{-u(x,\tau)\frac{\partial\varphi(x,\tau)}{\partial\tau} + \sum_{|\alpha|\leq m} A_\alpha(x,\tau,u(x,\tau),\ldots,D^m u(x,\tau))\}dxd\tau = 0 \quad (9)$$

for all $\varphi(x,t) \in \overset{\circ}{V}{}^{m,1}_{2,p,q}(Q_T)$ such that $\frac{\partial \varphi(x,t)}{\partial t} \in L_2(Q_T)$ and for all t_1, t_2, $0 < t_1 < t_2 < T$. Here $\overset{\circ}{V}{}^{m,1}_{2,p,q}(Q_T) = C(0,T; L_2(\Omega)) \cap L_p(0,T; \overset{\circ}{W}{}^m_p(\Omega)) \cap L_q(0,T; \overset{\circ}{W}{}^1_q(\Omega))$.

Using the Moser's method it is possible to prove the boundedness of solution analogously to [6]. So we will suppose that the estimate

$$\mathrm{ess\ sup}\{|u(x,t)| : (x,t) \in Q_T\} \leq M \tag{10}$$

is valid for some constant M.

Under these conditions we proved the following theorem.

Theorem 1. *Let $u(x,t) \in V^{m,1}_{2,p,q}(Q_T)$ be a generalized solution of equation (2) satisfying (10) and let $A_\alpha(x,t,\xi)$ satisfy the conditions 1), 2). Then $u(x,t)$ is locally Hölder continuous in Q_T and for each cylinder $Q_R(x_0,t_0) = B(x_0,R) \times (t_0 - R^q, t_0)$ such that $\overline{Q_R(x_0,t_0)} \subset Q_T$ there exist constants A and $\alpha \in (0,1)$ such that*

$$\mathrm{ess\ osc}\{u(x,t) : (x,t) \in Q_R(x_0,t_0)\} \leq AR^\alpha. \tag{11}$$

The constant α depends only on M and on the constants in conditions (4)-(8); the constant A depends only on the same parameters and on the distance from $Q_R(x_0,t_0)$ to $\Gamma_T = \{\partial\Omega \times (0,T)\} \cup \{\Omega \times \{0\}\}$.

Now we will formulate results about smoothness of solution of equation (2) by initial data

$$u(x,0) = u_0(x), \quad x \in \Omega \tag{12}$$

and Dirichlet or Neumann boundary conditions.

Let $g(x,t)$ be a function belonging to the space $V^{m,1}_{2,p,q}(Q_T)$ such that $G(x,t) \in L_{\rho_0,r_0}(Q_T)$ with the same ρ_0, r_0 as in (7) where

$$G(x,t) = |\frac{\partial g(x,t)}{\partial t}| + G_0(x,t) + \sum_{|\alpha|=1}^{m} [G_\alpha(x,t)]^{\frac{p_\alpha}{p_\alpha - 1}},$$

$$G_\alpha(x,t) = |A_\alpha(x, g(x,t), \ldots, D^m g(x,t)|$$

and we consider the problem concerning to find solution $u(x,t)$ of equation (0,1) by initial condition (12) if $u_0(x) \in L_2(\Omega)$ and Dirichlet condition

$$u(x,t) - g(x,t) \in \overset{\circ}{V}{}^{m,1}_{2,p,q}(Q_T) \tag{13}$$

The boundary $\partial\Omega$ is assumed to satisfy the condition

A) There exist numbers $\delta_0 \in (0,1)$, $R_0 > 0$ such that for every point $x_0 \in \partial\Omega$ and every ball $B(x_0, R)$ centered at x_0 with radius $R \leq R_0$ the inequality

$$\text{meas}\{\Omega \cap B(x_0, R)\} \leq (1 - \delta_0)\text{meas} B(x_0, R) \tag{14}$$

holds.

Theorem 2. *Let $u(x,t) \in V_{2,p,q}^{m,1}(Q_T)$ be a generalized solution of problem (2), (12), (13) and assume that conditions 1) - 2), condition A) and inequality (10) hold. Let $u_0(x)$ is Hölder continuous on $\overline{\Omega}$, $g(x,t)$ is Hölder continuous on $S = \partial\Omega \times [0,T]$ and $u_0(x) = g(x,o)$ for $x \in \partial\Omega$. Then $u(x,t)$ is Hölder continuous on $\overline{Q}_T$.*

By study of smoothness of solution of initial - Neumann boundary value problem we will suppose that the boundary $\partial\Omega$ belongs to class C^m.

We will tell that the function $u(x,t) \in V_{2,p,q}^{m,1}(Q_T)$ is a generalized solution of initial - Neumann boundary value problem for equation (2) with initial data $u_0(x)$ if the following integral identity

$$\int_{\Omega} u(x,t)\varphi(x,t)dx - \int_{\Omega} u_0(x)\varphi(x,0)dx+$$

$$+ \int_0^t \int_{\Omega} \{-u(x,\tau)\frac{\partial\varphi(x,\tau)}{\partial\tau} + \sum_{|\alpha|\leq m} A_\alpha(x,\tau,u(x,\tau),\ldots D^m u(x,\tau))D^\alpha \varphi(x,\tau)\}dxd\tau =$$

$$= \int_0^t \int_{\partial\Omega} \sum_{|\beta|\leq m-1} B_\beta(x,\tau,u(x,\tau),\ldots,D^{m-1}u(x,\tau))D^\beta \varphi(x,\tau)d\omega d\tau$$

$$\tag{15}$$

holds, for all $\varphi(x,t) \in V_{2,p,q}^{m,1}(Q_T)$ such that $\frac{\partial\varphi(x,t)}{\partial t} \in L_2(Q_T)$ and for all $t \in (0,T]$. Here $B_\beta(x,t,\eta)$ $|\beta| \leq m-1$, $(x,t) \in S_T = \partial\Omega \times [0,T]$, $\eta = \{\eta_\gamma : |\gamma| \leq m-1\}$, $\eta_\gamma \in R^1$ are Caratheodory's functions satisfying the inequalities

$$|B_\beta(x,t,\eta)| \leq C''' \sum_{|\gamma|=1}^{m-1} |\eta_\gamma|^{q_{\beta\gamma}} + C''' \tag{16}$$

where C''' is given positive constant, $q_{\beta\gamma}$ are defined by conditions

$$q_{\beta\gamma} = \tilde{p}_\gamma[1 - \frac{1}{\tilde{p}_\beta} - \frac{q_1 - p}{(m-1)q_1 q}] \text{ if } 1 \le |\beta| \le m - 1$$

$$q_{\beta\gamma} = 1 + p_\gamma[1 - \frac{|\gamma|}{(m-1)p} - \frac{m - |\gamma| - 1}{(m-1)q_1}] \text{ if } |\beta| = 0 \tag{17}$$

and $\tilde{p}_\gamma = p_\gamma$ for $|\gamma| > 1$, $\tilde{p}_\gamma = q_1$ for $|\gamma| = 1$, where p_γ, q_1 are the numbers from (6).

Theorem 3. *Let* $u(x,t) \in V_{2,p,q}^{m,1}(Q_T)$ *be a generalized solution of initial - Neumann boundary value problem for equation (2) and assume that conditions 1), 2), inequalities (10), (16), (17) hold and* $\partial\Omega \in C^m$. *Let* $u_0(x)$ *is Hölder continuous function on* $\overline{\Omega}$. *Then* $u(x,t)$ *is Hölder continuous on* $\overline{Q}_T$.

The theorem 1-3 will be followed from theorems of section 4 in order that solutions of equation (1) or corresponding initial - boundary value problems belong to classes $B_{q,s}$ and from imbedding theorems for classes $B_{q,s}$ in section 3.

2. Remarks about introduced class of equations

Introduced class of quasilinear parabolic equations with condition of parabolicity in form (4) is generalized corresponding class of elliptic equations from the paper [5] of the author.

In [5] was introduced the class of quasilinear elliptic equations

$$\sum_{|\alpha| \le m} (-1)^{|\alpha|} D^\alpha A_\alpha(x, u, \dots, D^m u) = 0 \tag{18}$$

with Caratheodory's coefficients $A_\alpha(x, \xi)$ satisfying the inequality of type (4). By such assumptions in [5] the results were proved about Hölder continuity of solutions of equation (18). In [5] the contrexamples were given showing that the condition $q > mp$ is necessary.

Let Ω be the unit ball with center at the origin. Straightforward calculations show that $u_1(x) = \ln|x|$ satisfies the equation

$$\sum_{k,l=1}^{n} \frac{\partial^2}{\partial x_k \partial x_l} \{ \sum_{i,j=1}^{n} (\frac{x_k x_l}{|x|^2} + \sigma_1 \delta_k^l)(\frac{x_i x_j}{|x|^2} + \sigma_1 \delta_i^j) \frac{\partial^2 u}{\partial x_i \partial x_j} +$$

$$+ \sigma_2 \frac{\partial^2 u}{\partial x_k \partial x_l} \} - 2\sigma_3 \sum_{i=1}^{n} \frac{\partial}{\partial x_i} [\sum_{j=1}^{n} (\frac{\partial u}{\partial x_j})^2 \frac{\partial u}{\partial x_i}] = 0 \tag{19}$$

if the constants σ_1, σ_2, σ_3 satisfy the relation

$$2\sigma_2(n-2) + 2\sigma_3 = [(n-2)\sigma_1 - 1][n - 3 - 2\sigma_1]. \tag{20}$$

It can be verified that σ_1 and sufficiently small positive σ_2, σ_3 can be chosen so that (20) will be satisfied. In our example $u_1(x) \in W_2^2(\Omega) \cap W_4^1(\Omega)$ if $n > 4$ and $u_1(x)$ is a generalized solution of equation (18). Coefficients of equation (18) satisfy all conditions of assumptions of type 1)-2) except $q > mp$ (here $q = mp = 4$). So this example shows that the inequality $q > mp$ is essential for smoothness of solution in elliptic case. And in the same time this contr-example shows that the conditions 1)-2) are essential in considered case of parabolic equations.

3. Classes $B_{q,s}$, $B_{q,s}^{(D)}$, $B_{q,s}^{(N)}$

We say that a mesurable function $u(x,t)$, $(x,t) \in Q_T$ belongs to the class $B_{q,s}(Q_T, M, \gamma, r, \delta, b, \kappa)$ if

$$u(x,t) \in V_{2,q}(Q_T) = C(0,T; L_2(\Omega)) \cap L_q(0,T; W_q^1(\Omega)) \tag{21}$$

$$\text{ess sup}\{|u(x,t)| : (x,t) \in Q_T\} \leq M \tag{22}$$

and for arbitrary points $(x_0, t_0) \in Q_T$ and positive numbers R and Θ such that

$$Q(R, \Theta) \equiv Q(x_0, t_0; R, \Theta) = B(x_0, R) \times (t_0 - \Theta, t_0) \subset \overline{Q(R, \Theta)} \subset Q_T$$

and for an arbitrary infinitely differentiable nondecreasing function $\eta(t)$ on R^1 the following inequalities hold:

$$\sup_{t_0 - \Theta \leq t \leq t_0} \int_{B(x_0, R - \sigma R)} [u(x,t) - k]_{\pm}^{s+1} \eta^q(t) dx +$$

$$+ \int_{t_0 - \Theta}^{t_0} \int_{B(x_0, R - \sigma R)} [u(x,\tau) - k]_{\pm}^{s-1} \left| \frac{\partial u(x,\tau)}{\partial x} \right|^q \eta^q(\tau) dx d\tau \leq$$

$$\leq \int_{B(x_0, R)} [u(x, t_0 - \Theta) - k]_{\pm}^{s+1} \eta^q(t_0 - \Theta) dx + \tag{$23_\pm$}$$

$$+ \gamma \left\{ \frac{1}{(\sigma R)^q} \iint_{Q(R,\Theta)} [u(x,\tau) - k]_{\pm}^{s+q-1} \eta^q(\tau) dx d\tau + \right.$$

$$+ \iint_{Q(R,\Theta)} [u(x,\tau) - k]_{\pm}^{s+1} \eta^{q-1}(\tau) \frac{d\eta(\tau)}{d\tau} dx d\tau + \left[\int_{t_0 - \Theta}^{t} [\text{mes} A_{k,R}^{\pm}(\tau)]^{\frac{r}{\rho}} d\tau \right]^{\frac{q}{r}(1+\kappa)} \right\}.$$

$$\sup_{t_0-\Theta\le t\le t_0} \int_{B(x_0,R-\sigma R)} [\ln \frac{H^\pm}{H^\pm \mp (u(x,t)-k)+\nu}]_+^{s+1} dx \le$$

$$\le \int_{B(x_0,R)} [\ln \frac{H^\pm}{H^\pm \mp (u(x,t_0-\Theta)-k)+\nu}]_+^{s+1} dx +$$

$$+\gamma\{\frac{1}{(\sigma R)^q} \iint_{Q(R,\Theta)} [\ln \frac{H^\pm}{H^\pm \mp (u(x,\tau)-k)+\nu}]_+^s [\frac{1}{H^\pm \mp (u(x,\tau)-k)+\nu}]^{2-q} dx\,d\tau +$$

$$+ \frac{1}{\nu^b}[1+(\ln \frac{H^\pm}{\nu})^s]\{\int_{t_0-\Theta}^{t_0} [\text{mes} A_{k,R}^\pm(\tau)]^{\frac{r}{\rho}}\}^{\frac{q}{r}(1+\kappa)}\}.$$

$$(24_\pm)$$

Here $s, \gamma, r, \rho, \delta, b, \kappa$ and σ are given positive numbers satisfying only the restrictions $k, \sigma \in (0,1)$, $b \le s$, $r, \rho > 1$ and

$$\frac{1}{r} + \frac{n}{\rho q} = \frac{n}{q^2} \tag{25}$$

and the possible values of r and ρ are limited by the conditions

$$\rho \in (q,\infty], \ r \in [q^2,\infty) \text{ if } n = 1,$$

$$\rho \in [q, \frac{nq}{n-q}], \ r \in [q,\infty] \text{ if } n > q > 1,$$

$$\rho \in [q,\infty), \ r \in (\frac{q^2}{n},\infty] \text{ if } 1 < n \le q.$$

The following notations are also used in $(23)_\pm$, $(24)_\pm$:

$$[u(x,t)-k]_\pm = \max\{\pm(u(x,t)-k),0\}, \tag{26}$$

$$A_{k,R}^\pm(t) = \{x \in B(x_0,R) : \pm(u(x,t)-k) > 0\} \tag{27}$$

and $[\ln \frac{H^\pm}{H^\pm\mp(u(x,t)-k)+\nu}]_+$ is understood by analogy with (26). In $(23)_\pm$, $(24)_\pm$ k is arbitrary real number satisfying the condition

$$\text{ess sup}\{(u(x,t)-k)_\pm : (x,t) \in Q(R,\Theta)\} \le \delta \tag{28}$$

and $H^\pm$ and ν are positive numbers such that

$$\text{ess sup}\{(u(x,t)-k)_\pm : (x,t) \in Q(R,\Theta)\} \le H^\pm \le \delta, \ \nu \le \min\{H^\pm,1\} \tag{29}$$

The following theorem is a major importance for proving Theorem 1.

209

Theorem 4. *Assume that $u(x,t)$ is an arbitrary function of class $B_{q,s}(Q_T, M, \gamma, r, \delta,$ $b, k)$, $q \geq 2$. Then the inequality (11) is valid for each cylinder $Q_R(x_0, t_0)$ such that $\overline{Q_R(x_0, t_0)} \subset Q_T$. By this the positive constant α depends only on $q, s, M, \gamma, r, \delta, b, \kappa$ and A depends from the same parameters and, in addition, on the distance between $Q_R(x_0, t_0)$ and Γ_T.*

For study of behaviour of solutions of equation (1) near Γ_T we will introduce some analogs of class $B_{q,s}$.

We will tell that a function $u(x,t)$, $(x,t) \in Q_T$ belongs to the class $B_{q,s}^{(D)}(Q_T \cup \Gamma_T, M, \gamma, r, \delta, b, \kappa)$ if the conditions (21), (22) are valid and for arbitrary point $(x_0, t_0) \in Q_T \cup \Gamma_T$, positive numbers R, Θ, infinitely differentiable nondecreasing on R^1 function $\eta(t)$ such that $0 \leq \eta(t) \leq 1$ the inequalities $(23)_\pm$, $(24)_\pm$ hold if $\Theta \leq t_0 \leq T$ and $s, \gamma, r, \rho, \delta, b, \kappa, \sigma, k, H^\pm, \nu$ are limited by the same conditions as in (25)-(29) and, in addition, k is such that

$$[u(x,t) - k]_\pm = 0, \text{ for } (x,t) \in Q(R, \Theta) \cap S_T \tag{30}$$

Theorem 5. *Assume that the condition A is fulfilled and $u(x,t)$ is an arbitrary function from class $B_{q,s}^{(D)}(Q_T \cup \Gamma_T, M, \gamma, r, \delta, b, \kappa)$, $q \geq 2$ which satisfies the inequality*

$$|u(x_1, t_1) - u(x_2, t_2)| \leq B[|x_1 - x_2| + |t_1 - t_2|^{\frac{1}{q}}]^\beta \tag{31}$$

with $B > 0$, $\beta \in (0, 1]$ for every pair of points (x_1, t_1), $(x_2, t_2) \in \Gamma_T$. Then there exist positive numbers A, α depending only on $q, s, M, \gamma, r, \delta, b, \kappa, B, \beta, R_0, \sigma_0$ such that the inequality (11) holds for every pair of points (x_1, t_1), $(x_2, t_2) \in \overline{Q}_T$.

We will tell that a measurable function $u(x,t)$, $(x,t) \in Q_T$ belongs to the class $B_{q,s}^{(N)}(Q_T \cup \Gamma_T, M, \gamma, r, \delta, b, \kappa)$ if the conditions (21)-(29) are valid for arbitrary point $(x_0, t_0) \in Q_T \cup \Gamma_T$ with the same as in (25)-(29) values of parameters and with the same function $\eta(t)$.

Theorem 6. *Assume that $\partial \Omega$ is C^1 manifold and $u(x,t)$ is an arbitrary function from class $B_{q,s}^{(N)}(Q_T \cup \Gamma_T, M, \gamma, r, \delta, b, \kappa)$, $q \geq 2$ which satisfies the inequality*

$$|u(x_1, 0) - u(x_2, 0)| \leq B|x_1 - x_2|^\beta, \ B > 0, \ \beta \in (0, 1] \tag{32}$$

for every pair of points $x_1, x_2 \in \overline{\Omega}$. Then there exist positive numbers A, α depending only on $q, s, M, \gamma, r, \delta, b, \kappa, B, \beta, \Omega$ such that the inequality (11) holds for every pair of points $(x_1, t_1), (x_2, t_2) \in \overline{Q}_T$.

Proofs of Theorem 4-5 are based (see [4]) on the modification of De Giorgi method [1] and by this we essentially used the idea of Di Benedetto [2] to work with cylinders whose dimensions are connected with oscilation of solutions.

4. Belonging of solutions to class $B_{q,s}$

The results of Theorems 1-3 about smoothness of solutions will be followed from Theorems of preceeding section and Theorem which we will now formulate.

Theorem 7. Let $A_\alpha(x, t, \xi)$ satisfy the conditions 1), 2) and $u(x, t) \in V_{2,p,q}^{m,1}(Q_T)$ be a generalized solution of equation (1) satisfying the inequality (10). Then $u(x, t)$ belongs to $B_{q,s}(Q_T, M, \gamma, r, \delta, b, \kappa)$ with constants $s, \gamma, r, \delta, b, \kappa$ depending only on the parameters in conditions (4)-(8) and $\|F(x, t)\|_{L_{\rho_0, r_0}(Q_T)}$.

Theorem 8. Let $u(x, t) \in V_{2,p,q}^{m,1}(Q_T)$ be a generalized solution of problem (2), (12), (13) satisfying the inequality (10) and let $A_\alpha(x, t, \xi)$ satisfy the conditions 1), 2). Then $u(x, t)$ belongs to class $B_{q,s}^{(D)}(Q_T \cup \Gamma_T, M, \gamma, r, \delta, b, \kappa)$ with constants $s, \gamma, r, \delta, b, \kappa$ depending only on the parameters in conditions (4)-(8), $\|F(x, t)\|_{L_{\rho_0, r_0}(Q_T)}$ and $\|G(x, t)\|_{L_{\rho_0, r_0}(Q_T)}$.

Theorem 9. Let $u(x, t) \in V_{2,p,q}^{m,1}(Q_T)$ be a weak solution of initial-Neumann boundary value problem for equation (2) satisfying the inequality (10), $A_\alpha(x, t)$ satisfy the conditions 1), 2), $B_\beta(x, t, \eta)$ satisfy the inequality (16), (17) and let $\partial\Omega \in C^m$. Then $u(x, t)$ belongs to class $B_{q,s}^{(N)}(Q_T \cup \Gamma_T, M, \gamma, r, \delta, b, \kappa)$ with constants $s, \gamma, r, \delta, b, \kappa$ depending only on Ω, the parameters in conditions (4)-(8), (16)-(17) and $\|F(x, t)\|_{L_{\rho_0, r_0}(Q_T)}$.

By proofs of Theorems 7-9 we used the methods of paper [6] and also by proof of Theorem 9 we constructed some continuations of function $u(x, t)$ and another auxiliary functions outside of Q_T.

References

1. E. De Giorgi, *Sulla differenziabilita a l'analitica delle estremali degli integrali multipli regolari*, Met. Acc. Sci. Torino, Cl. Sci. Fis. Mat. Natur. **3** (1957), no. 3, 24–43.
2. E. Di Benedetto, *On the local behaviour of solutions of degenerate parabolic equations with measurable coefficients*, Ann. Scuola Norm. Super. Pisa, Cl. Sci. Ser. IV **13** (1986), no. 3, 487–535.
3. O.A.Ladyzhenskaya, V.A.Solonnikov, N.N.Ural'tzeva, *Linear and quasilinear equations of parabolic type, Trans. Math., Monograps, 23*, Amer. Math. Soc., Providence, R.I., 1968.
4. I.V.Skrypnik, *On the quasilinear parabolic equations of higher order with hölder solutions*, Diff. equations **23** (1993), no. 3, 501–514 (Russian).
5. I.V.Skrypnik, *On Hölderity of functions from class $B_{q,s}$*, Ukrainian math. journal **45** (1993), no. 7 (Russian).
6. I.V.Skrypnik, *On the quasilinear elliptic equations of higher order with continuous generalized solutions*, Diff. equations **14** (1978), no. 6, 1104–1119 (Russian).

I.V.SKRYPNIK, INSTITUTE FOR APPLIED MATHEMATICS AND MECHANICS, ROSA LUXEMBURG ST., 74, DONETSK 340114, UKRAINE

212

M A VIVALDI

Existence and uniqueness results for degenerate-elliptic integro-differential problems

Degenerate elliptic differential equations have been studied by many authors: Edmunds and Peletier, Kruskov, Murty and Stampacchia, Trudinger... (see [4], [8], [9], [12] and [13]) more recently by Fabes, Jerison, Kenig and Serapioni (see [5], [6]) and in the larger context of the Dirichlet forms of diffusion type by Biroli and Mosco ([2], [3]).

On the other hand uniformly elliptic-integro-differential operators are investigated by Bensousson and Lions (see [1]) in connection with stochastic control problems, Wiener-Poisson process and diffusion process with interior jumps.

I refer to this book [1] for references, probabilistic interpretations and an exhaustive discussion on the motivations of this type of problems.

More recently parabolic integro-differentials operators not in divergence form, with Hölder continuous coefficients are studied by Garroni and Menaldi (see [7]): the authors construct the fundamental solution and the Green function, investigate the main properties and give many applications.

In this paper we study the Dirichlet problem and the obstacle problems for a class of degenerate elliptic integro differential operators A-$\mathcal{I}$ (see (1.3), (1.4) (1.6)): the previous mentioned results do not apply to operators A-$\mathcal{I}$ that are both degenerate and non local and to my knowledge there are no results for this type of degenerate and nonlocal operators.

1. Problems, notations and results

Consider the following problems

$$\begin{cases} \text{Find } u \in \overset{\circ}{H^1_2}(\Omega, w) \text{ s.t.} \quad \forall v \in \overset{\circ}{H^1_2}(\Omega, w) \\ a(u, v) - \int_\Omega \mathcal{I}uv \, dx = \int_\Omega fvw \, dx \end{cases} \quad (1.1)$$

$$\begin{cases} \text{Find } u \in \mathbb{K} \text{ s.t.} \quad \forall v \in \mathbb{K} \\ a(u, v - u) - \int_\Omega \mathcal{I}u(v - u) \, dx \geq \int_\Omega f(v - u)w \, dx \, . \end{cases} \quad (1.2)$$

Ω is a bounded connected open set of $\mathbb{R}^N$, $a(\cdot, \cdot)$ is the bilinear form associated to the operator A

$$a(u, v) = \int (a_{ij} u_x u_{xj} + a_0 uv) \, dx$$

i) $Au \equiv A_0 u + a_0 u = -(a_{ij} u_{x_i})_{x_j} + a_0 u$, $a_{ij} = a_{ji}$

ii) $\lambda_0 w(x) |\xi|^2 \leq a_{ij}(x) \xi_i \xi_j \leq \Lambda w(x) |\xi|^2, \lambda_0, \Lambda \in \mathbb{R}, \lambda_0 > 0$ (1.3)

iii) $\alpha_0 w(x) \leq a_0(x) \leq \overline{\alpha}_0 w(x)$, $\alpha_0, \overline{\alpha}_0 \in \mathbb{R}, \alpha_0 > 0$

here and in the following the summation symbol is understood. The weight $w(x)$ belongs to the Muckenhoupt's class A_2 i.e. $w(x)$ is a nonnegative function, w and w^{-1} belong to $L^1_{\text{loc}}(\mathbb{R}^N)$ and the following estimate holds

$$\sup_B \left(\int_B w \, dx \cdot \int_B w^{-1} \, dx \right) \leq k_0 \qquad (1.4)$$

The supremum is taken over all euclidean balls.

From now on we use the notations for any measurable set E

$$\int_E g \, dx = \frac{\int_E g \, dx}{|E|}, \quad |E| = \int_E dx, \quad w(E) = \int_E w \, dx$$

$$B_r(x) = \{ y \in \mathbb{R}^N : \| x - y \|_{\mathbb{R}^N} \leq r \}, \quad r > 0$$

Remark 1.1. If w belongs the Muckenhoupt's class A_2 then there exists $k_1 \in \mathbb{R}$ s.t. $\forall x \in \mathbb{R}^N, \forall \rho > 0$

$$w(B_{2\rho}(x)) \leq k_1 w(B_\rho(x)) \qquad (1.5)$$

$k_1 = k_1(N, k_0)$.

I refer to [5] and [6] for the proof of (1.5).

$\mathcal{I}$ is the integral operator

$$\mathcal{I} u = \int_{\mathbb{R}^N} m(dz) \int_0^1 z \nabla u(x + \theta z) w^{1/2}(x) w^{1/2}(x + \theta z) c(x, z) \, d\theta$$

$c(x, z)$ is a measurable function defined in $\mathbb{R}^N \times \mathbb{R}^N$ s.t.

$$0 \leq c(x, z) \leq 1$$

m is a singular measure s.t.

$$\int_{\mathbb{R}^N} |z| m(dz) = k_2 < +\infty$$

(1.6)

From now on any element of $\overset{\circ}{H}{}^1_2(\Omega, w)$ is supposed extended by zero outside of Ω.

Set $\|f\|_{0,\infty} = \operatorname{esssup}_\Omega |f|$. Denote by $L^p(\Omega, w)$ the weighted Lebesgue class with norm

$$\|f\|_{0,p} = \left(\int_\Omega |f|^p w \, dx \right)^{\frac{1}{p}} \quad p \in [1, +\infty) \ .$$

$H^1_2(\Omega, w)$ the completion of $\operatorname{Lip}(\Omega)$ w.r. to the norm

$$\|f\|_{1,2} = \{\|f\|^2_{0,2} + \|\nabla f\|^2_{0,2}\}^{\frac{1}{2}}$$

and $\overset{\circ}{H}{}^1_2(\Omega, w)$ the closure of $C_0^\infty(\Omega)$ in $H^1_2(\Omega, w)$.

Let K be a compact subset of Ω define:

$$w - cap(K, \Omega) = \inf \left\{ \int_\Omega |\nabla v|^2 w \, dx, \ v \in C_0^1(\Omega), \ v \geq 1 \text{ in } K \right\}$$

we can extend naturally to any open set A and any arbitrary set E by the usual passage to the supremum and to the infimum.

We will say that a set F has w-capacity zero if there exist a ball B_R s.t. $\overline{F} \subset B_R$ and w-cap $(F, B_R) = 0$ we will say that a property $P(x)$ holds in w-capacity almost everywhere (w-q.e.) in E if there exist a set E^0 with w-capacity zero s.t. $P(x)$ holds $\forall x \in E - E^0$.

Assume

$$f \in L^2(\Omega, w) \ . \tag{1.7}$$

Finally set

$$\mathbb{K} = \{v \in \overset{\circ}{H}{}^1_2(\Omega, w) : \psi_1 \leq v \leq \psi_2 \quad w\text{-q.e. in } \Omega\} \tag{1.8}$$

$$\psi_1, \psi_2 : \ \Omega \to [-\infty, +\infty]$$

In section 2 we use the following Sobolev-Poincaré-inequality, see [5] [6] and also [3] for the proof, details and comments:

$$\forall u \in \overset{\circ}{H}{}^1_2(\Omega, w)$$

$$w(\Omega)^{-\frac{1}{s}} \|u\|_{0,s} \leq k_3 (\operatorname{diam} \Omega) w(\Omega)^{-\frac{1}{2}} \|\nabla u\|_{0,2} \tag{1.9}$$

$$s = \frac{2k_4}{k_4 - 2} \quad k_4 \geq \max \left\{ 2, \frac{\log k_1}{\log 2} \right\}$$

We are now in a position to state results

215

Theorem 1.1. *Assume previous hypotheses and notations (1.3), (1.4), (1.6) and (1.7) then problem (1.1) admits one and only one solution.*

$$f \in L^p(\Omega, w) \,, \quad p > p_0^{\cdot} = \max\left\{2, \frac{s}{s-2}\right\} \quad \text{and} \quad L^{\infty}(\Omega) \cap \mathbb{K} \neq \emptyset \qquad (1.10)$$

Theorem 1.2. *Assume previous hypotheses and notations (1.3), (1.4), (1.6), (1.7) and (1.10) then problem (1.2) admits one solution.*

Remark 1.2. The integral operator $\mathcal{I}$ behaves as an operator of order one (see [1] and [7]); in fact the following estimates hold:

$$\left| \int_{\Omega} (\mathcal{I}\varphi_1 - \mathcal{I}\varphi_2)(\varphi_1 - \varphi_2) \, dx \right| \leq k_2 \|\varphi_1 - \varphi_2\|_{0,2} \|\nabla(\varphi_1 - \varphi_2)\|_{0,2} \quad \forall \varphi_1, \varphi_2 \in \overset{\circ}{H}{}^1_2(\Omega, w)$$
$$(1.11)$$

and

$$\int_{\Omega} \mathcal{I}\varphi(\varphi - k)^+ \leq k_2 \|\nabla(\varphi - k)^+\|_{0,2} \|(\varphi - k)^+\|_{0,2} \quad \forall \varphi \in \overset{\circ}{H}{}^1_2(\Omega, w)$$
$$(1.12)$$
$$\text{and } \forall \text{ constant } k \geq 0 \,.$$

Actually Theorem 1.1 and Theorem 1.2 hold for any linear operator $\mathcal{I}$ that satisfies inequalities (1.11) and (1.12); we can in fact easily verify that only estimates (1.11) and (1.12) are crucial in the proofs.

2. Proofs of the theorems

We split the proof of theorem 1.1 in five steps:

Step 1. *There exists λ_1 s.t. $\forall \lambda \geq \lambda_1$ the following problem $(2,1)\lambda$ admits one and only one solution u_λ*

$$\text{Find } u_\lambda \in \overset{\circ}{H}{}^1_2(\Omega, w) \text{ s.t.} \quad \forall v \in \overset{\circ}{H}{}^1_2(\Omega, w)$$
$$a(u_\lambda, v) + \int_{\Omega} (\lambda uw - \mathcal{I}u)v \, dx = \int_{\Omega} fvw \, dx \qquad (2.1)_\lambda$$

Proof of step 1. From estimate (1.10) of Remark 1.2 we derive for any choice of functions φ_1 and φ_2 in the space $\overset{\circ}{H}{}^1_2(\Omega, w)$ the following inequality

$$a(\varphi_1 - \varphi_2, \varphi_1 - \varphi_2) - \int_{\Omega} \mathcal{I}(\varphi_1 - \varphi_2)(\varphi_1 - \varphi_2) \, dx \geq$$
$$\overline{\lambda}\|\varphi_1 - \varphi_2\|^2_{1,2} - \lambda_1 \|\varphi_1 - \varphi_2\|^2_{0,2} \qquad (2.2)$$

216

where we can choose $\overline{\lambda} = \min\left\{\frac{\lambda_0}{2}, \alpha_0\right\}$ and $\lambda_1 = \frac{k_2^2}{2\lambda_0}$ k_2 is the constant in estimate (1.6) and (1.11).

The known results on the coercive bilinear forms (see e.g. [10], [11]) provides the existence and the uniqueness of the solution of problem $(2.1)_\lambda$ and conclude the proof of step 1.

Step 2. *Assume*

$$f \in L^p(\Omega, w) , \quad p > p_0 = \max\left\{2, \frac{s}{s-2}\right\} \tag{2.3}$$

where s is the Sobolev exponent in $\overset{\circ}{H}{}_2^1(\Omega, w)$, (see estimate (1.9)). Then for any solution u of problem (1.1) the following estimate holds

$$\sup_\Omega |u| \le c(\operatorname{diam}\Omega)^2 w(\Omega)^{-\frac{1}{p}}\|f\|_{0,p} + N\|u\|_{0,2} \tag{2.4}$$

c depends only on the constant k_3 in (1.9).

Proof of step 2. Here we use the Stampacchia's method of truncation (see [10]). Choose as test function $v = \zeta$ where $\zeta = \max\{u - k, 0\}$ and set $A(k) = \{x \in \Omega : u \ge k\}$, $k > 0$. By the Sobolev-Poincaré inequality (1.9) and the definition of the integral operator $\mathcal{I}$ (see (1.10) of Remark 1.2):

$$\overline{\lambda}\|\zeta\|_{1,2}^2 \le$$

$$\le k_3(\operatorname{diam}\Omega)\left(\int_{A(k)}(|f|^{s'} + \lambda_1^{s'}|u|^{s'})w\,dx\right)^{\frac{1}{s'}} \cdot \left(\int_\Omega |\nabla\zeta|^2 w\,dx\right)^{\frac{1}{2}} \cdot w(\Omega)^{\frac{1}{s}-\frac{1}{2}} \le$$

$$\le \frac{k_3^2}{\overline{\lambda}}(\operatorname{diam}\Omega)^2 w(\Omega)^{\frac{2}{s}-1}(\|f\|_{0,p}^2 w(A(k))^{\frac{2}{s'}-\frac{2}{p}} +$$

$$+ \lambda_1^2\|u\|_{0,r}^2 w(A(k))^{\frac{2}{s'}-\frac{2}{r}}) + \frac{\overline{\lambda}}{4}\int_\Omega |\nabla\zeta|^2 w\,dx$$

Now we choose $m = \frac{2s}{s-2}$, we use Sobolev-Poincaré inequality and we have $\forall h > k > 0$:

$$(h-k)^2 w(A(h))^{\frac{2}{s}} \le \left(\int_{A(k)}((u-k)^+)^s w\,dx\right)^{\frac{2}{s}} \le$$

$$\le 4k_3^4(\operatorname{diam}\Omega)^4 w(\Omega)^{-\frac{4}{m}}\left\{\|f\|_{0,p}^2 w(A(k))^{\frac{2}{s'}-\frac{2}{p}} +\right.$$

$$\left.+ \lambda_1^2\|u\|_{0,r}^2 w(A(k))^{\frac{2}{s'}-\frac{2}{r}}\right\}$$

217

then

$$w(A(h)) \leq (4k_3^4(\operatorname{diam}\Omega)^4 w(\Omega)^{-\frac{4}{m}}(\|f\|_{0,p}^2 + \lambda_1^2\|u\|_{0,r}^2)) \cdot \frac{w(A(k))^\gamma}{(h-k)^s} \tag{2.6}$$

where

$$\gamma = \min\left\{\frac{s}{s'} - \frac{s}{p}, \frac{s}{s'} - \frac{s}{r}\right\} \tag{2.7}$$

Since $p > \frac{m}{2}$ so we have that $\frac{s}{s'} - \frac{s}{p} > 1$.

Now we proceed as in the proof of theorem 4.1 of [10] we start by choosing $r = 2$ in (2.6) and from Lemma 4.1 case (iii) of [10] we deduce $u \in L^{t_1}(\Omega, w)$ with $t_1 > 2$ then we choose $r = t_1$ in (2.6) and by an iterative procedure we obtain $u \in L^r(\Omega, w)$ with $r \geq p > \frac{m}{2}$ then $\gamma > 1$ in (2.7) and from Lemma 4.1 case (i) of [10] we deduce $u \in L^\infty(\Omega)$ and $w(A(d)) = 0$ with

$$d = k_4^2(\operatorname{diam}\Omega)^2 w(\Omega)^{-\frac{1}{p}}\{\|f\|_{0,p} + \lambda_1\|u\|_{0,2}\} \ .$$

In an analogous way we set $\eta = \max\{-u - k, 0\}$ and $A(k) = \{x \in \Omega : -u \geq k\}$ and by repeating the previous procedure we conclude the proof of step 2.

Remark 2.1. Of course the claim of step 2 still holds for u_λ; more precisely we have

$$\|u_\lambda\|_{0,\infty} \leq c(\operatorname{diam}\Omega)^2 w(\Omega)^{-\frac{1}{p}}\|f\|_{0,p} \quad p > p_0 \tag{2.8}$$

Step 3. *Let f be a bounded function then the following estimate holds for the solution u_λ of auxiliary problem $(2.1)_\lambda$:*

$$\|u_\lambda\|_{0,\infty} \leq \frac{\|f\|_{0,\infty}}{\lambda + \alpha_0} \ ,$$

α_0 *is the constant in (1.3) (iii).*

Proof of step 3. Let $k > \frac{\|f\|_{0,\infty}}{\lambda + \alpha_0}$ and choose as "test" function in $(2.1)_\lambda$ $v = \max\{u_\lambda - k, 0\}$. By using estimate (1.12) we obtain

$$\lambda_0\|\nabla v\|_{0,2}^2 + (\alpha_0 + \lambda)\|v\|_{0,2}^2 \leq a(v,v) + \lambda\|v\|_{0,2}^2 \leq$$
$$\leq \int_\Omega (\mathcal{I}u_\lambda + (f - (\lambda + \alpha_0)k)w)v\,dx \leq$$
$$\leq \frac{\lambda_0}{2}\|\nabla v\|_{0,2}^2 + \frac{k_2^2}{2\lambda_0}\|v\|_{0,2}^2 \tag{2.9}$$

So if we choose $\lambda + \alpha_0 \geq \frac{k_2^2}{2\lambda_0}$ we derive $v \equiv 0$ i.e. $u_\lambda \leq k$. In an analogous way we choose as test function in $(2.1)_\lambda$ $v = \max\{-u_\lambda - k, 0\}$ to obtain $-u_\lambda \leq k$; hence step 3 is achieved.

Step 4. *Let f belongs to $L^p(\Omega, w)$ with $p > \max\left\{2, \frac{s}{s-2}\right\}$ then there exists one and only one solution of problem (1.1).*

Proof of step 4. Define the sequence $\{\tilde{u}_n\}_{n \in \mathbb{N} \cup \{0\}}$ by the following procedure: $\tilde{u}_0 \in \overset{\circ}{H^1_2}(\Omega, w)$ is the (unique) solution of problem (2.10)

$$
\begin{cases}
a(\tilde{u}_0, v) + \int_\Omega (-\mathcal{I}\tilde{u}_0 + \lambda w \tilde{u}_0) v \, dx = \int_\Omega f v w \, dx \\
\tilde{u}_0 \in \overset{\circ}{H^1_2}(\Omega, w) \quad \forall v \in \overset{\circ}{H^1_2}(\Omega, w)
\end{cases}
\tag{2.10}
$$

and for $n \geq 1$ $\tilde{u}_n$ is the (unique) solution of

$$
a(\tilde{u}_n, v) + \int_\Omega (-\mathcal{I}\tilde{u}_n + \lambda \tilde{u}_n w) v \, dx = \int_\Omega (f + \lambda \tilde{u}_{n-1} w) v \, dx
$$
$$
\tilde{u}_n \in \overset{\circ}{H^1_2}(\Omega, w) \quad \forall v \in \overset{\circ}{H^1_2}(\Omega, w)
\tag{2.11}
$$

Set $\tilde{u}_{-1} \equiv 0$ from (2.10) and (2.11) we obtain $\forall n \in \mathbb{N} \cup \{0\}$

$$
a(\tilde{u}_{n+1} - \tilde{u}_n, v) + \int_\Omega (-\mathcal{I}(\tilde{u}_{n+1} - \tilde{u}_n) + \lambda(\tilde{u}_{n+1} - \tilde{u}_n) w) v \, dx =
$$
$$
= \int_\Omega \mathbf{v}\lambda(\tilde{u}_n - \tilde{u}_{n-1}) w \, dx \quad \forall v \in \overset{\circ}{H^1_2}(\Omega, w)
\tag{2.12}
$$

By estimate (2.8) we derive $\tilde{u}_0 \in L^\infty(\Omega)$ and then from (2.8) and (2.12) $\tilde{u}_1 - \tilde{u}_0$ belongs to $L^\infty(\Omega)$.

By induction also $\tilde{u}_n - \tilde{u}_{n-1} \in L^\infty(\Omega)$, by step 3

$$
\|\tilde{u}_{n+1} - \tilde{u}_n\|_{0,\infty} \leq \frac{\lambda}{\lambda + \alpha_0} \|\tilde{u}_n - \tilde{u}_{n-1}\|_{0,\infty} \leq \left(\frac{\lambda}{\lambda + \alpha_0}\right)^n \|\tilde{u}_1 - \tilde{u}_0\|_{0,\infty}
\tag{2.13}
$$

and hence:

$$
\overline{\lambda}\|\tilde{u}_{n+1} - \tilde{u}_n\|_{1,2}^2 \leq \lambda \|\tilde{u}_n - \tilde{u}_{n-1}\|_{0,2}^2 \leq \lambda \left(\frac{\lambda}{\lambda + \alpha_0}\right)^{2(n-1)} w(\Omega)\|\tilde{u}_1 - \tilde{u}_0\|_{0,\infty}
$$

We conclude that the sequence $\{\tilde{u}_n\}_{n \in \mathbb{N}}$ converges (strongly) in $\overset{\circ}{H^1_2}(\Omega, w)$ and in $L^\infty(\Omega)$ towards a function u that solves problem (1.1).

To prove the uniqueness suppose that u^* and u are two solutions of problem (1.1): from step 2 we obtain that u^* and u belong to $L^\infty(\Omega)$ from step 3 we have

$$
\|u - u^*\|_\infty \leq \frac{\lambda}{\lambda + \alpha_0} \|u - u^*\|_\infty \quad \text{i.e.} \quad u = u^*.
$$

Proof of theorem 2.1. Set T the inverse operator of $A - \mathcal{I} + \lambda$ ($\lambda \geq \lambda_1$); T is compact from $\overset{\circ}{H}{}^1_2(\Omega, w) \rightarrow \overset{\circ}{H}{}^1_2(\Omega, w)$ from step 4 we derive that problem (1.1) with $f \equiv 0$ admits only $u \equiv 0$ as solution by Fredholm-alternative theorem we conclude that for any $v \in \overset{\circ}{H}{}^1_2(\Omega, w)$ there exists a unique solution $u \in \overset{\circ}{H}{}^1_2(\Omega, w)$ of $(I - \lambda T)u = v$ then for any $f \in L^2(\Omega, w)$ problem (1.1) admits a unique solution u (see e.g. [11]).

Remark 2.2. Actually if in the recurrence procedure of step 4 we assume $f \in L^2(\Omega, w)$ then $\tilde{u}_0$ belongs to $L^s(\Omega, w)$ (see Poincaré inequality (1.9)). By repeating the iterative argument of step 2 and using Lemma 4.1 of [10] we obtain that $\tilde{u}_n - \tilde{u}_{n-1}$ belongs to $L^\infty(\Omega)$ for $n \geq n_0 > \frac{1}{s-2}$ and estimates (2.13) and (2.14) still hold if we replace n by $n - n_0$. Hence the extra sommability assumption on f (2.3) can be removed just by repeating the procedure of step 2 and step 4.

Remark 2.3. Consider the proof of theorem 1.1: step 1 and step 2 work also for obstacle problems, a term depending on the obstacles appears in the claim of step 2, on the contrary step 3 does not work. To show Theorem 1.2 we shall follow the procedure used by Troianiello for the *differential nondegenerate* case (see [11] Theorem4.17 and 4.30). By the presence of the non local term $\mathcal{I}$ i am not able to prove the "weak maximum principle" (see Theorem 2.4 of [11]) and hence we obtain a weaker result. Actually i succeeded in proving uniqueness results only by doing some "extra" coerciveness assumptions of the type:

$$\sqrt{2\lambda_0 \alpha_0} \geq k_2 \ .$$

Proof of theorem 1.2. Consider first the one obstacle problem with the upper constraint, i.e. suppose $\psi_1 \equiv -\infty$ in the convex $\mathbb{K}$.

Denote by $\hat{u}$ the (unique) solution of problem (1.1) as f belongs to $L^p(\Omega, w)$ with $p > p_0$ then $\hat{u} \in L^\infty(\Omega)$, see step 2 of the proof of theorem 1.1.

Set $d_2 = \min\{0, \operatorname*{essinf}_\Omega(\psi_2 - \hat{u})\}$.

Fix λ as in step 1 and define by recurrence $u_0 = \hat{u}$ and u_n the (unique) solution of problem

$$\begin{cases} a(u_n, v - u_n) + \int_\Omega (-\mathcal{I}u_n + \lambda u_n w)(v - u_n)\, dx \geq \int_\Omega (f + \lambda u_{n-1})(v - u_n) w\, dx \\ u_n \in \mathbb{K}, \quad \forall v \in \mathbb{K}. \end{cases}$$

$$(2.16)_n$$

As the bilinear form in (2.16) is coercive we claim that:

$$\hat{u} + d_2 \leq u_n \leq u_{n-1} \leq \hat{u} \tag{2.17}$$

Choose as test function in equation (1.1) $v = \zeta = \max\{u_1 - u_0, 0\}$ and in problem $(2.16)_1$ $v = \min\{u_1, u_0\} \equiv u_1 - \zeta$.

We add the inequalities and we use the coerciveness of the form in $(2.1)_1$ to find

$$\|\zeta\|_{1,2}^2 \leq 0 \quad \text{i.e.} \quad u_1 \leq u_0 \ .$$

Set $\zeta_n = \max\{u_n - u_{n-1}, 0\}$ and choose $v = \max\{u_n, u_{n-1}\} = u_{n-1} + \zeta_n$ in problem $(2.16)_{n-1}$ and $v = \min\{u_n, u_{n-1}\} = u_n - \zeta_n$ in problem $(2.16)_n$ by recurrence we prove:

$$\|\zeta_n\|_{1,2}^2 \leq 0 \quad \text{i.e.} \quad u_n \leq u_{n-1} \ .$$

Finally set $\eta_m = \max\{\hat{u} + d_2 - u_n, 0\}$ and choose $v = \max\{u_n, \hat{u} + d_2\} = u_n + \eta_n$ in problem $(2.16)_n$ and $v = \eta_n$ in equation (1.1) (by recurrence) we prove

$$\|\eta_m\|_{1,2}^2 \leq 0 \quad \text{i.e.} \quad u_n \geq \hat{u} + d_2 \ .$$

Hence the sequence $\{u_n\}_{n \in \mathbb{N}}$ is monotone decreasing and bounded in $L^2(\Omega, w)$.

By fixing $v_0 \in \mathbb{K}$ we see that the sequence $\{u_n\}$ is also bounded in $\overset{\circ}{H}{}_2^1(\Omega, w)$:

$$\|u_n\|_{1,2} \leq c(\|f\|_{0,2} + \|\hat{u}\|_{0,2} + d_2 + \|v_0\|_{1,2})$$

thus the whole sequence converges weakly in $\overset{\circ}{H}{}_2^1(\Omega, w)$ and strongly in $L^2(\Omega, w)$ toward $\underline{u} \leq \hat{u}$ solution of the "unilateral" problem (1.2).

Of course if we suppose $\psi_1 \neq -\infty$ and $\psi_2 \equiv +\infty$ by the same procedure we can construct a solution $\overline{u} \geq \hat{u}$ of the unilateral problem with constraint below.

Consider now the bilateral problem, introduce the bounded closed convex $\mathcal{K} \neq \emptyset$ of $L^2(\Omega, w)$

$$\begin{cases} \mathcal{K} = \{v \in L^2(\Omega, w) : \ \Psi_1 \leq v \leq \Psi_2\} \\ \text{where} \\ \Psi_1 = \max\{\underline{u}, \psi_1\} \text{ and } \Psi_2 = \min\{\overline{u}, \psi_2\} \end{cases} \tag{2.18}$$

and define a continuous mapping $S : L^2(\Omega, w) \to \overset{\circ}{H}{}_2^1(\Omega, w)$ by $S(u)$ the (unique) solution of the problem

$$\begin{cases} a(S(u), v - S(u)) + \\ \quad + \int_\Omega (-\mathcal{I}S(u) + \lambda S(u)w)(v - S(u)) \, dx \geq \int_\Omega (f + \lambda u)(v - S(u))w \, dx \quad (2.19) \\ S(u) \in \mathbb{K}, \quad \forall v \in \mathbb{K}. \end{cases}$$

We claim that S maps $\mathcal{K}$ into itself. The functions $v = \max\{S(u), \underline{u}\}$ and $v = \min\{S(u), \overline{u}\}$ are admissible in (2.19) indeed, the bilinear form in (2.19) is coercive and we can proceed as in the previous part to show that

$$\underline{u} \le S(u) \le \overline{u} \ .$$

Again by the coerciveness S maps bounded subsets of $L^2(\Omega, w)$ in bounded subsets of $\overset{\circ}{H}{}^1_2(\Omega, w)$, hence into relatively compact subsets of $L^2(\Omega, w)$. Thus the Schauder theorem yields the existence of a fixed point $u = S(u) \in \mathcal{K}$ hence a solution of the two obstacle problem (1.2).

Remark 2.4. Hypothesis (1.10) is "technical": it is indeed clear from the previous proof that theorem 1.2 holds, in the "natural" hypothesis $\mathbb{K} \ne \emptyset$ if we can associate to the obstacles ψ_1 and ψ_2 a bounded closed convex set $\mathcal{K}$ of $L^2(\Omega, w)$ that is for instance if ψ_1 and ψ_2 belong to $L^2(\Omega, w)$. Moreover we need only inequalities in $L^2(\Omega, w)$ but, due to the presence of the nonlocal operator "$\mathcal{I}$", the comparison results seem work only with the "constants" if we do not assume extra coerciveness assumptions (see also Remark 2.3). The question of the uniqueness is strictly connected to this situation.

REFERENCES

[1] Bensoussan A. and Lions J.L., *Contrôle impulsionnel et inéquations quasi variationnelles*, Dunod, Paris 1982.

[2] Biroli M. and Mosco U., Wiener criterion and potential estimates for obstacle problems relative to degenerate elliptic operators, *Ann. Mat. Pura Appl.*, (IV) CLIX (1991), 255-281.

[3] Biroli M. and Mosco U., A Saint-Venant principle for Dirichlet forms on discontinuous media, *Publ. Lab. Anal. Num. Univ. Paris* VI 93005 (1993).

[4] Edmunds D.E. and Peletier L.A., A Harnack inequality for weak solutions of degenerate quasilinear elliptic equations, *J. London Math. Soc.* 5 (1972), 21-31.

[5] Fabes E.B., Jerison D.S. and Kenig C.E., The Wiener test for degenerate elliptic equations, *Ann. Inst. Fourier* 32 (1982), 151-182.

[6] Fabes E.B., Kenig C.E. and Serapioni R.P., The local regularity of solutions of degenerate elliptic equations, *Comm. P.D.E.* 7 (1982), 77-116.

[7] Garroni M.G. and Menaldi J.L., *Green functions for second order parabolic integro-differential problems*, Longman, London 1992.

[8] Kruskov S.N., Certain properties of solutions to elliptic equations, *Soviet Mathematics* **4** (1963), 686-695.

[9] Murthy M.K.V. and Stampacchia G., Boundary value problems for some degenerate elliptic operators, *Ann. Mat. Pura. Appl.* **80** (1968), 1-122.

[10] Stampacchia G., Le problème de Dirichlet pour les équations elliptiques du second ordre à coefficients discontinus, *Ann. Inst. Fourier*, Grenoble **15** (1965), 189-258.

[11] Troianiello G.M., *Elliptic differential equations and obstacle problems*, Plenum Press, New York 1987.

[12] Trudinger N., On the regularity of generalized solutions of linear nonuniformly elliptic equations, *Arch. Rat. Mech. Anal.* **42** (1971), 51-62.

[13] Trudinger N., Linear elliptic operators with measurable coefficients, *Ann. Sc. Norm. Sup. Pisa* **27** (1973), 265-308.

M.A. VIVALDI
Dipartimento di Metodi e Modelli Matematici per le Scienze Applicate
via A. Scarpa, 10
Università di Roma "La Sapienza"
00185 - Roma, Italia

A I VOLPERT AND V A VOLPERT

Application of the Leray–Schauder degree to investigation of travelling wave solutions of parabolic equations

We study travelling wave solutions of the semilinear parabolic system of equations

$$\frac{\partial u}{\partial t} = a\Delta u + r(x')\frac{\partial u}{\partial x_1} + F(u) \tag{1}$$

in the infinite cylinder $\Omega \subset R^m$, with the boundary condition

$$\frac{\partial u}{\partial n}\Big|_{\partial\Omega} = 0. \tag{2}$$

Here $\Omega = D \times R^1$, D is a bounded domain in R^{m-1} with a smooth boundary, x_1 is a coordinate along the axis of the cylinder, $x' = (x_2, ..., x_m)$, $w = (w_1, ..., w_n)$, $F = (F_1, ..., F_n)$, a is a constant symmmetric positive-definite matrix, $r(x')$ is a scalar function. Travelling wave solution of this system is a solution of the form

$$u(x,t) = w(x_1 - ct, x_2, ..., x_n).$$

Here c is a constant, the wave velocity. The function $w(x)$ is a solution of the following problem

$$a\Delta w + (r(x') + c)\frac{\partial w}{\partial x_1} + F(w) = 0, \quad \frac{\partial w}{\partial n}\Big|_{\partial\Omega} = 0. \tag{3}$$

We assume also that it has the limits at the infinity:

$$\lim_{x_1\to\pm\infty} w(x) = w_\pm, \quad w_+ \neq w_-. \tag{4}$$

Existence of multidimensional waves was studied in [1], [6], [8] (see the complete bibliography in [17], [23]).

The contents of the paper is as follows. In Section 1 we define the Leray-Schauder degree for the elliptic operators in unbounded domains which correspond to the problem (3), (4). Section 2 is devoted to the problem of wave existence. We study wave trains in Section 3. Applications of the obtained results to chemical kinetics are discussed in Section 4.

1. Degree construction.

It is well known (see, for example, [11]) that if (3) is considered in a bounded domain then the corresponding vector field can be reduced to a compact one. Indeed, let A

be the operator corresponding to the left hand side of (3), acting in the space C^α with the domain $C^{2+\alpha}$. Due to the compact imbedding of $C^{2+\alpha}$ into C^α it can be represented in the form

$$A = L + B = L(I + L^{-1}B).$$ (5)

where L has a compact inverse, and the equality $Lu = 0$ implies that $u = 0$, B is a bounded operator. Thus we can consider the completely continuous vector field $I + L^{-1}B$ and apply the usual definition of the Leray- Schauder degree. If we consider unbounded domains then there is no the compact imbedding of $C^{2+\alpha}$ into C^α, and this approach can not be used. Nevertheless the reduction to compact vector fields can be fulfilled even for unbounded domains. In [4] a weighted Sobolev space with strong weights, for example $exp(x^2)$, is considered. Such weights lead to a compact imbedding of H^{k+1} into H^k, where H^m is a space of functions for which m derivatives are integrable with a square. As above the representation (5), where L^{-1} is compact, can be obtained. It is important to note that this construction is possible only in the case when the weight function growth is more fast than the exponential one. Hence, exponentially decreasing functions do not belong to these spaces, and it is a strong restriction on the application of this approach. In particular, it can not be used to the study of the problem formulated above.

We also consider weighted Sobolev spaces but with weak weights [15], [16]. In this case exponentially decreasing functions belong to them but there is no compact imbedding of the spaces which leads to (5) with a compact operator L^{-1}. Hence we can not use the Leray-Schauder theory known for the completely continuous vector fields, and the aim of this work is to construct the degree for the operators corresponding to the problem (3), (4). To do this we obtain certain lower estimates of the operators which give a possibility to apply the method of Skrypnik of the degree construction [12] (see also [3]). In this method the degree for the original operator is defined as the degree for some finite-dimensional operators, and it is possible due to the compactness of the set of zeros of the operator. For the operators under consideration this compactness follows from the estimates mentioned above.

We introduce the function space and the operator which correspond to the problem (3), (4).

We consider the weighted Sobolev space $W^1_{2,\mu}(\Omega)$ of the vector-valued functions, defined in the cylinder Ω, with the inner product

$$[u,v]_\mu = \int_\Omega (\sum_{k=1}^m (\frac{\partial u}{\partial x_k}, \frac{\partial v}{\partial x_k}) + (u,v))\mu dx,$$

where $u, v : \Omega \to R^p, (\partial u/\partial n)\,|_{\partial\Omega} = 0$. The norm in this space is denoted by $\| \cdot \|_\mu$.

The weight function μ depends on x_1 only, and it is supposed to satisfy the following conditions:

1. $\mu(x_1) \geq 1, \mu(x_1) \to \infty$ as $|x_1| \to \infty$,
2. μ'/μ and μ''/μ are continuous functions which tend to zero as $|x_1| \to \infty$.

For example we can take $\mu(x_1) = 1 + x_1^2$ or $\mu(x_1) = \ln(1 + x_1^2)$.

We emphasize again that this weight is weak in the sense that exponentially decreasing at infinity functions belong to $W^1_{2,\mu}(\Omega)$. Contrary to the weight spaces with a strong weight there is no a compact imbedding of $W^1_{2,\mu}(\Omega)$ into a space $L_{2,\mu}(\Omega)$ of square-summable functions with the weight μ.

We define the operator $A(u)$ acting from $W^1_{2,\mu}(\Omega)$ into the conjugate space $(W^1_{2,\mu}(\Omega))^*$ by the following formula:

$$\langle A(u), v \rangle = \int_\Omega \sum_{k=1}^m (a\frac{\partial u}{\partial x_k}, \frac{\partial(v\mu)}{\partial x_k}) dx - \qquad (6)$$

$$\int_\Omega (a\frac{\partial^2 \psi}{\partial x_1^2} + (r+c)\frac{\partial}{\partial x_1}(u+\psi) + F(u+\psi, x), v)\mu dx,$$

where $u, v \in W^1_{2,\mu}(\Omega)$, the notation $\langle f, v \rangle$ means the action of the functional $f \in W^1_{2,\mu}(\Omega)$ on the element v, ψ is a twice continuously differentiable function of x_1 of the form

$$\psi(x_1) = w_-\omega(x_1) + w_+(1 - \omega(x_1)).$$

Here $\omega(x_1)$ is a monotone sufficiently smooth finction which is equal to zero for $x_1 \geq 1$ and to unity for $x_1 \leq -1$. We consider here a more general case where the nonlinearity F can depend on x.

We suppose that $r(x')$ is a bounded function, and the function $F(w, x)$ satisfies the following conditions:

1.$F(\psi, x) \in W^1_{2,\mu}(\Omega)$.

For example, if the function F does not depend on x explicitly, and

$$F(w_+) = F(w_-) = 0,$$

then, obviously, this condition is satisfied.

2.The function $F(w, x)$ and the matrices $F'(w, x)$ and $F''_i(w, x), i = 1, ..., n$ are unifomly bounded for all $w \in R^n$ and $x \in \Omega$.

3.There exist uniform limits

$$b_\pm = \lim_{x_1 \to \pm\infty} F'(w_\pm, x), \ x' \in D.$$

where $b_\pm$ are constant matrices.

We note that for twice continuously differentiable functions u and compactly supported functions v (6) can be rewritten in the form

$$\langle A(u), v \rangle = \qquad (7)$$

$$- \int_\Omega (a\Delta(u+\psi) + (r+c)\frac{\partial}{\partial x_1}(u+\psi) + F(u+\psi, x), v)\mu dx.$$

If we put $w = u + \psi$ then (3) has the form

$$a\Delta(u+\psi) + (r+c)\frac{\partial}{\partial x_1}(u+\psi) + F(u+\psi, x) = 0, \qquad (8)$$

and the connection between (7) and (8) is clear.

Any solution $u \in W_{2,\mu}^1(\Omega)$ of the equation

$$A(u) = 0 \tag{9}$$

has continuous second derivatives and satisfies (8). Conversely, every solution u of (8) having continuous second derivatives, satisfying the boundary conditions in (3) and belonging to $W_{2,\mu}^1(\Omega)$ is a solution of (9).

We should make some remarks about the case when the function F does not depend on x explicitly. In this case some additional difficulties appear. First of all, along with a solution $w(x)$ of (3) there are also solutions $w(x + h)$, where h is an arbitrary number. It means that for each solution $u(x)$ of

$$a\Delta(u + \psi) + (r + c)\frac{\partial}{\partial x_1}(u + \psi) + F(u + \psi) = 0, \tag{10}$$

there is one-parameter family of solutions

$$u_h(x) = u(x_1 + h, x_2, ..., x_m) + \psi(x_1 + h) - \psi(x_1). \tag{11}$$

This nonisolatedness of solutions complicates further investigations. Moreover, we know from the investigations in the one-dimensional case [15] that (10) can have solutions for isolated values of c. It means that a small changing of the system (10) can lead to the disappearance of the solution. There is no a contradiction with the homotopy invariance of the degree since the system linearized on the solution has zero eigenvalue, and the index of the stationary point can be equal to zero. But the construction of the degree does not have sense in this case.

Thus if the function F does not depend on x explicitly the constant c should not be considered as given. We have the following formulation of the problem: to find c for which (10) has a solution. So the constant c is unknown along with the function $u(x)$, and we have to consider the operator defined on the space $W_{2,\mu}^1(\Omega) \times R$. To avoid this complication and nonisolatedness of solutions we apply the method of a functionalization of a parameter. It means that instead of unknown constant c we consider a functional $c(u)$. It is supposed to be given and to satisfy the following conditions:

1. For each $u \in W_{2,\mu}^1(\Omega)$ the function $c(u_h)$, where u_h is defined by (11), is monotone in h,

2. $c(u_h) \to \mp\infty$ as $h \to \pm\infty$.

In this case (10) is equivalent to the equation

$$a\Delta(u + \psi) + (r + c(u))\frac{\partial}{\partial x_1}(u + \psi) + F(u + \psi) = 0 \tag{12}$$

in the following sense. Let the constant $c = c_0$ and the family $u_h(x)$, $-\infty < h < +\infty$ be a solution of (10). We choose $h = h_0$ to satisfy the equality $c(u_h) = c_0$. It follows from the conditions 1 and 2 that such h exists for any c_0, and it is unique. Obviously, the function $u(x) = u_{h_0}(x)$ satisfies (12). Conversely, let $u(x)$ be a solution of (12). Then the constant $c = c(u)$ and the family $u_h(x)$, defined by (11), satisfy (10).

We now construct a functional $c(u)$ which satisfies the conditions above. Let $\sigma(x_1)$ be a monotone increasing function such that $\sigma(x_1) \to 0$ as $x_1 \to -\infty$, $\sigma(x_1) \to 1$ as $x_1 \to +\infty$,

$$\int_{-\infty}^{0} \sigma(x_1)dx_1 < \infty.$$

We put

$$\rho(u) = (\int_{\Omega} | u + \psi - w_+ |^2 \sigma(x_1)dx)^{1/2} \tag{13}$$

and

$$c(u) = \ln \rho(u). \tag{14}$$

Assertion 1 *The functional $c(u)$ defined on $W^1_{2,\mu}(\Omega)$ by (13),(14) satisfies Lipschitz condition on every bounded set of $W^1_{2,\mu}(\Omega)$, and has the following properties: $c(u_h)$ is a monotone decreasing function of h, $c(u_h) \to \mp\infty$ as $h \to \pm\infty$. Here u_h is defined by (11), $u \in W^1_{2,\mu}(\Omega)$.*

The properties of the operator A which are formulated below are the same for the cases when c is a constant and a functional, and when F depends on x explicitly and does not depend. So, if the contrary is not pointed out, we consider all these cases.

Assertion 2 *The operator $A(u)$ satisfies Lipschitz condition on every bounded set of $W^1_{2,\mu}(\Omega)$.*

Condition 1 *All eigenvalues of the matrices $b_\pm - a\xi^2$ are in the left half-plane for all real ξ.*

We note that this condition is connected with the location of the continuous spectrum of the linearized operator [9], [18].

Theorem 1 *Let Condition 1 be satisfied. Then there exists a linear bounded symmetric positive definite operator S, acting in the space $W^1_{2,\mu}(\Omega)$, such that the inequality*

$$\langle A(u) - A(u_0), S(u - u_0)\rangle \geq \|u - u_0\|_\mu^2 + \phi(u, u_0) \tag{15}$$

takes place for any $u, u_0 \in W^1_{2,\mu}(\Omega)$. Here $\phi(u_n, u_0) \to 0$ as $u_n \to u_0$ weakly.

From Theorem 1 it follows that a condition similar to Condition α of Skrypnik [12] is valid:
if u_n is a sequence in $W^1_{2,\mu}(\Omega)$ which converges weakly to an element $u_0 \in W^1_{2,\mu}(\Omega)$ and if

$$\lim_{n\to\infty} \langle A(u_n), S(u_n - u_0)\rangle \leq 0 \tag{16}$$

then u_n converges to u_0 strongly in $W^1_{2,\mu}(\Omega)$.

From this we conclude that the Leray-Schauder degree (the rotation of a vector field) $\gamma(A, \mathcal{M})$ can be constructed similar to [12] for any bounded set $\mathcal{M}$ in $W^1_{2,\mu}(\Omega)$. The degree does not depend on the arbitrariness in the choice of the operator S satisfying the conditions of the Theorem 1.

The Leray-Schauder degree is proved to possess two usual properties: the principle of nonzero rotation and the homotopy invariance. The principle of nonzero rotation means that if $\gamma(A, \mathcal{M}) \neq 0$ then (9) has a solution in $\mathcal{M}$.

We define now the homotopy of the operators under consideration. We consider families of matrices a_τ, functions $F_\tau(u, x)$, $r_\tau(x')$ and constants c_τ, in case if they are considered as given, depending on the parameter $\tau \in [0, 1]$. The following conditions are supposed to be satisfied:

1. The matrices a_τ are symmetric positive definite and continuous in τ,
2. The functions $F_\tau(\psi_\tau, x)$ are continuous in τ in the $W^1_{2,\mu}(\Omega)$ norm,
3. The matrices $F'_\tau(w, x)$ and $F''_{u_i,\tau}(w, x), i = 1, ..., n$ are uniformly bounded for all $w \in R^p, x \in \Omega, \tau \in [0, 1]$. The matrix $F'_\tau(w, x)$ satisfies Lipschitz condition in $w \in R^p, \tau \in [0, 1]$ uniformly in $x \in \Omega$.
4. For each $\tau \in [0, 1]$ there exist uniform limits

$$b_\pm(\tau) = \lim_{x_1 \to \pm\infty} F'(w_\pm(\tau), x), x' \in D.$$

All eigenvalues of the matrices

$$b_\pm(\tau) - a_\tau \xi^2$$

lie in the left half plane for all real ξ.

5. The functions $r_\tau(x')$ are bounded and continuous in τ uniformly in $x' \in D$. If c_τ is a given constant then it is continuous in τ.

Here $w_+(\tau)$ and $w_-(\tau)$ are given vector-valued functions which are supposed to be continuous,

$$\psi_\tau(x_1) = w_-(\tau)\omega(x_1) + w_+(\tau)(1 - \omega(x_1)).$$

In the case when F_τ does not depend on x explicitly, and c_τ is a functional, it is defined by the formulas

$$\rho_\tau(u) = (\int_\Omega | u + \psi_\tau - w_+(\tau) |^2 \sigma(x_1)dx)^{1/2} \tag{17}$$

$$c_\tau(u) = \ln \rho_\tau(u). \tag{18}$$

The operator $A_\tau(u) : W^1_{2,\mu}(\Omega) \to (W^1_{2,\mu}(\Omega))^*$ is defined by the equality

$$\langle A_\tau(u), v \rangle = \int_\Omega \sum_{k=1}^m (a_\tau \frac{\partial u}{\partial x_k}, \frac{\partial(v\mu)}{\partial x_k})dx - \tag{19}$$
$$\int_\Omega (a_\tau \psi''_\tau + (r_\tau + c_\tau)(\frac{\partial u}{\partial x_1} + \psi'_\tau) + F_\tau(u + \psi_\tau, x), v)\mu dx,$$

Theorem 2 *Let $\mathcal{M}$ be a bounded domain in the space $W_{2,\mu}^1(\Omega)$ with the boundary Γ, and $A_\tau(u) \neq 0$ for $u \in \Gamma, \tau \in [0,1]$. If the conditions 1-5 are satisfied then*

$$\gamma(A_0, \mathcal{M}) = \gamma(A_1, \mathcal{M}).$$

Let $u_0 \in W_{2,\mu}^1(\Omega)$ be an isolated stationary point of the operator $A(u)$:

$$A(u_0) = 0,$$

and $A(u) \neq 0$ for $u \neq u_0$ in some neighbourhood of the point u_0. Then the index of the stationary point u_0 is defined in the usual way: it is the rotation of the vector field $A(u)$ on a sphere with the center u_0 of sufficiently small radius.

We linearize the operator $A(u)$ at the stationary point u_0. The linearized operator $A'(u_0) : W_{2,\mu}^1(\Omega) \to (W_{2,\mu}^1(\Omega))^*$ is defined by the equality

$$\langle A'(u_0)u, v \rangle = \int_\Omega \sum_{k=1}^m (a\frac{\partial u}{\partial x_k}, \frac{\partial(v\mu)}{\partial x_k})dx \tag{20}$$

$$- \int_\Omega (c'(u)\frac{\partial(u_0 + \psi)}{\partial x_1} + (r(x') + c(u_0))\frac{\partial u}{\partial x_1} + F'(u_0 + \psi)u, v)\mu dx,$$

where

$$c'(u) = \frac{\int_\Omega (u_0(x) + \psi - w_+, u(x))\sigma(x_1)dx}{\int_\Omega | u_0 + \psi - w_+ |^2 \sigma(x_1)dx}$$

If c is a given constant then the term with $c'(u)$ in (20) should be omitted.

Theorem 3 *Let Condition 1 be satisfied. Then there exists a linear symmetric positive definite bounded operator S acting in the space $W_{2,\mu}^1(\Omega)$ such that for any $u \in W_{2,\mu}^1(\Omega)$*

$$\langle A'(u_0)u, Su \rangle \geq \|u\|_\mu^2 + \theta(u),$$

where $\theta(u)$ is a functional defined on $W_{2,\mu}^1(\Omega)$ and satisfying the condition : $\theta(u_n) \to 0$ as $u_n \to 0$ weakly in $W_{2,\mu}^1(\Omega)$.

From this theorem it follows, in particular, that the operators are Fredholm. We introduce an operator $J : W_{2,\mu}^1(\Omega) \to (W_{2,\mu}^1(\Omega))^*$ by the equality

$$\langle Ju, v \rangle = \int_\Omega (u, v)\mu dx$$

Theorem 4 *Let Condition 1 be satisfied. Then for all $\lambda \geq 0$ the operator $A'(u_0) + \lambda J$ is Fredholm. For all $\lambda \geq 0$, except perhaps a finite number, it has a bounded inverse defined on the whole $(W_{2,\mu}^1(\Omega))^*$.*

We use this theorem to investigate the isolatedness of a stationary point and to calculate its index.

230

Theorem 5 *Let u_0 be a stationary point of the operator $A(u)$, and suppose that the equation*

$$A'(u_0)u = 0$$

has no solutions except zero. Then the stationary point u_0 is isolated, and the absolute value of its index is equal to 1.

We show that the sign of the index is connected with the multiplicity of eigenvalues similarly to completely continuous vector fields (see, for example, [10]). To do this we map $(W^1_{2,\mu}(\Omega))$ into $(W^1_{2,\mu}(\Omega))^*$ by means of the operator J, and denote $(W^1_{2,\mu}(\Omega))^*_0 = J(W^1_{2,\mu}(\Omega))$. We consider the operator $A_* = A'(u_0)J^{-1}$ acting in $(W^1_{2,\mu}(\Omega))^*$ with the domain $(W^1_{2,\mu}(\Omega))^*_0$. It follows from Theorem 4.1 that the operator A_* has no more than a finite number of negative eigenvalues, and all other negative numbers are its regular points. We note that the real eigenvalues λ of the operator A_* satisfy the equality

$$\langle A'(u_0)u, v \rangle = \lambda \int_\Omega (u,v)\mu dx$$

for some $u \neq 0$, $u \in W^1_{2,\mu}(\Omega)$ and for all $v \in W^1_{2,\mu}(\Omega)$. Here Ju is an eigenfunction of the operator A_* which corresponds to the eigenvalue λ. In fact, we are speaking about the usual definition of eigenvalues and eigenfunctions for differential operators in the class of generalized solutions in $W^1_{2,\mu}(\Omega)$.

Theorem 6 *Under the conditions of Theorem 5 the index of a stationary point u_0 is equal to $(-1)^\nu$, where ν is the sum of the multiplicities of the negative eigenvalues of the operator A_*.*

Proofs of Theorems 1 - 6 can be found in [15], [16].

2. Existence of waves.

When the degree is constructed we can apply the Leray-Schauder method to prove existence of travelling wave solutions for certain classes of systems. We consider the locally monotone systems which are determined by the following condition:

If $F_i(u_0) = 0$ for some i and u_0, then

$$\frac{\partial F_i}{\partial u_j} > 0, \quad j \neq i, \quad j = 1, ..., n. \tag{21}$$

From now on we consider the case of one spatial variable assuming also that the matrix a is diagonal.

We remind further that existence of waves depends on stability of the points w_+ and w_- as stationary points of the system $du/dt = F(u)$. There are three basic cases: *bistable case* where all eigenvalues of the matrices $F'(w_\pm)$ lie in the left half-plane, *monostable case* where one of the matrices $F'(w_+)$ and $F'(w_-)$ has all eigenvalues in the left half-plane and the other has at least one eigenvalue with a positive real part, *unstable case* where each of the matrices $F'(w_\pm)$ have at least one eigenvalue in the right half-plane.

The following theorem gives the main result on wave existence for the bistable case [15].

Theorem 7 *Let the function $F(u)$ have a finite number of zeros $w^{(1)}, ..., w^{(k)}$ in the interval (w_+, w_-) (i.e. for $w_+ < w < w_-$). Suppose that the matrices $F'(w_\pm)$ have all eigenvalues in the left half-plane, and each of the matrices $F'(w^{(i)})$, $i = 1, ..., k$ has at least one eigenvalue in the right half-plane.*

Then there exists a monotone travelling wave solution, i.e. the function $w(x)$ with the limits (4) and a constant c which satisfy the equation

$$aw'' + cw' + F(w) = 0. \tag{22}$$

If the inequality (21) is satisfied for all u, then the wave solution is unique up to translation.

We note that under some additional conditions the strict inequality (21) can be replaced by the nonstrict one [17].

We present for completness the results on wave existence for the monostable and unstable cases [23]. For the monostable case we suppose that the inequality (21) is satisfied everywhere, for the unstable case we don't need this condition.

To formulate the existence theorem for the monostable case we introduce the functional ω^* which is used to calculate the minimal velocity of the wave. Let K be the class of functions $\rho \in C^2(-\infty, \infty)$ which decrease monotonically and satisfy the following conditions at infinity:

$$\lim_{x \to \pm\infty} \rho(x) = w_\pm. \tag{23}$$

We put

$$\psi^*(\rho) = sup_{x,k} \frac{a_k \rho_k''(x) + F_k(\rho(x))}{-\rho_k'}, \tag{24}$$

$$\omega^* = inf_{\rho \in K} \psi^*(\rho). \tag{25}$$

Here a_k are the diagonal elements of the matrix a.

Theorem 8 *Suppose that there exists a vector $p \geq 0$, $p \neq 0$ such that*

$$F(w_+ + sp) \geq 0 \quad for \quad 0 < s \leq s_0, \tag{26}$$

where s_0 is a positive number. Suppose further that in the interval $[w_+, w_-]$ (i.e. for $w_+ \leq w \leq w_-$) there are no other zeros of the function $F(w)$ except for w_+ and w_-. Then there is a monotone decreasing solution of the problem (1.2), (1.3) for all $c \geq \omega^*$. Such solutions do not exist for $c < \omega^*$.

We discuss finally the unstable case. We begin with the following definition.

Definition. We say that the function $F(u)$ satisfies the condition of positivity in a positive neighbourhood of the point u_0 if there is a vector $p \geq 0$ such that

$$(p, F(u) - F(u_0)) > 0$$

for all $u \geq u_0$, $u \neq u_0$ in some neighbourhood of the point u_0. In this case we use the notation $F \in P(u_0)$.

Similarly, the function $F(u)$ satisfies the condition of negativity in a negative neighbourhood of the point u_0 if there is a vector $p \geq 0$ such that

$$(p, F(u) - F(u_0)) < 0$$

for all $u \leq u_0$, $u \neq u_0$ in some neighbourhood of the point u_0. In this case we denote $F \in N(u_0)$.

Obviously, if there is a vector $p \geq 0$ such that $pF'(u_0) > 0$ then $F \in P(u_0)$ and $F \in N(u_0)$. Here F' denotes the matrix $\partial F/\partial u$.

Theorem 9 *If $F \in P(w_+)$ and $F \in N(w_-)$ then a monotone wave solution with the limits (4) does not exist.*

3. Wave trains.

In the previous Section we have presented the results on wave existence in the bistable case assuming that there are no stable zeros in the interval $w_+ < w < w_-$. If however there are stable zeros in it, then the $[w_+, w_-]$-wave (i.e. the wave with the limits (4)) may not exist.

We explain the situation on a simple example which is a particular case of the results obtained in [5] for the scalar equation. We consider a function F which has three intermediate zeros, one stable and two unstable. More precisely it satisfies the following conditions:

$$F(w_0) = F(w_1) = F(w_2) = 0, \ w_+ < w_1 < w_0 < w_2 < w_-,$$

$$F'(w_\pm), F'(w_0) < 0,$$

$$F(w) \neq 0, \ w \in (w_+, w_-), \ w \neq w_0, w_1, w_2.$$

As it is known, under these conditions there exist $[w_+, w_0]$-wave and $[w_0, w_-]$-wave, i.e. the waves with the limits

$$\lim_{x \to -\infty} w(x) = w_0, \ \lim_{x \to +\infty} w(x) = w_+,$$

and
$$\lim_{x \to -\infty} w(x) = w_-, \quad \lim_{x \to +\infty} w(x) = w_0,$$

respectively. We denote by c_1 and c_2 the velocities of these waves. The result on the wave existence can be formulated in the following way:

If $c_2 > c_1$ then $[w_+, w_-]$-wave exists, if $c_2 \leq c_1$ it does not exist.

In the last case there are two waves which propagate one after another with different velocities. The solution of the Cauchy problem for large time has two "steps" and the distance between them increases in time. This structure which contains several "steps" is called wave trains. In the most simple case a wave train consists of a single wave.

We present here some of our recent results devoted to wave trains for the monotone systems [24] for which the inequality (21) is supposed to be satisfied everywhere.

We assume that there is a finite number of zeros of the function F in the interval (w_+, w_-) : $w^0, ..., w^p$ and $\tilde{w}^0, ..., \tilde{w}^k$. We assume further that the matrices $F'(w^i)$, $i = 0, 1, ..., p$ have all eigenvalues in the left half-plane while each of the matrices $F'(\tilde{w}^i)$, $i = 0, 1, ..., k$ has at least one eigenvalue in the right half-plane.

We begin with the most simple case when there is only one stable stationary point in the interval (w_+, w_-), i.e. $p = 0$.

Theorem 10 *Let $p = 0$ and c^+ and c^- be the velocities of the $[w_+, w^0]$-wave and $[w^0, w_-]$-wave, respectively. Then the $[w_+, w_-]$-wave exists if $c^+ < c^-$. It does not exist if $c^+ \geq c^-$.*

We consider now the general case.

Theorem 11 *Let $[w_+, w^i]$-wave and $[w^j, w_-]$-wave exist, and c^+ and c^- be their velocities, respectively. Suppose that $w^i \leq w^j$.*
If $c^+ \geq c^-$, then the $[w_+, w_-]$-wave does not exist.
If the $[w_+, w_-]$-wave exists and c is its velocity, then $c^+ \leq c \leq c^-$.

Theorem 12 *Let the points $w^0, ..., w^p$ satisfy the inequality $w^+ < w^0 < ... < w^p < w^-$, and c^+, $c^1, ..., c^p$, c^- be the velocities of $[w^+, w^0]$-, $[w^0, w^1]$-,..., $[w^{p-1}, w^p]$-, $[w^p, w^-]$- waves, respectively.*
If $c^+ < c^1 < ... < c^p < c^-$, then the $[w_+, w_-]$-wave exists.

In these two theorems the stable stationary points are ordered in the sense of the inequality $w^+ < w^0 < ... < w^p < w^-$. It is always the case for the scalar equation. However for systems it can be different. The next theorem considers the "nonordered" case where the multiplicity of wave trains occurs.

234

Theorem 13 *Let $p = 1$ and the points w^0 and w^1 are such that none of the inequalities*

$$w^0 \leq w^1 \quad \text{or} \quad w^1 \leq w^0$$

are satisfied. Let further $c_0^{\pm}$ and $c_1^{\pm}$ be the velocities of the $[w_+, w^0]$-, $[w^0, w_-]$-, $[w_+, w^1]$-, $[w^1, w_-]$- waves.
If the both of the inequalities

$$c_0^+ < c_0^-, \quad c_1^+ < c_1^-$$

are satisfied, then the $[w_+, w_-]$-wave exists. If at least one of them is not satisfied, then it does not exist.

The results on wave existence are obtained by the Leray-Schauder method [24].

4. Waves in combustion and chemical kinetics.

Travelling waves in chemical kinetics and combustion are well studied in the case of one-step chemical reaction and for some simple example of the multistep kinetics. Waves with complex kinetics are less studied because the mathematical analysis becomes essentially more complicated. We discuss here some results and approaches to the study of chemical waves with complex kinetics.

We consider a chemical reaction of the general form

$$\sum_{j=1}^{m} \alpha_{ij} A_j \to \sum_{j=1}^{m} \beta_{ij} A_j, \quad i = 1, ..., n. \tag{27}$$

Here $A_1, ..., A_m$ are concentrations of the reactants, α_{ij}, β_{ij} the stoichometric coefficients. Under the usual approximation of constant density the distribution of the temperature and the concentrations can be described by the reaction-diffusion system

$$\frac{\partial T}{\partial t} = \kappa \frac{\partial^2 T}{\partial x^2} + \sum_{i=1}^{n} q_i W_i, \tag{28}$$

$$\frac{\partial A_j}{\partial t} = d \frac{\partial^2 A_j}{\partial x^2} + \sum_{i=1}^{n} \gamma_{ij} W_i, \quad j = 1, ..., m. \tag{29}$$

where T is the temperature, κ and d are the coefficients of the heat and mass diffusion, respectively, q_i is the adiabatic heat release of the i-th reaction, $\gamma_{ij} = \beta_{ij} - \alpha_{ij}$, W_i is the rate of the i-th reaction,

$$W_i = K_i(T) A_1^{\alpha_{i1}} \times ... \times A_1^{\alpha_{im}}.$$

The functions $K_i(T)$ determine the temperature dependence of the reaction rate and usually have the form of the Arrhenius exponent,

$$K_i(T) = k_i^0 e^{-E_i/RT},$$

where k_i^0 is a constant, E_i is the activation energy of the i-th reaction, R is the gas constant.

Our main approach to study travelling waves in chemical kinetics is to reduce the reaction-diffusion system (28), (29) to the locally monotone or monotone system. If it can be done, then we can apply the results on wave existence (Sections 2, 3), stability and velocity (see [17], [23]) to chemical waves. We consider briefly some classes of reactions for which this reduction can be done [17], [19]-[22].

1. *Reactions with an open graph.*

By the definition reaction with an open graph is such reaction that there exists a vector σ for which $\sigma\Gamma < 0$ [13]. Here Γ is the matrix with elements γ_{ij}. In particular it can be linearly independent reactions for which the columns of the matrix Γ are linearly independent, or reactions with positive heat releases q_i.

The basic property of reactions with an open graph can be explained as follows. If we consider the kinetic system of equations

$$\frac{dA}{dt} = \Gamma W, \tag{30}$$

$A = (A_1, ..., A_m)$, $W = (W_1, ..., W_n)$, and multiply it from the right by the vector σ, then we obtain $d(\sigma, A)/dt < 0$ when the concentrations are positive. Thus there are linear combinations of the concentrations which decrease in time. If r is the rank of the matrix Γ, we introduce r such linear combinations $u_1, ..., u_r$ and rewrite the kinetic system in the form

$$\frac{du}{dt} = KW,$$

where $u = (u_1, ..., u_r)$, K is a numerical matrix with nonpositive elements. Thus the nonlinear term in the reaction-diffusion system becomes negative (in the region where the concentrations are positive) and this gives a particular case of the locally monotone systems. We can't apply here directly the result of the Section 2 on wave existence because the reaction-diffusion systems have some specific features not taken into account in the general theory. However we can apply the same approach based on the Leray-Schauder method (see [17], [19]).

There are different realizations of this approach. In [2], [7] existence of solutions is proved first on a finite interval and then the passage to the limit is fulfilled as the length of the interval goes to infinity. In [17], [19] from the very beginning we study the problem on the whole axis. We consider a model system for which existence of solutions is known, and then deform it to the system under consideration. The difference in realization of the Leray-Schauder method gives some difference in the results. In all cases a travelling wave connects stationary points of the kinetic system (30). However in [2], [7] these stationary points are not determined, while in [17], [19] is proved existence of waves connecting a priori given stationary points.

There is also some difference in the results connected with the restrictions on the parameters. In [7] is considered the case of exhothermic reactions, in [17], [19] reactions can be exhothermic or endothermic but the case of heat releases of the alternating signs is studied under the additional condition $\kappa = d$, in [2] this additional condition is omitted in the case of the graph of reactions with no circuit.

Reduction to the locally monotone system allows to prove existence of travelling waves. If we reduce the reaction-diffusion system to the monotone system, we can obtain not only existence of waves but also their stability and the minimax representation of the velocity. The general conditions of reducibility to the monotone systems are given in [17], [20]-[22]. They have a pure algebraic form but nevertheless are rather complicated. For simplicity we present here only some examples of reactions for which this reduction can be done. We note that if the reactions are not isothermic, we assume that $\kappa = d$.

2. *Exhothermic reactions without parallel stages.*
It can be for example independent reaction:

$$A_1 + A_2 \to ..., \ A_3 + A_4 \to ..., \ A_5 + A_6 \to ...,$$

sequential reactions:

$$A_1 \to A_2 \to A_3 \to ...$$

or

$$A_1 + A_2 \to A_3, \ A_3 + A_4 \to A_5, \ A_5 + A_6 \to A_7,$$

nonbranching chain reactions

$$A_2 + B \to AB + A, \ B_2 + A \to AB + B,$$

and many others.

We note that if the reaction contains parallel stages or some of the elementary reactions are endothermic, then we obtain some additional conditions on the parameters under which the system can be reduced to the monotone system. This means that we obtain analytical conditions of uniqueness and stability of a wave. This approach can be also applied if the reaction contains reversible stages [17], [20]-[22].

References

[1] Berestycki H., Nirenberg L. Travelling fronts in cylinders. Annales de l'IHP. Analyse non lineaire. **9** (1992), No. 5, pp. 497-572.

[2] Bonnet A. Travelling waves for plannar flames with complex chemistry reaction network. Commun. on Pure and Appl. Math., **45** (1992), No. 12.

[3] Browder F.E. Degree theory for nonlinear mappings. Proceedings of Symposia in Pure Mathematics. **45** (1986), Part I, pp. 203-226.

[4] M.Escobedo, O.Kavian. Variational problems related to self-similar solutions of the heat equation. Nonlinear Analysis TMA, Vol. 11 (1987), No. 10, 1103- 1133.

[5] Fife P.C., McLeod J.B. The approach to solutions of nonlinear diffusion equations to travelling front solutions. Arch. Ration. Mech. and Anal. **65** (1977), pp. 335- 361.

[6] Gardner R. Existence of multidimensional travelling wave solutions of an initial-boundary value problem. J. Diff. Eqns **61** (1986), pp. 335-379.

[7] Heinze S. Traveling waves in combustion processes with complex chemical networks. Trans. Amer. Math. Soc. **304** (1987), No. 1, pp. 405-416.

[8] Heinze S. Traveling waves for semilinear parabolic partial differential equations in cylindrical domains. Preprint No. 506, Heidelberg, 1989, 46 p.

[9] D.Henry. Geometrical theory of semilinear parabolic equations. Lecture notes in mathematics, Vol. 840, Springer-Verlag, Berlin - New York, 1981.

[10] Krasnoselskii M.A., Zabreiko P.P. Geometrical methods of nonlinear analysis. Springer-Verlag, Berlin - New York, 1984.

[11] O.A.Ladyzhenskaya, N.N.Uraltseva. Linear and quasilinear elliptic equations. Academic Press, New York, 1968.

[12] Skrypnik I.V. Nonlinear elliptic equations of higher order. Naukova Dumka, Kiev, 1973 (in Russian). Nonlinear elliptic boundary value problems. Teubner-Texte Zur Mathematik, Vol. 91 (1986), BSB B.G. Teubner Verlagsgesellschaft, Leipzig, 232 p.

[13] Volpert A.I., Hudjaev S.I. Analysis in classes of discontinuous functions and equations of mathematical physics. "Nauka", Moscow, 1975. English translation: Martinus Nijhoff, Dordrecht, 1985.

[14] Volpert A.I., Volpert V.A. Construction of the rotation of the vector field for operators describing wave solutions of parabolic systems. Soviet Math. Dokl., Vol. 36 (1988), No. 3, 452-455.

[15] Volpert A.I., Volpert V.A. Application of the rotation theory of vector fields to the study of wave solutions of parabolic equations. Trans. Moscow Math. Soc., Vol. 52 (1990), 59-108.

[16] Volpert A.I., Volpert V.A. Construction of the Leray-Schauder degree for the elliptic operators in unbounded domains. Annales de l'inst. H. Poincare. Analyse non lineaire, 1994, **11**, No. 3, pp. 245-247

238

[17] A.I.Volpert, V.A.Volpert, V.A.Volpert. Travelling wave solutions of parabolic systems. American Math. Society, Providence, 1994.

[18] Volpert V.A. The spectrum of an elliptic operator in an unbouded cylindrical domain. Dokl. Akad. Nauk Ukrain. SSR, Ser. A, 1981, No. 9, pp. 9-12.

[19] Volpert V.A., Volpert A.I. Existence and stability of traveling waves in chemical kinetics. In: Dynamics of Chemical and Biological Systems, Nauka, Novosibirsk, 1989, pp. 56-131 (in Russian).

[20] Volpert V.A., Volpert A.I. Waves of chemical transformation having complex kinetics. Dokl. Phys. Chemistry **309** (1989), No. 1-3, pp. 877-879.

[21] Volpert V.A., Volpert A.I. Existence and stability of waves in chemical kinetics. Khimicheskaya Fizika **9** (1990), No. 2, pp. 238-245 (in Russian).

[22] Volpert V.A., Volpert A.I. Some mathematical problems of wave propagation in chemical active media. Khimicheskaya Fizika **9** (1990), No. 8, pp. 1118-1127 (in Russian).

[23] Volpert V.A., Volpert A.I. Traveling waves described by monotone parabolic systems. Preprint No. 146, CNRS URA 740, 1993, 46 p.

[24] Volpert V.A., Volpert A.I. Wave trains described by monotone parabolic systems. Submitted to Comm. in PDE.

Department of Mathematics
Technion, 32000 Haifa, Israel

Laboratoire d'analyse numérique
Université Lyon-1, URA 740 CNRS
69622 Villeurbanne, France

H F WEINBERGER

Degenerate elliptic models for perfectly plastic flows

1. Introduction.

In this work we shall discuss some questions which arise in the modelling of the drawing or extrusion of metal through a die. This process is illustrated in the figure below. Its purpose is to turn a cylindrical ingot of some given cross section such as a rectangle into a cylindrical bar with a different cross section, such as an I-beam.

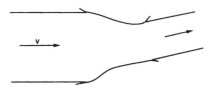

It is assumed that the two ends are infinitely long, and that the flow is steady.

2. A class of models for perfectly plastic flow.

The model which is frequently used for such a process is that of a perfectly plastic (or rigid-plastic) flow. Such a flow is described as follows. Let $\mathbf{v}(\mathbf{x})$ be a velocity field in the flow region. The associated rate of strain is the symmetric matrix D defined by

$$D_{ij} := \frac{1}{2}\left(\frac{\partial v_i}{\partial x_j} + \frac{\partial v_j}{\partial x_i}\right).$$

There is also a stress field $\sigma(\mathbf{x})$, which is a symmetric-matrix valued function in the material.

The properties of the metal are specified by prescribing a fixed closed convex set S of symmetric matrices. The stress is required to lie in this set. As long as $\sigma(\mathbf{x})$ is in the interior of S, the metal is rigid, which means that the rate of strain D at $\mathbf{x}$ is zero. When $\sigma(\mathbf{x})$ is on the boundary of S, the material yields sufficiently to prevent the stress from leaving S. There are various ways of related the resulting rate of strain to the stress. We shall assume that the rates of strain is normal to the surface S at the stress σ. The stress and the rate of strain are then related in the following manner:

For D a fixed symmetric matrix, define the support function

$$H(D) = \sup_{\sigma \in S} \operatorname{tr}(\sigma D). \tag{1}$$

The stress-strain law states that two symmetric matrices σ and $D \neq 0$ can be the stress and rate of strain at any particular point $\mathbf{x}$ if and only if $\sigma \in S$ and $\operatorname{tr}(\sigma D) =$

240

$H(D)$, so that σ is a maximizer in the problem (1). When H is differentiable at D, the stress-rate of strain relation is given by

$$\sigma_{ij} = \frac{\partial H}{\partial D_{ij}}.$$

Because the function $H(D)$ defined by (1) is positive homogeneous of degree 1, we see that if a rate of strain D has the associated stress σ, then any positive multiple of D has the same associated stress.

The quantity $H(D)$ represents the rate of energy dissipation per unit volume due to friction. The equation of motion is obtained from the variational principle: Minimize

$$\int H(D) d\mathbf{x} \tag{2}$$

under the boundary conditions that the velocity is tangential on the lateral boundary including the die, and that $\mathbf{v}$ takes on prescribed constant values on prescribed cylinders upstream and downstream. The integral is to be taken over the flow region. We see from the definition (1) that $D = 0$ implies that $H(D) = 0$. Since the velocity is to be constant outside a bounded set, the integral in (2) may be taken over this bounded set.

When H is a smooth function of D, equating the first variation of the integral in (2) to zero gives the Euler equation

$$\sum_{j=1}^{3} \frac{\partial}{\partial x_j} \left(\frac{\partial H}{\partial D_{ij}} \right) = 0 \tag{3}$$

and the boundary condition

$$\sum_{j=1}^{3} \frac{\partial H}{\partial D_{ij}} n_j = \gamma n_i$$

for some scalar function γ on the lateral boundary, including the die. In view of the above formula for σ, (3) is just the equilibrium force condition, which states that the divergence of the stress tensor is zero. The boundary condition states that no tangential force is applied.

The formal Fourier transform of the Euler equation (3) has the form

$$\sum_{j,k,\ell=1}^{3} \frac{\partial^2 H}{\partial D_{ij} \partial D_{k\ell}} \xi_j \xi_\ell w_k = 0. \tag{4}$$

Because the function H is convex, the quadratic form

$$\sum_{i,j,k,\ell=1}^{3} \frac{\partial^2 H}{\partial D_{ij} \partial D_{k\ell}} Q_{ij} Q_{k\ell}$$

241

of the coefficient matrix of this system is positive semidefinite over the symmetric matrices Q. Because the function H is homogeneous of degree one, this coefficient matrix has the null vector $D_{k\ell}$. Thus the system is at least degenerate elliptic, and it is not strongly elliptic. We shall investigate whether or not this system has real characteristics.

By taking the dot product of (4) with the vector $\mathbf{w}$, we see that if (4) holds, then the quadratic form of the coefficient matrix vanishes on the symmetric matrix $\frac{1}{2}\{\boldsymbol{\xi}\otimes\mathbf{w}+\mathbf{w}\otimes\boldsymbol{\xi}\}$. Because this quadratic form is positive semidefinite, we conclude that (4) is satisfied if and only if the symmetric part of the rank one matrix $\boldsymbol{\xi}\otimes\mathbf{w}$ lies in the null space of the coefficient matrix. That is, **the Euler equation (3) has a real characteristic if and only if the null space of the coefficient matrix** $\partial^2 H/\partial D_{ij}\partial D_{k\ell}$ **contains the symmetric part of a matrix of rank one.**

The following lemma (see, e.g., Kohn [4]) gives a sufficient condition for the existence of real characteristics. For the sake of completeness, we give a proof.

LEMMA 1. *A 3×3 matrix can be written as the symmetric part of a rank one matrix if and only if its intermediate eigenvalue is zero.*

Proof. Suppose the middle eigenvalue of the matrix D is zero. Choose coordinates which diagonalize the matrix. Then $D=\mathrm{diag}(\lambda_1,0,\lambda_3)$, and hence

$$D = \frac{1}{2}\{(\sqrt{\lambda_1},0,\sqrt{-\lambda_3})\otimes(\sqrt{\lambda_1},0,-\sqrt{-\lambda_3})+(\sqrt{\lambda_1},0,-\sqrt{-\lambda_3})\otimes(\sqrt{\lambda_1},0,\sqrt{-\lambda_3})\},$$

which is the symmetric part of a rank one matrix.

If, on the other hand, $D = \frac{1}{2}[\mathbf{a}\otimes\mathbf{b}+\mathbf{b}\otimes\mathbf{a}]$, then its quadratic form vanishes on the union of the orthogonal complements of $\mathbf{a}$ and $\mathbf{b}$, while the null space of the matrix is the intersection of these two complements. Therefore, if $\mathbf{a}$ and $\mathbf{b}$ are linearly independent, the quadratic form is indefinite, so that 0 is the intermediate eigenvalue. If $\mathbf{a}$ and $\mathbf{b}$ are linearly dependent, the matrix has two zero eigenvalues one of which can then be thought of as the intermediate eigenvalue.

This Lemma yields the following Corollary.

Corollary 1. If $H(D)$ is smooth at a D which has the eigenvalues $\lambda_1 \geq 0$, 0, and $\lambda_3 \leq 0$, then in coordinates in which D is diagonal, the normal cone of the Euler equation system (1) at this D contains the vectors which have the form $(\sqrt{\lambda_1},0,\pm\sqrt{-\lambda_3})$.

Proof. Because $\partial H/\partial D$ is homogeneous of degree 0, the matrix D is in the null space of $\partial^2 H/\partial D_{ij}\partial D_{k\ell}$. We see from the Lemma that if in the coordinates which diagonalize D we let $\mathbf{a} := (\sqrt{\lambda_1},0,\sqrt{-\lambda_3})$ and $\mathbf{b} := (\sqrt{\lambda_1},0,-\sqrt{-\lambda_3})$, then

$$\sum_{k,\ell=1}^{3}\frac{\partial^2 H}{\partial D_{ij}\partial D_{k\ell}}a_k b_\ell = 0.$$

Therefore, if ξ is either the vector **a** or the vector **b**, the characteristic matrix $\sum \partial^2 H/\partial D_{ij} \partial D_{k\ell}\xi_j\xi_\ell$ is singular.

We remark that if S is invariant under the transformations $\sigma \to U^t\sigma U$ for all orthogonal U, then the function $H(D)$ is also invariant under these transformations. This occurs when the material is isotropic.

Because it is found experimentally that the yield behavior of most plastic materials is not affected significantly by the addition of a a hydrostatic stress, it is usually assumed that the set S is invariant under the translations $\sigma \to \sigma + \alpha I$. Then $H(D) = \infty$ unless $\text{tr}(D)=0$. Therefore, minimization of (2) implies the constraint $\text{tr}(D)=0$, so that the flow is incompressible. The maximizer σ is only determined to within an additive scalar matrix. That is,

$$\sigma_{ij} = \frac{\partial H}{\partial D_{ij}} - p\delta_{ij}$$

where the scalar p is a hydrostatic pressure. The Euler equation states that the divergence of this matrix is zero, and we add to this the fact that the divergence of **v** is zero. The formal Fourier transform of the resulting system is

$$\sum_{j,k,\ell=1}^{3} \frac{\partial^2 H}{\partial D_{ij}\partial D_{k\ell}}\xi_j\xi_\ell\hat{v}_k - \hat{p}\xi_i = 0$$

$$\sum_{k=1}^{3} \xi_k\hat{v}_k = 0.$$

By taking the scalar product of $\hat{\mathbf{v}}$ with the first equation, we again see that the matrix $\hat{\mathbf{v}} \otimes \boldsymbol{\xi} + \boldsymbol{\xi} \otimes \hat{\mathbf{v}}$ must lie in the null space of the second derivative matrix of H. We note that if D is singular, then because its trace is zero, it can be written in the form $D = \text{diag}(\lambda, 0, -\lambda)$ with $\lambda > 0$. Therefore the vector pairs $\boldsymbol{\xi} = (\sqrt{\lambda}, 0, \pm\sqrt{\lambda})$, $\hat{\mathbf{v}} = (\sqrt{\lambda}, 0, \mp\sqrt{\lambda})$ satisfy this equation with $\hat{p} = 0$. That is, Corollary 1 is still valid:

Corollary 2. The incompressible Euler equation has real characteristics at points where D is singular, and if $D = \text{diag}(\lambda, 0, -\lambda)$ with $\lambda > 0$, then the vectors $(1, 0, \pm1)$ lie in the normal cone.

If the flow is two-dimensional, the rate of strain has a zero eigenvalue in the third direction. Thus, Corollaries 1 and 2 imply that two-dimensional problems are automatically hyperbolic, with double characteristics. It is this kind of problem which one finds in most of the examples in the classical text books [1,2].

The situation is quite different in three dimensions. Corollaries 1 and 2 give sufficient conditions for the existence of some real characteristics, but they neither predict nor preclude the existence of further real characteristics. We shall use the

243

two most common models of plasticity to show that other characteristics may or may not occur.

The von Mises model : The constraint set is $S = \{\sigma : \text{tr}([\sigma - \frac{1}{3}\text{tr}(\sigma)]^2) \leq 1\}$.

We find that $H(D) = \sqrt{\text{tr}(D^2)}$ when $\text{tr}(D)=0$ and ∞ otherwise, so that the formal Fourier transform of the Euler equation is of the form

$$\{\text{tr}(D^2)\}^{-3/2} \left\{ \frac{1}{2}\text{tr}(D^2)|\boldsymbol{\xi}|^2 I - D\boldsymbol{\xi} \otimes D\boldsymbol{\xi} \right\} \hat{\mathbf{v}} - \hat{p}\boldsymbol{\xi} = 0$$

$$\boldsymbol{\xi} \cdot \hat{\mathbf{v}} = 0.$$

If local coordinates along the principal axes of D are used and the eigenvalues of D are denoted by $\lambda_1 \geq \lambda_2 \geq \lambda_3$, the characteristic polynomial, which is the determinant of the coefficient matrix, becomes

$$-\frac{1}{4}\{\text{tr}(D^2)\}^{-7/2}|\boldsymbol{\xi}|^2 \left\{ (\lambda_1^2 + \lambda_2^2 + \lambda_3^2)(\xi_1^2 + \xi_2^2 + \xi_3^2)^2 - \sum_{i,j=1}^{3}(\lambda_i - \lambda_j)^2\xi_i^2\xi_j^2 \right\}.$$

To determine the behavior of this function, we obtain a bound for the sum. Because the sum is a convex function of λ_2, its value is bounded by the larger of values obtained by replacing λ_2 by λ_1 and by λ_3. Therefore,

$$\sum_{i,j=1}^{3}(\lambda_i - \lambda_j)^2\xi_i^2\xi_j^2 \leq 2(\lambda_3 - \lambda_1)^2[\xi_1^2\xi_3^2 + \xi_2^2 \max(\xi_1^2, \xi_3^2)]$$

$$= 2(\lambda_3 - \lambda_1)^2 \max(\xi_1^2, \xi_3^2)[|\boldsymbol{\xi}|^2 - \max(\xi_1^2, \xi_3^2)]$$

$$\leq \frac{1}{2}(\lambda_3 - \lambda_1)^2|\boldsymbol{\xi}|^4.$$

Thus the factor in braces in the characteristic polynomial is bounded below by $\{\frac{1}{2}(\lambda_1 + \lambda_3)^2 + \lambda_2^2\}|\boldsymbol{\xi}|^4$. By the trace condition, $\lambda_1 + \lambda_3 = -\lambda_2$, so that this lower bound can be written as $\frac{3}{2}\lambda_2^2|\boldsymbol{\xi}|^4$.

We conclude that the system is elliptic unless $\lambda_2 = 0$. When $\lambda_2 = 0$, the factor in braces in the characteristic polynomial becomes

$$-\lambda_1^4|\boldsymbol{\xi}|^2\{(\xi_1^2 - \xi_3^2)^2 + \xi_2^2(\xi_1^2 + \xi_3^2) + \xi_2^4\}.$$

Thus the system has the real normal cone $\xi_2 = 0$, $\xi_1 = \pm\xi_3$ at those points where $\lambda_2 = 0$, or otherwise stated, where the determinant of D vanishes. That is, the von Mises system has precisely the real characteristics required by Corollary 2, and no others.

This result was found by T.Y.Thomas [8].

The Tresca model. We let

$$S = \{\sigma : |\sigma \mathbf{n}|^2 - (\mathbf{n} \cdot \sigma \mathbf{n})^2 \leq |\mathbf{n}|^2 \qquad \text{for all vectors } \mathbf{n} \text{ with } |\mathbf{n}| = 1\}.$$

That is, it is required that the tangential component of the force per unit area on any surface in the material does not exceed the constant 1.

It is easily shown that $H(D) = ||D||$ when the trace of D is zero and infinity otherwise. Here $||D||$ is the operator norm of the linear transformation $\mathbf{a} \to D\mathbf{a}$ on the Euclidean space R^3. As is well known, $||D||$ is the largest of the absolute values of the eigenvalues of D.

The function $||D||$ is smooth when the eigenvalue of largest absolute value is unique, that is, when the determinant of D is not zero. In this case, the Euler equation is most easily derived in the following way.

Let $\mathbf{w}$ be a vector-valued function which points in the direction of the eigenvector which corresponds to the eigenvalue λ of D which has the largest magnitude. That is,

$$D\mathbf{w} = \lambda \mathbf{w}. \tag{5}$$

The usual derivative formula for the eigenvalue shows that

$$\frac{\partial \lambda}{\partial D_{ij}} = \frac{w_i w_j}{|\mathbf{w}|^2}.$$

Since $H(D) = |\lambda|$, we find that

$$\sigma_{ij} = \text{sgn}(\lambda) \frac{w_i w_j}{|\mathbf{w}|^2} - p\delta_{ij}.$$

The eigenvector $\mathbf{w}$ is, of course, only defined to within a scalar multiple. We now define this multiple by choosing $|\mathbf{w}| = e^{-\text{sgn}(\lambda)p}$, so that

$$\sigma_{ij} = \text{sgn}(\lambda) \left\{ \frac{w_i w_j}{|\mathbf{w}|^2} + \ln |\mathbf{w}| \delta_{ij} \right\}.$$

The Euler equation states that the divergence of this matrix field must vanish, which yields the system

$$\frac{\text{div } \mathbf{w}}{|\mathbf{w}|^2} w_i + \frac{1}{|\mathbf{w}|^2} \sum_{j=1}^{3} \left(\frac{\partial w_i}{\partial x_j} + \frac{\partial w_j}{\partial x_i} \right) w_j - \frac{2}{|\mathbf{w}|^4} \sum_{j,k=1}^{2} \frac{\partial w_k}{\partial x_j} w_j w_k \ w_i = 0. \tag{6}$$

By taking the dot product of this expression with $\mathbf{w}$, one sees that div $\mathbf{w}=0$. One thus obtains a system of the form

$$\frac{1}{2} \sum_{j=1}^{3} \left(\frac{\partial w_i}{\partial x_j} + \frac{\partial w_j}{\partial x_i} \right) w_j - \gamma w_i = 0$$

$$\text{div } \mathbf{w} = 0. \tag{7}$$

By taking the dot product of the first equation with $\mathbf{w}$ and using the second equation, we see that this equation is equivalent to (6).

The equation (7) is to be used in conjunction with the eigenvalue equation (5) and the incompressibility condition:

$$\frac{1}{2}\sum_{j=1}^{3}\left(\frac{\partial v_i}{\partial x_j}+\frac{\partial v_j}{\partial x_i}\right)w_j-\lambda w_i=0$$

$$\text{div } \mathbf{v}=0.$$

(8)

We have thus replaced the usual Euler equation, which is of second order in $\mathbf{v}$ and first order in λ by the system (7), (8), which is of first order in the variables $(\mathbf{w}, \mathbf{v})$ and of order zero in the variables (γ, λ).

In order to find the characteristics, we freeze the coefficient variable $\mathbf{w}$ in these equations. We then find that the equations (7) for $(\mathbf{w}, \gamma)$ and the equations (8) for $(\mathbf{v}, \lambda)$ are uncoupled and identical. Consequently, the real characteristics of the Euler system are just the characteristics of (8), but with their multiplicity doubled.

To find the normal cone, we use the reasoning of the derivation of Lemma 1 and its Corollaries. We see from this reasoning that ξ is in the characteristic cone if and only if there is a vector $\mathbf{b}$ such that the matrix $\xi \otimes \mathbf{b} + \mathbf{b} \otimes \xi$ has trace zero and lies in the null space of $\partial^2|\lambda|/\partial D_{ij}\partial D_{k\ell}$. Because λ is independent of the two lower eigenvalues, we see that this null space contains all linear combinations of matrices for which $\mathbf{w}$ is a null vector and the matrix D. This class of matrices is characterized by the fact that $\mathbf{w}$ is an eigenvector.

The conditions that $\mathbf{w}$ is an eigenvector of $\xi \otimes \mathbf{b} + \mathbf{b} \otimes \xi$ and that this matrix has trace zero are

$$\mathbf{b}\cdot\mathbf{w}\,\xi+\xi\cdot\mathbf{w}\,\mathbf{b}=\frac{2\mathbf{b}\cdot\mathbf{w}\,\xi\cdot\mathbf{w}}{|\mathbf{w}|^2}\mathbf{w}$$

and

$$\xi\cdot\mathbf{b}=0.$$

We distinguish two cases. If

$$\xi\cdot\mathbf{w}=0,$$

the two conditions are satisfied by $\mathbf{b}=\xi\times\mathbf{w}$, so that the normal cone contains the orthogonal complement of $\mathbf{w}$.

If, on the other hand, $\xi\cdot\mathbf{w}\neq 0$, the first condition shows that $\mathbf{b}$ must be a multiple of the reflected vector $2(\xi\cdot\mathbf{w}/|\mathbf{w}|^2)\mathbf{w}-\xi$. The orthogonality of this vector to ξ is satisfied if and only if

$$(\xi\cdot\mathbf{w})^2=\frac{1}{2}|\mathbf{w}|^2|\xi|^2.$$

The normal cone thus contains this right circular cone of half-angle $\pi/4$ about the w-direction as well as the plane perpendicular to w. It is easily seen that the characteristic polynomial of the system (7), (8) is of degree six. Since all characteristics are double, we conclude that all characteristics are real.

We have thus shown that at points where $\det(D) \neq 0$ the Euler system for the Tresca model has only real characteristics, so that it can be considered a hyperbolic system with double characteristics. We recall that the corresponding equations for the von Mises model are elliptic when $\det(D) \neq 0$.

Unfortunately, the Tresca model behaves badly at points where D is singular. At such points the norm $||D||$ is equal to the absolute values of both the positive and the negative eigenvalue. As a consequence, the functional $\int ||D|| dx$ is differentiable in the sense of Gâteaux but not in the sense of Fréchet at such points. That is, the vanishing of the first variation cannot be characterized in terms of a differential equation at these points.

We remark that Thomas [8] found the characteristics of a different system of flow equations based on the Tresca yield condition. Following von Mises, he required that the stress deviator be proportional to the strain rate, with the constant of proportionality determined by the Tresca yield condition. For these equations the normal cone at every point consists of only two directions.

3. An optimization problem.

An interesting problem in connection with the die flow is to determine a die profile which takes a given ingot shape into a prescribed product shape, and which makes the rate of dissipation $\int H(D) dx$ as small as possible. This problem for two-dimensional flows (sheet drawing) with the Tresca model was solved by a Richmond and Devenpeck [6] by imposing an extra condition on the solution. The idea was extended to axially symmetric flows by Richmond [5] and Richmond and Morrison [7]. Hill [3] showed how the extra condition should look in three dimensions.

Our purpose here is to derive the three-dimensional version of the Richmond condition from the optimum problem. We are interested in minimizing the integral (2) with $H(D) = ||D||$ not only over choices of incompressible v but also over choices of the die boundary.

We first observe that since $||D||$ is at least as large as the largest eigenvalue of D, the Poincaré principle shows that

$$||D|| \geq \frac{\mathbf{v} \cdot D\mathbf{v}}{|\mathbf{v}|^2} \tag{9}$$

for any vector field v, and in particular for the velocity field. Because this v is divergence free and because of the definition of D, we can write

$$\frac{\mathbf{v} \cdot D\mathbf{v}}{|\mathbf{v}|^2} = \operatorname{div}(\log |\mathbf{v}| \ \mathbf{v}).$$

We substitute this identity into the inequality (9) and integrate over a bounded part of the flow region which includes the set where the flow is plastic. Since $\mathbf{v} \cdot \mathbf{n} = 0$ on the lateral boundaries, we thus find the lower bound

$$\int H(D)dx \geq Q \ln(|\mathbf{v}_d|/|\mathbf{v}_u|)$$

where Q is the known flux rate of the material, and $\mathbf{v}_d$ and $\mathbf{v}_u$ are the prescribed constant downstream and upstream velocities.

We have thus obtained a lower bound for the dissipation rate which is independent of the die shape and depends only on prescribed quantities. (Note that because the flow is incompressible, the ratio of speeds is the inverse of the ratio of cross-sectional areas.) Equality will hold if and only if equality holds in (9). This is the case precisely when $\mathbf{v}$ is the eigenvector which corresponds to the largest eigenvalue of D, and this eigenvalue λ is the positive number $||D||$. Thus if we can find a velocity field which satisfies this condition in addition to the equations (7), (8), we have an optimal die.

At first glance, it seems unlikely that one can make a solution of the system (7), (8) satisfy an additional differential equation by leaving the die boundary undetermined. However, a miracle, which was observed by Richmond and Devenpeck in the two-dimensional case, occurs. Namely, this additional condition is equivalent to setting $\mathbf{v} = \mathbf{w}$ in the system (7), (8). Because the system (8) is obtained from (7) by replacing $\mathbf{v}$ by $\mathbf{w}$, we can then replace the whole system (7), (8) by the simpler system

$$\frac{1}{2} \sum_{j=1}^{3} \left(\frac{\partial v_i}{\partial x_j} + \frac{\partial v_j}{\partial x_i} \right) v_j - \lambda v_i = 0$$

$$\text{div } \mathbf{v} = 0.$$

(10)

This system has the characteristic polynomial $\boldsymbol{\xi} \cdot \mathbf{v}[|\mathbf{v}|^2|\boldsymbol{\xi}|^2 - 2(\boldsymbol{\xi} \cdot \mathbf{v})^2]$.

The system is now hyperbolic with simple characteristics. The streamlines constitute one family of characteristics, and the streamline direction is the axis of the right circular part of the normal cone. Thus the streamline direction is timelike.

The normal component of the velocity vanishes on the lateral boundary, including the part of the die in the flow.

The optimization problem is thus reduced to the following problem in control theory: Given an upstream profile and velocity and a compatible downstream profile and velocity, find a solution of the hyperbolic system (10) which connects them.

Richmond and Devenpeck [6] were able to solve this problem in two dimensions because the timelike and spacelike directions can be interchanged in a two-space. Richmond [5] and Richmond and Morrison [7] showed that the same idea still works in axially symmetric flows.

248

Work on the three-dimensional case is still in progress. At the moment even the existence of the solutions of the Euler system (7), (8) has not been established. It is, however, quite possible that the optimization problem is easier than this existence problem.

If one is to use the result, one needs not just an existence theorem, but a method for approximating the solution. One of the difficulties is that the die boundary, which is to be determined, is characteristic surface. One can get around this fact with the following artifice.

It is easily verified that the system (10) is equivalent to the system

$$\mathbf{v} \times \text{curl}\left(\frac{\mathbf{v}}{|\mathbf{v}|^2}\right) = 0$$
$$\text{div } \mathbf{v} = 0. \tag{11}$$

The first of these equations can be written as

$$\text{curl}\left(\frac{\mathbf{v}}{|\mathbf{v}|^2}\right) = \alpha\mathbf{v}$$

for some scalar-valued function α. We take the divergence of this equation to find that $\mathbf{v} \cdot \text{grad } \alpha = 0$, so that α is constant along each streamline. Since $\mathbf{v}$ is constant on the upstream side, we conclude that $\alpha = 0$ on all streamlines.

If we assume that the flow domain is simply connected, it follows that we may write

$$\frac{\mathbf{v}}{|\mathbf{v}|^2} = \text{grad } \tau$$

for some scalar-valued function $\tau(\mathbf{x})$. The equations (11) are then equivalent to the single condition

$$\text{div}\left(\frac{\text{grad } \tau}{|\text{grad } \tau|^2}\right) = 0. \tag{12}$$

This second order scalar equation is easily seen to be hyperbolic. Its normal cone consists of the right circular conical part of the normal cone of (10). Because the plane part of the cone of (10) is absent, the lateral boundary is not characteristic, which will make computations more stable.

The boundary condition $\mathbf{v} \cdot \mathbf{n} = 0$ becomes $\partial\tau/\partial n = 0$. Thus the optimization problem is reduced to finding a die boundary with the property that the solution of the initial-boundary value problem for (12) with prescribed constant gradient upstream and Neumann data on the boundary has a constant gradient downstream as well.

Acknowledgement: This work was begun as a result of lectures by O. Richmond and R. Mallett of ALCOA at the Seminar on Industrial Problems of the Institute for Mathematics and its Applications. It has been partly supported by grants from the ALCOA Foundation.

REFERENCES

1. W. Prager and P.G.Hodge, Theory of Perfectly Plastic Solids. Wiley, 1951.
2. R. Hill, The Mathematical Theory of Elasticity. Oxford U. Press, 1950.
3. R. Hill, *Ideal forming operations for perfectly plastic solids.* J. of Mech. and Phys. of Solids 15 (1967), pp. 223-227.
4. R. V. Kohn, *The relaxation of a double-well energy.* Continuum Mechanics and Thermodynamics 3 (1991), pp. 193-235.
5. O. Richmond, *Theory of streamlined dies for drawing and extrusion.* in Mechanics of the Solid State, ed. F. P. J. Rimrott and J. Schwaighofer, U. of Toronto Press, 1968, pp. 154-167.
6. O. Richmond and M. L. Devenpeck, *A die profile for maximum efficiency in strip drawing.* Proc. of the Fourth U .S National Congress of Applied Mechanics, Vol.2 (1962) pp. 1053-1057.
7. O. Richmond and H. L. Morrison, *Streamlined wire drawing dies of minimum length.* J. of Mech. and Phys. of Solids 15 (1967) pp. 195-203.
8. T. Y. Thomas, *On the characteristic surfaces in the von Mises plasticity equations.* Journal of Rational Mechanics and Analysis 1 (1952), pp. 343-357.

School of Mathematics
University of Minnesota
Mineapolis, MN 55455, U.S.A.

E ZUAZUA

Some recent results on the large time behavior for scalar parabolic conservation laws

Dedicated to the memory of Pierre Grisvard

In this paper we present some recent results on the large time behavior of solutions of multidimensional scalar conservation laws of the form:

$$u_t - \Delta u + \sum_{i=1}^{N} \frac{\partial f_i(u)}{\partial x_i} = 0 \quad \text{in} \quad \mathbf{R}^N \times (0, \infty), \tag{0.1}$$

$$u(x, 0) = u_0(x) \quad \text{in} \quad \mathbf{R}^N, \tag{0.2}$$

where $N \geq 1$, $f_i \in C^1([0, \infty)) \cap C^2((0, \infty))$, $f(0) = 0$ and $u_0 \in L^1(\mathbf{R}^N)$.

It is easy to see that the Cauchy problem (0.1)-(0.2) has a unique solution u in $C([0, \infty); L^1(\mathbf{R}^N) \cap L^\infty_{loc}(0, \infty; L^\infty(\mathbf{R}^N))$. Moreover, solutions are smooth for $t > 0$ (we refer to [EZ1] for this type of existence and uniqueness result).

Solutions of (0.1)-(0.2) with L^1-initial data satisfy the two following properties:

$$\int_{\mathbf{R}^N} u(x, t) dx = \int_{\mathbf{R}^N} u_0(x) dx, \ \forall t \geq 0 \quad \text{(conservation of mass)} \tag{0.3}$$

$$\|u(t)\|_\infty \leq C \left(\int_{\mathbf{R}^N} u_0 \right) t^{-\frac{N}{2}}, \ \forall t > 0 \quad (L^\infty \text{ - decay}) \tag{0.4}$$

In the sequel by $\int$ we denote the integral over $\mathbf{R}^N$.

We are interested on the asymptotic behavior of solutions as $t \to \infty$ and more precisely on how the constant mass is distributed in space. In other words, we want to obtain the asymptotic profile of the solutions. As we will see, in order to answer to these problems it is necessary to determine the rate of decay of solutions which, due to the effect of the convection, may be faster than that of (0.4).

Up to now three different types of asymptotic behavior have been observed: weakly non-linear, self-similar and strongly non-linear.

Before describing them we will briefly discuss these questions in the frame of the heat equation.

If u solves the linear heat equation

$$u_t - \Delta u = 0$$

with initial data u_0, then

$$u = G(\cdot, t) * u_0(\cdot)$$

where G is the fundamental solution of the heat equation:

$$G(x,t) = (4\pi t)^{-N/2}\exp(-|x|^2/4t).$$

It is then easy to see that when $\int u_0 = M$, then

$$t^{\frac{N}{2}(1-\frac{1}{r})}\|u(t) - MG(t)\|_{L^r(\mathbf{R}^N)} \to 0,$$

as $t \to \infty$ for every $r \in [1, \infty]$. In other words, the asymptotic behavior of solutions is given by the fundamental solution with the appropriate mass. This result is easy to prove (see for instance [EZ1] and [DZ]).

Suppose now that u solves the linear heat equation with a convection term:

$$u_t - \Delta u + a \cdot \nabla u = 0$$

with $a \in \mathbf{R}^N$ (by $\cdot$ we denote the scalar product in $\mathbf{R}^N$). Then $v(x,t) = u(x + at, t)$ solves the linear heat equation

$$v_t - \Delta v = 0.$$

The result above applies to v and this gives the asymptotic behavior of u:

$$t^{\frac{N}{2}(1-\frac{1}{r})}\|u(.,t) - MG(. - at, t)\|_{L^r(\mathbf{R}^N)} \to 0.$$

We note that the asymptotic profile is that of the gaussian MG but now the center of mass moves with velocity a as t increases.

In view of this, in order to understand the behavior for general equations of the form (0.1) we can assume without loss of generality that $f_i(0) = f_i'(0) = 0$ for $i \in \{1, ..., N\}$.

Notice that in view of the uniform decay rate (0.4), when studying the large time behavior of solutions, only the behavior of the nonlinearity at $u = 0$ matters. In particular, if the nonlinearities are very flat at the origin we may expect solutions to behave as in the linear heat equation. This is described in the next section. In the second one we describe the self-similar behavior which corresponds to a critical case where convection and diffusion are of the same order a typical example being the one-dimensional viscous Burgers equation. Finally, in section 3 we describe some strongly nonlinear results where convection dominates diffusion and we formulate some open problems.

1. Weakly nonlinear behavior

This case was considered in [EZ1](Theorem 5 in Section 6) where the following result was proved.

Theorem 1. *Suppose that*

$$\lim_{s \to 0} \frac{f_i(s)}{|s|^{1/N}s} = 0, \quad \text{for all } i \in \{1, ..., N\}$$

and assume that u solves (0.1)–(0.2) with $\int u_0 = M$. Then

$$\lim_{t \to \infty} t^{\frac{N}{2}(1-\frac{1}{r})}\|u(t) - MG(t)\|_{L^r(\mathbf{R}^N)} = 0$$

for every $r \in [1, \infty]$ where $G(x,t)$ is the heat kernel.

This result can be easily proved by working on the integral equation associated with (0.1)-(0.2) and applying well known estimates on the heat kernel as well as (0.3)-(0.4).

This result has been extended recently by G. Duro [D] for scalar conservation laws with variable diffusivity of the form

$$u_t - \text{div}(a(x)\nabla u) + \sum_{i=1}^{N} \frac{\partial f_i(u)}{\partial x_i} = 0 \tag{1.1}$$

with $a = 1 + b$, $b \in W^{1,1}(\mathbf{R}^N)$, Hölder continuous and such that $\|b\|_{L^\infty(\mathbf{R}^N)} < 1$.

In this context we can go farther. For instance, in [Z2], the second term of the asymptotic development was given for constant diffusivity. It is important to notice that the effect of the nonlinearity does afect the second term of the asymptotic development. For instance, when $f_1(s) = |s|^{q-1}s$ and $f_i \equiv 0$ for $i = 2, ..., N$, the second term of the asymptotic behavior depends on whether $q > 1+2/N$, $q = 1+2/N$ or $1 + 1/N < q < 1 + 2/N$. Similar results for Navier-Stokes equations have been proved by A. Carpio [C1].

2. Self-similar behavior

It was proved in [AEZ] that if $q = 1 + \frac{1}{N}$ then, for every $M \in \mathbf{R}$ and every $a \in \mathbf{R}^N$ there is a unique self-similar solution $\omega_{M,a}$ of the problem

$$\begin{cases} \omega_t - \Delta\omega + a \cdot \nabla|\omega|^{q-1}\omega = 0 & \text{in } \mathbf{R}^N \times (0, \infty) \\ \omega(t, x) = t^{-N/2}g(x/\sqrt{t}); \quad \int g = M. \end{cases} \tag{2.1}$$

Notice that self-similar solutions present the same scaling as the linear heat kernel.

We denote by $g_{M,a}$ the similarity profile of mass M associated to the equation with convection direction a. The profile $g_{M,a}$ satisfies the elliptic equation

$$-\Delta g - \frac{x \cdot \nabla g}{2} + a \cdot \nabla|g|^{q-1}g - \frac{N}{2}g = 0 \quad \text{in } \mathbf{R}^N \tag{2.2}$$

and the constraint

$$\int g = M. \tag{2.3}$$

In [AEZ] solutions g of (2.2)-(2.3) were obtained by a fixed point argument. Among other properties, solutions g are of constant sign and they are such that $g_{M_1,a} > g_{M_2,a}$ when $M_1 > M_2$. Of course, when $a \neq 0$ solutions are not radially symmetric. When $a = 0$ we have $g_{M,0} = M(4\pi)^{-N/2}\exp(-|x|^2/4)$, i.e., $\omega_{M,0} = MG$. We refer to [AEZ] and S. Kawashima [K] for further properties of the profiles g.

In [EZ1] we proved the following result of self-similar large time behavior:

Theorem 2. *Suppose that the nonlinearities satisfy*

$$\lim_{s \to 0} \frac{f_i(s)}{|s|^{1/N} s} = a_i, \quad \text{for all } i \in \{1, ..., N\}$$

with $a = (a_1, ..., a_N) \neq 0$, and assume that u is the solution of (0.1)–(0.2) with $\int_{\mathbf{R}^N} u_0 = M$. Then

$$\lim_{t \to \infty} t^{\frac{N}{2}(1-\frac{1}{r})}\|u(t) - \omega_{M,a}(t)\|_{L^r(\mathbf{R}^N)} = 0 \tag{2.4}$$

for every $r \in [1, \infty]$ where $\omega_{M,a}$ is the self-similar solution of (2.1).

In space dimension $N = 1$, $(N + 1)/N = 2$. Thus, a simple example of this self-similar behavior is the viscous Burgers equation

$$u_t - u_{xx} + (|u|u)_x = 0.$$

In this case, self-similar profiles can be computed explicitly by means of the Cole-Hopf trasformation.

Theorem 2 was proved working in the similarity variables and in the frame of the weighted Sobolev spaces introduced by Escobedo and Kavian [EK]. When working in the similarity variables the similarity profiles become stationary solutions of the corresponding convection-diffusion equation in which (2.2) is the equilibrium equation. In [EZ1] we applied La Salle's invariance principle with a suitable Lyapunov function. The construction of this Lyapunov function was based on the following strict L^1-contraction property: If u and v are two solutions of (0.1) with initial data u_0 and v_0, $u_0 \neq v_0$, of the same mass then,

$$\int |u(x,t) - v(x,t)|dx$$

is strictly decreasing in time.

In [Z1] this problem was reconsidered. It was shown that the existence and uniqueness of self-similar solutions $\omega_{M,a}$ is also a consequence of the strict L^1-contraction property. In view of [Z1] Theorem 2 can be proved by dynamical arguments without having the necessity of analyzing the elliptic problem (2.2)-(2.3).

254

It is important to note that self-similar solutions $\omega_{M,a}$ are also fundamental solutions of

$$\omega_t - \Delta\omega + a \cdot \nabla|\omega|^{q-1}\omega = 0, \tag{2.5}$$

i.e.,

$$\omega_{M,a}(t) \to M\delta, \quad \text{as } t \to 0 \tag{2.6}$$

where δ is the Dirac delta at the origin in the weak sense of measures:

$$\int \omega_{M,a}(x,t)\phi(x)dx \to M\phi(0), \quad \text{as } t \to 0$$

for every bounded and continuous ϕ.

In [EVZ1, 2] the uniqueness of fundamental solutions was proved in the class of constant sign solutions. In [C2,3] this uniqueness result was improved removing the sign condition. The fact that fundamental solutions are unique allows to give a new proof Theorem 2 by simple scaling arguments in the spirit of what we will explain in the next section.

The right scaling for this problem in this self-similar case is that of the linear heat equation. Indeed if u solves

$$u_t - \Delta u + a \cdot \nabla|u|^{q-1}u = 0 \tag{2.7}$$

with $q = (N+1)/N$ then

$$u_\lambda(x,t) = \lambda^N u(\lambda x, \lambda^2 t) \tag{2.8}$$

solves the same equation. This shows the invariance of (2.7) under the scaling transformation (2.8). On the other hand, the initial data of u_λ, $u_\lambda(x,0) = \lambda u_0(\lambda x)$ satisfies

$$\lambda u_0(\lambda x) \to M\delta, \quad \text{as} \quad \lambda \to \infty.$$

Therefore, it is natural to expect the convergence of u_λ to $\omega_{M,a}$, the latter being the fundamental solution of (2.5)–(2.6).

Using the estimates (0.3) and (0.4) and the regularizing effect of the heat kernel we can prove indeed that

$$u_\lambda(\cdot,t_0) \to \omega_{M,a}(\cdot,t_0), \quad \text{in} \quad L^r(\mathbf{R}^N), \quad \text{as} \quad \lambda \to \infty \tag{2.9}$$

for every $r \in [1,\infty]$ and $t_0 > 0$. On the other hand, it is easy to see that (2.9) is equivalent to (2.4).

Theorem 2 has been extended by G. Duro [D] to the equation with partial diffusivity (1.1) when $a = 1 + b$ with $b \in W^{1,1}(\mathbf{R}^N)$, Hölder continuous and such that $\|b\|_{L^\infty(\mathbf{R}^N)} < 1$ and $|b(x)| \le C/(1+|x|^\alpha)$ for some $\alpha > 0$.

3. Strongly nonlinear behavior

Let us consider first the one-dimensional problem $(N = 1)$:

$$u_t - u_{xx} + (|u|^{q-1}u)_x = 0 \tag{3.1}$$

with $1 < q < 2$.

In this range of q's the large time behavior is determined by the fundamental entropy solutions of the hyperbolic scalar conservation law

$$u_t + (|u|^{q-1}u)_x = 0. \tag{3.2}$$

Such solutions are known to exist, to be unique and of self-similar form (see T. P. Liu and M. Pierre [LP]). If ω_M is the fundamental entropy solution of mass M then it has the self-similar form

$$\omega_M = t^{-1/q}g_M(x/t^{1/q}) \tag{3.3}$$

with a profile g_M that can be computed explicitly to be of mass M:

$$g_M(x) = x^{1/(q-1)}1_{I_M}$$

where 1_{I_M} denotes the characteristic function of the interval

$$I_M = (0, (qM/(q-1))^{(q-1)/q}).$$

Note that the profiles g are of compact support and that they present a discontinuity at $x = (qM/(q-1))^{(q-1)/q}$.

From (3.3) we see that ω_M decays in L^∞ as $t^{-1/q}$ which is a faster decay rate than that of solutions of the heat equation (i.e., $t^{-1/2}$) since $q < 2$.

The following result was proved in [EVZ1]:

Theorem 3. *Suppose that $1 < q < 2$ and let u be a solution of (3.1) with initial data u_0 of mass M. Then,*

$$\lim_{t \to \infty} t^{\frac{1}{q}(1-\frac{1}{r})}||u(t) - \omega_M(t)||_{L^r(\mathbf{R}^N)} = 0 \tag{3.4}$$

for every $r \in [1, \infty)$ where ω_M is the fundamental entropy solution of (3.2) with mass M.

When proving Theorem 3 the main difficulty is to prove that solutions of (3.1) satisfy the following decay estimate:

$$||u(t)||_{L^\infty(\mathbf{R})} \le Ct^{-1/q} \tag{3.5}$$

which is stronger than (0.4) since $1 < q < 2$. By comparison it is sufficient to prove (3.5) for positive solution. Now if u is a positive solution of (3.1), we have

$$(u^{q-1})_x \leq 1/qt. \tag{3.6}$$

To prove (3.6) it is sufficient to write the equation satisfied by $q(u^{q-1})_x$ and to observe that $W(t) = 1/t$ is a supersolution that takes $+\infty$ as initial data. Thus by comparison (3.6) holds. From (3.6), taking into account that the mass of u is constant in time and applying Taylor's expansion we obtain (3.5) easily. Indeed, let us fix some $t > 0$ and let x_0 be such that $u(x_0, t) = \|u(t)\|_{L^\infty(\mathbf{R}^N)}$. By Taylor's expansion and (3.6), if $x < x_0$, we have:

$$u^{q-1}(x,t) = A^{q-1} + (u^{q-1})_x(y,t)(x - x_0)$$

for some $y \in [x, x_0]$. But then by (3.6):

$$u^{q-1}(x,t) \geq \frac{A^{q-1}}{2^{q-1}}, \forall x \in I = [x_0 - A^{q-1}(1 - 2^{1-q})qt, x_0].$$

Therefore

$$u(x,t) \geq \frac{A}{2}, \forall x \in I$$

Integrating this inequality in the interval I we deduce that

$$\frac{A}{2}|I| = \frac{A}{2}A^{q-1}(1 - 2^{1-q})qt \leq \int_I u(x,t)dx \leq M$$

and therefore

$$A \leq \left(\frac{2M}{(1 - 2^{1-q})q}\right)^{1/q} t^{-1/q}$$

which implies (3.5).

Once (3.5) is proved we define the rescaled family:

$$u_\lambda(x,t) = \lambda u(\lambda x, \lambda^q t).$$

We observe that the behavior of $u(\cdot, t)$ as $t \to \infty$ can be deduced from the behavior of $u_\lambda(\cdot, 1)$ as $\lambda \to \infty$. Indeed, (3.4) holds if and only if

$$\|u_\lambda(., 1) - \omega_M(., 1)\|_{L^r(\mathbf{R}^N)} \to 0$$

as $\lambda \to \infty$. On the other hand, u_λ satisfies

$$\partial_t u_\lambda + \partial_x(|u_\lambda|^{q-1} u_\lambda) = \lambda^{q-2} \partial_x^2 u_\lambda \tag{3.7}$$

and

$$u_\lambda(x,0) = \lambda u_0(\lambda x). \tag{3.8}$$

From (3.7)-(3.8) we see that (at least formally), since $q-2 < 0$, u_λ has to converge to the entropy fundamental solution of mass $M = \int u_0$ of the hyperbolic conservation law (3.2). The proof of the convergence can be made rigorous by using the classical compensated compactness results by L. Tartar [T].

Following these arguments we proved Theorem 3 in [EVZ1] in the frame of the more general conservation law:

$$u_t - u_{xx} + \partial_x(f(u)) = 0$$

provided $f(0) = f'(0) = 0$ and that there exists some $q \in (1,2)$ such that the following limit exists

$$\lim_{s \to 0} \frac{f_j''(s)}{|s|^{q-3}s}.$$

In this case the limiting behavior is given by the entropy fundamental solutions of the problem

$$u_t + a\partial_x(|u|^{q-1}u) = 0$$

where $a = \lim_{s\to 0} \frac{f(s)}{|s|^{q-1}s}$.

Let us consider now the multidimensional problem:

$$u_t - \Delta u + \partial_1(|u|^{q-1}u) = 0 \quad \text{in} \quad \mathbf{R}^N \times (0,\infty). \tag{3.9}$$

We denote by ∂_1 the derivative with respect to x_1. We will use the notation $x = (x_1, y)$ with $y = (x_2, ..., x_N)$.

Let us assume that $1 < q < 1 + 1/N$. In this case the asymptotic behavior of solutions is given by the fundamental solutions of the following *reduced equation*:

$$u_t - \Delta_y u + \partial_1(|u|^{q-1}u) = 0 \quad \text{in} \quad \mathbf{R}^N \times (0,\infty). \tag{3.10}$$

In this equation Δ_y denotes the laplacian in the variables $y = (x_2, ..., x_N)$. Thus, equation (3.10) is parabolic in the directions y and hyperbolic in x_1.

In order to guarantee the uniqueness of solutions of (3.10) we have to add an entropy condition in the spirit of Kruzhkov (see [EVZ3]):

$$|u - \psi|_t - \Delta_y|u - \psi| + \partial_1||u|^{q-1}u - |\psi|^{q-1}\psi| \le \text{sign}(u - \psi)\Delta_y\psi \tag{3.11}$$

in $\mathcal{D}'((0,\infty) \times \mathbf{R}^N)$ for all $\psi \in \mathcal{D}(\mathbf{R}^{N-1})$; $\psi = \psi(x_2, \cdots, x_N)$. We refer to Bénilan and Touré [BT] for an abstract approach to entropy solutions of this type of problems based on the theory of semigroups in L^1.

In [EVZ2] we proved that for every $M \ge 0$ there exists a unique non-negative entropy fundamental solution u of (3.10) with mass M in $BC((0,\infty); L^1(\mathbf{R}^N))$ (by

BC we denote the space of continuous and bounded functions). More recently, in [C2, 3] the sign restriction was removed so that we know now that the fundamental entropy solution is unique.

As a consequence of [EVZ2] and [C2, 3] we have the following result on the asymptotic behavior of solutions of (3.9):

Theorem 3. *Suppose that $1 < q < 1 + 1/N$ and assume that u is the solution of (3.8) with initial data u_0 of mass $\int_{\mathbf{R}^N} u_0 = M$. Then*

$$\lim_{t \to \infty} t^{\frac{N+1}{2q}(1-\frac{1}{r})}||u(t) - \omega_M(t)||_{L^r(\mathbf{R}^N)} = 0$$

for every $r \in [1, \infty)$ where ω_M is the entropy fundamental solution of the reduced equation (3.10) which is self-similar, i.e.,

$$\omega_M(x, t) = t^{-(N+1)/2q} f_M(x_1/t^\alpha, y/\sqrt{t})$$

with $\alpha = (N + 1 + q - Nq)/2q$.

Note that the self-similar profiles f_M have compact support in the direction x_1.

When proving Theorem 3 and once the uniqueness of the fundamental entropy solution of the reduced equation (3.10) is known, as in space dimension N=1, the main difficulty is to prove the decay estimate:

$$||u(t)||_{L^\infty(\mathbf{R}^N)} \le Ct^{-(N+1)/2q}. \tag{3.11}$$

Notice that since $q < (N + 1)/N$, $(N + 1)/2q > N/2$ and therefore the decay rate of (3.11) is faster than (0.4).

To prove (3.12) we follow the one-dimensional argument. If u is non-negative we show that

$$\partial_1(u^{q-1}) \le 1/qt. \tag{3.13}$$

This can be proved easily. It is sufficient to observe that $w = q\partial_1(u^{q-1})$ satisfies the heat equation

$$w_t - \Delta w + w^2 + \gamma\frac{|\nabla z|^2}{z^2}w + (z - 2\gamma\frac{w}{z})\frac{\partial w}{\partial x_1} - 2\frac{\gamma}{z}\sum_{j=2}^{N}\frac{\partial z}{\partial x_j}\frac{\partial w}{\partial x_j} = 0$$

where $z = qu^{q-1}$ and $\gamma = (2 - q)/(q - 1)$. By comparison an taking into account that $W = 1/t$ is a super-solution of this equation we obtain (3.13).

In order to apply the one-dimensional argument we also need and L^∞-estimate for

$$v(y, t) = \int_{\mathbf{R}} u(x_1, y, t)dx_1.$$

But this is easy to do since v satisfies the $N - 1$-dimensional heat equation

$$v_t - \Delta_y v = 0 \quad \text{in} \quad \mathbf{R}^{N-1} \times (0, \infty)$$

and takes the initial data $v_0(y) = \int_{\mathbf{R}} u_0(x_1, y)dx_1$ which belongs to $L^1(\mathbf{R}^{N-1})$. Thus

$$\|v(t)\|_{L^\infty(\mathbf{R}^{N-1})} \le Ct^{-(N-1)/2}. \tag{3.14}$$

Combining (3.13) and (3.14) we deduce (3.12).

Once (3.12) is proved we introduce the re-scaled family of solutions:

$$u_\lambda(t, x_1, y) = \lambda^{\frac{N+1}{2q}} u(\lambda^\alpha x_1, \sqrt{\lambda} y, \lambda t) \tag{3.15}$$

whith α as in Theorem 3. The function u_λ satifies

$$\frac{\partial u_\lambda}{\partial t} - \Delta_y u_\lambda + \frac{\partial |u_\lambda|^{q-1} u_\lambda}{\partial x_1} = \lambda^{1-2\alpha} \frac{\partial^2 u_\lambda}{\partial x_1^2}, \tag{3.16}$$

and the initial condition

$$u_\lambda(x, 0) = \lambda^{\frac{N+1}{2q}} u_0(\lambda^\alpha x_1, \sqrt{\lambda} y). \tag{3.17}$$

Observe that

$$\alpha > \frac{1}{2} \Longleftrightarrow q < 1 + \frac{1}{N}. \tag{3.18}$$

Therefore, formally, u_λ converges to the fundamental entropy solution of the reduced equation (3.10) as $\lambda \to \infty$. When u is non-negative this can be made rigorous with classical estimates. Indeed, in view of (3.13) we have

$$\int |\partial_{x_1} u^q| dx \le \frac{1}{(q-1)t} \int u dx = \frac{M}{(q-1)t}$$

and this provides an uniform bound for $\partial_{x_1} u_\lambda^q(t)$ in $L^1(\mathbf{R}^N)$ for $\lambda \ge 1$ and t belonging to any compact interval $[t_0, t_1]$ with $0 < t_0 < t_1 < \infty$. On the other hand the parabolic character of (3.16) allows to obtain estimates for $\partial_t u_\lambda$ and $\partial_{x_j} u_\lambda$ for $j = 2, ..., N$. Compactness results in the spirit of [S] allow to pass to the limit in (3.16). The entropy inequality (3.11) is also easy to get by means of Kato's inequality.

When dealing with changing sign solutions the obtention of the compactness of u_λ is more delicate since we do not have the entropy estimate (3.13). This problem was solved by A. Carpio in [C2, 3] by using the kinetic methods developed by Lions, Perthame and Tadmor [LPT].

Theorem 3 can be extended to more general situations in which the nonlinearities f_i satisfy the following conditions:

$$\begin{cases} \exists q \in (1, 1 + \dfrac{1}{N}) \text{ and } j \in \{1, ..., N\} \text{ such that } f_i \equiv 0 \text{ if } i \ne j, \\[2mm] f_i(0) = f_i'(0) = 0 \text{ and the following limit exists} \\[2mm] \lim_{s \to 0} \dfrac{f_j''(s)}{|s|^{q-3} s}. \end{cases}$$

260

Under this condition, if

$$\lim_{s \to 0} \frac{f_j(s)}{|s|^{q-1}s} = c$$

the asymptotic behavior of solutions of (0.1)-(0.2) is given by the fundamental entropy solutions of the reduced equation

$$u_t - \Delta_y u + c\partial_{x_j}(|u|^{q-1}u) = 0.$$

More recently, in [EZ2], we have addressed the problem of the strongly nonlinear large time behavior when the convection points in two directions. For instance, when the equation is of the form

$$u_t - \Delta u + \partial_{x_1}(|u|^{q-1}u) + \partial_{x_2}(|u|^{p-1}u) = 0 \quad \text{in} \quad \mathbf{R}^N \times (0, \infty). \tag{3.19}$$

We have shown that when $1 < q < (N+1)/N$ and $p > 1 + q/(N+1)$ the asymptotic behavior is that of the problem with one single convection:

$$u_t - \Delta u + \partial_{x_1}(|u|^{q-1}u) = 0.$$

Therefore it is given by the fundamental entropy solutions of the reduced equation

$$u_t - \Delta_y u + \partial_{x_1}(|u|^{q-1}u) = 0.$$

The proof uses the same techniques. However, in this case, the proof of the decay rate (3.12) is much more technical since, due to the second convection term, the obtention of (3.13) and (3.14) is far from being obvious. At this level the restriction $p > 1 + q/(N+1)$ appears naturally.

Note that $1 + q/(N+1) > q$ if and only if $q < (N+1)/N$. Thus, the asymptotic behavior of solutions of (3.17) is open for $1 < q < (N+1)/N$ and $q < p \leq 1 + q/(N+1)$.

Applying the scaling (3.15) to (3.19) we get:

$$\frac{\partial u_\lambda}{\partial t} - \Delta_y u_\lambda + \frac{\partial |u_\lambda|^{q-1}u_\lambda}{\partial x_1} = \lambda^{1-2\alpha}\frac{\partial^2 u_\lambda}{\partial x_1^2} - \lambda^{\frac{1}{2} - \frac{(N+1)(p-1)}{2q}}\frac{\partial |u_\lambda|^{p-1}u_\lambda}{\partial x_2}. \tag{3.20}$$

We observe that

$$\frac{1}{2} - \frac{(N+1)(p-1)}{2q} < 0 \quad \text{if and only if} \quad p > 1 + q/(N+1).$$

When $p = 1 + q/(N+1)$ the (formal) limit equation of (3.20) is

$$u_t - \Delta_y u + \partial_{x_1}(|u|^{q-1}u) + \partial_{x_2}(|u|^{p-1}u) = 0. \tag{3.21}$$

261

This suggests that when $1 < q < (N+1)/N$ and $p = 1 + q/(N+1)$ the aymptotic behavior of solutions of (3.19) will be given by the fundamental entropy solutions of (3.21). But this is an open problem. The existence of fundamental entropy solutions of (3.21) is unknown.

When $1 < q < (N+1)/N$ and $q < p < 1 + q/(N+1)$ one has to expect a different asymptotic behavior. In this case the decay rate is probably faster than $t^{-(N+1)/2q}$. But, up to now, this problem has not been solved.

Acknowledgements. The author was supported by the grant PB90-0245 of the DGICYT (Spain) and by the Eurhomogenization project SC1*-CT91-0732 of the EC.

References.

[AEZ] J. Aguirre, M. Escobedo and E. Zuazua, Self-similar solutions of a convection diffusion equation and related semilinear problems, Comms. Part. Differ. Eqs. **15** (1990), 139–157.

[B] Ph. Bénilan and H. Touré, Solution entropique pour une équation parabolique-hyperbolique non linéaire, Ann. Fac. Sci. Toulouse, **III** (1) (1994), 63–80.

[C1] A. Carpio, Asymptotic behavior in incompresible Navier-Stokes equations, SIAM J. Math. Anal., to appear.

[C2] A. Carpio, Unicité et comportement asymptotique pour des équations de convection-diffusion scalaires, C. R. Acad. Sci. Paris, to appear.

[C3] A. Carpio, Large time behavior in convection-diffusion equations, preprint.

[DZ] J. Duoandikoetxea and E. Zuazua, Moments, masses de Dirac et développements de fonctions, C. R. Acad. Sci. Paris, **315** (6) (1992), 693–698.

[D] G. Duro, Ph. D. Thesis, Universidad Complutense de Madrid, in preparation.

[EK] M. Escobedo and O. Kavian, Variational problems related to self-similar solutions of the heat equation, Nonlinear Anal. TMA, **11** (10) (1987), 1103–1133.

[EVZ1] M. Escobedo, J.L. Vázquez and E. Zuazua, Source-type solutions and asymptotic behaviour for a diffusion-convection equation, Arch. Rat. Mech. Anal. **124** (1993), 43–66.

[EVZ2] M. Escobedo, J.L. Vázquez and E. Zuazua, A diffusion-convection equation in several space dimensions, Indiana Univ. Math. J., **42**(4) (1993), 1413–1440.

[EVZ3] M. Escobedo, J.L. Vázquez and E. Zuazua, Entropy solutions for diffusion-convecti- on equations with partial diffusivity, to appear in Trans. of the A.M.S.

[EZ1] M. Escobedo and E. Zuazua, Large-time behaviour for solutions of a convection diffusion equation in $\mathbf{R}^N$, Jour. Funct. Analysis **100** (1991),119–161.

[EZ2] M. Escobedo and E. Zuazua, Long time behavior for a convection-diffusion equation in higher dimension, preprint.

[K] S. Kawashima, Self-similar solutions of a convection-diffusion equation, Nonlinear PDE - Japan Symposium 2, Lecture Notes in Num. Appl. Anal., 12, K. Masuda, M. Mimura and T. Nishida eds., (1993), 123–136.

[LPT] P.L. Lions, B. Perthame and E. Tadmor, A kinetic formulation of multidimensional scalar conservation laws and related equations, Journal of the Amer. Math. Soc., 7 (1) (1994), 169–189.

[LP] T. P. Liu and M. Pierre, Source solutions and asymptotic behavior in conservation laws, J. Diff. Eqs, 51 (1984), 419–441.

[S] J. Simon, Compact sets in the space $L^p(0, T; B)$, Annali Mat. Pura Appl. CXLVI (1987), 65–96.

[T] L. Tartar, Compensated compactness and applications to partial differential equations, in Nonlinear Analysis and Mechanics, Heriot-Watt Symp., IV, R. J. Knops ed., Pitman, London, 1979, 136–212.

[Z1] E. Zuazua, A dynamical system approach to the self similar large time behavior in scalar convection-diffusion equations, J. Diff. Eqs. 108 (1994), 1–35.

[Z2] E. Zuazua, Weakly non-linear large time behavior for scalar convection-diffusion equations, Diff. Int. Eqs, 6(6) (1993), 1481–1492.

Enrique ZUAZUA

Departamento de Matemática Aplicada
Universidad Complutense
E–28040 Madrid.
zuazua@mat.ucm.es

Printed and bound by CPI Group (UK) Ltd, Croydon, CR0 4YY

23/10/2024

01778224-0007